Student Problem Set with Solutions

Second Edition

ELECTRIC CIRCUIT ANALYSIS

Student Problem Set with Solutions

Second Edition

ELECTRIC CIRCUIT ANALYSIS

David E. Johnson
Birmingham-Southern College

Johnny R. Johnson
University of North Alabama

John L. Hilburn
President, Microcomputer Systems Inc.

Prentice Hall, Englewood Cliffs, New Jersey 07632

Editorial Production/ Supervision and Interior Design: **Benjamin D. Smith**
Supplement Acquisitions Editor: **Alice Dworkin**
Acquisitions Editor: **Elizabeth Kaster**
Manufacturing Buyers: **Linda Behrens/ David Dickey**

A Paramount Communications Company
Englewood Cliffs, New Jersey 07632

Printed in the United States of America

10 9 8 7 6 5 4

ISBN 0-13-251091-X

Prentice-Hall International (UK) Limited, *London*
Prentice-Hall of Australia Pty. Limited, *Sydney*
Prentice-Hall Canada Inc. *Toronto*
Prentice-Hall Hispanoamericana, S.A., *Mexico*
Prentice-Hall of India Private Limited, *New Delhi*
Prentice-Hall of Japan, Inc., *Tokyo*
Simon & Schuster Asia Pte. Ltd., *Singapore*
Editora Prentice-Hall do Brasil, Ltda., *Rio de Janeiro*

Contents

	Preface	vii
1	Introduction	1
2	Resistive Circuits	6
3	Dependent Sources	16
4	Analysis Methods	28
5	Network Theorems	45
6	Independent Equations	60
7	Energy Storage Elements	65
8	Simple RC and RL Circuits	78
9	Second-Order Circuits	94
10	Sinusoidal Excitation and Phasors	106
11	AC Steady-State Analysis	119
12	AC Steady-State Power	134
13	Three-Phase Circuits	143
14	Complex Frequency and Network Functions	155
15	Frequency Response	168
16	Transformers	180
17	Fourier Series	192
18	Fourier Transforms	202
19	Laplace Transforms	210
20	Laplace Transform Applications	218

PREFACE

This manual was written for use by students who are studying introductory circuit analysis from our textbook, ELECTRIC CIRCUIT ANALYSIS 2/E (Prentice Hall, 1992), but it could serve as a study guide for students using other calculus-based beginning circuits books, as well. The problems presented in this manual are not found in our textbook. The primary purpose of this problem set is to provide additional problem material for further student practice.

The problems are arranged by section in the textbook, so that a student needing more practice in a certain area could easily locate the problems in that area. Detailed solutions are presented following the statements of each set of problems. The ideal way for students to use the manual is to read the problem and attempt to solve it on their own before looking at the solution. It is very easy to follow someone else's solution of a problem and be lulled into falsely thinking that the problem is thoroughly understood.

One of the most effective ways to study circuit theory is, of course, to work large numbers of problems. We are almost tempted to say that it is impossible to work too many. In this spirit we hope this manual will be extremely useful to students of circuit analysis.

David E. Johnson
Johnny R. Johnson
John H. Hilburn

Student Problem Set
with Solutions

Second Edition

ELECTRIC CIRCUIT ANALYSIS

Chapter 1

Introduction

1.1 Definitions and Units

1.1 Let $f(t)$ in the graph shown be the charge $q(t)$ in coulombs that has entered the positive terminal of an element as a function of time. Find (a) the total charge that has entered the terminal between 1 and 3s, (b) the total charge that has entered between 4 and 7s, and (c) the current entering the terminal at 1s, 4s and 6s.

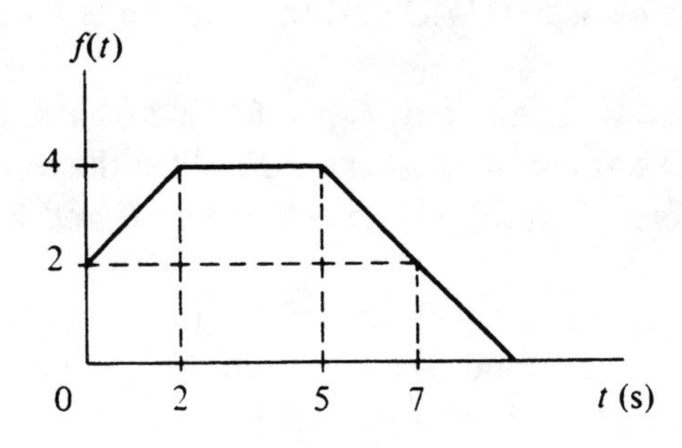

PROBLEM 1.1

1.2 Let $f(t)$ in the graph shown be the current $i(t)$ in milliamperes entering an element terminal as a function of time. Find (a) the charge that enters the terminal between 0 and 10s and (b) the rate at which the charge is entering at $t = 2$s and $t = 4$s.

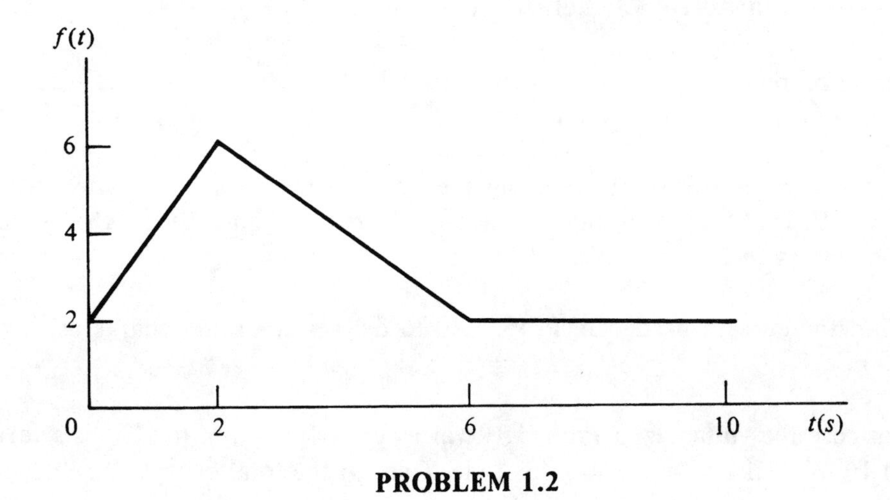

PROBLEM 1.2

1.2 Charge and Current

1.3 Let $f(t)$ in Prob. 1.1 be the current $i(t)$ in milliamperes entering an element terminal as a function of time. (a) Find the charge that enters the terminal between 0 and 7s and (b) the rate at which the charge is entering at $t = 5$s.

1.4 Let $f(t)$ in Prob. 1.2 be the charge $q(t)$ in coulombs that has entered the positive terminal of an element as a function of time. Find (a) the total charge that has entered the terminal between 1 and 2s, (b) the total charge that has entered between 6s and 10s, and (c) the current entering the terminal at 1s, 4s, and 8s.

1.3 Voltage, Energy, and Power

1.5 Find the power delivered to an element at $t = \pi$ seconds if the charge entering the positive terminal is $q = 8\sin(t/4)$C and the voltage is $v = 6\cos(t/4)$V.

1.6 The power delivered to an element is $p = 6e^{-4t}$W and the charge entering the positive terminal of the element is $q = -20e^{-2t}$mC. Find (a) the voltage across the element and (b) the energy delivered to the element between 0 and 250ms.

1.7 The charge entering the positive terminal of an element is $q = -2e^{-4t}$C. Find the power delivered to the element as a function of time and the energy delivered to the element between 0 and 1s if (a) $v = 6i$, (b) $v = 3di/dt$, (c) $v = 2\int_0^t i dt - 4$.

1.8 The current entering the positive terminal of an element is $i = -2e^{-t}A$. Find the power delivered to the element as a function of time and the energy delivered to the element between 0 and 1s if (a) $v = 6i$, (b) $v = 3di/dt$, and (c) $v = 2\int_0^t i dt + 4$. (*Note:* The element voltage v is in volts if the current i is in amperes.)

1.4 Passive and Active Elements

1.9 If the current entering the positive terminal of an element is $i = \frac{1}{6}t - 20$ mA and the voltage is $v = 12$V, find the energy delivered to the element in 5 min.

1.10 If the current $i = 3$A. is entering the positive terminal of a battery with voltage $v = 12$V, find (a) the energy supplied to the battery, and (b) the charge delivered to the battery in 2 h (hours).

1.11 Find the current needed in Prob. 1.10 to deliver the same charge as in part (b) in 10 min.

1.12 Suppose the voltage v in Prob. 1.10 varies linearly from 4 to 12V as t varies from 0 to 10 min. If $i = 3$A during this time, find (a) the total energy supplied and (b) the total charge delivered to the battery.

1.13 Let the current entering the positive terminal of an element be $i = 0$ for $t < 0$, and $i = 3\cos(4t)$ for $t > 0$. If the voltage is $v = 2di/dt$ V, find the energy delivered to the element for all time.

1.14 In Prob. 1.13 find the total charge delivered to the element at $t = \pi/4$ s and the power absorbed at $\pi/8$ s.

1.15 The voltage across an element is 6V and the charge q entering the positive terminal is as shown. Find the total charge and total energy delivered to the element between 1 and 10ms.

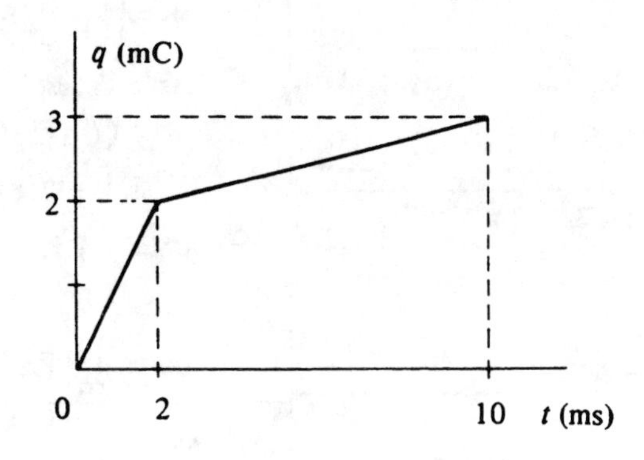

PROBLEM 1.15

1.16 If the graph of Prob. 1.15 is the current i A versus t s, and $v = 10di/dt$ V, find p at 1s and at 4s.

1.17 If the graph of Prob. 1.15 is the voltage v V versus t s, and $p = v^2/12$, find p in terms of time on the intervals $0 < t < 2$s and $2 < t < 10$s.

1.18 If a current $i = 2$A is entering the positive terminal of a battery with voltage $v = 6$V, then the battery is in the process of being charged. (It is absorbing rather than delivering power.) Find (a) the energy supplied to the battery, and (b) the charge delivered to the battery in 2 h (hours). Note the consistency of the units, 1V = 1 J/C.

1.19 Let the current entering the positive terminal of an element be $i = 0$ for $t < 0$, and $i = 6\sin(2t)$A for $t > 0$. (a) If the voltage is $v = 4di/dt$ V, show that the energy delivered to the element is nonnegative for all time. (The element is passive.) (b) Repeat part (a) if $v = 2\int_0^t i dt$ V.

1.1 From the graph;

(a) $q_T = q(3) - q(1) = 4 - 3 = \underline{1C}$

(b) $q_T = q(7) - q(4) = 2 - 4 = \underline{-2C}$

(c) $i = \frac{dq}{dt}$ = slope of q curve;

$i(1) = \frac{2-0}{2} = \underline{1A}$;

$i(4) = \underline{0A}$;

$i(6) = \frac{4-2}{5-7} = \underline{-1A}$.

1.2 (a) $q_T = \int_0^{10} i(t)dt$ = area under $i(t)$

$= 2\times10 + \frac{1}{2}(6)(4) = \underline{32\,mC}$

(b) rate = $i(t)$, from the graph

$i(2) = \underline{6mA}$; $i(4) = \frac{1}{2}(2+6) = \underline{4mA}$

1.3 (a) $q_T = \int_0^7 i(t) = 2\times7 + 2\times5 = \underline{24mC}$

(b) $i(t)$ = rate;

$i(5) = \underline{4mA}$

1.4 (a) From the graph

$q_T = q(2) - q(1) = 6 - 4 = \underline{2C}$

(b) $q_T = q(10) - q(6) = 2 - 2 = \underline{0C}$

(c) $i = \frac{dq}{dt}$ = slope of q curve;

$i(1) = \frac{6-2}{2} = \underline{2A}$;

$i(4) = \frac{2-6}{6-2} = \underline{-1A}$; $i(8) = \underline{0A}$

1.5 $i(t) = \frac{dq}{dt} = \frac{8}{4}\cos(t/4)$ A

$p(\pi) = vi = (6\cos(\pi/4))(2\cos(\pi/4))$

$= \frac{12}{2} = \underline{6W}$

1.6 (a) $i = \frac{dq}{dt} = 40e^{-2t}$ mA;

$v = \frac{P}{i} = \frac{6e^{-4t}\,W}{40e^{-2t}mA} = \underline{150e^{-2t}V}$

(b) $w = \int_0^{250ms} p\,dt = \int_0^{0.25} 6e^{-4t}dt$

$= \underline{1.5(1-e^{-1})\,J}$

1.7 $i = \frac{dq}{dt} = 8e^{-4t}$ A

(a) $p = vi = 6i^2 = 6(8e^{-4t})^2 = \underline{384e^{-8t}W}$

$w = \int_0^1 p\,dt = \int_0^1 384e^{-8t}dt = \underline{48(1-e^{-8})\,J}$.

1.7 cont.

(b) $v = 3\frac{di}{dt} = 3(-32e^{-4t}) = -96e^{-4t}V$,

$p = (-96e^{-4t})(8e^{-4t}) = \underline{-768e^{-8t}W}$;

$w = \int_0^1 -768e^{-8t}dt = \underline{96(e^{-8}-1)\,J}$.

(c) $v = 2\int_0^t 8e^{-4t}dt - 4 = -4e^{-4t}V$,

$p = (-4e^{-4t})(8e^{-4t}) = \underline{-32e^{-8t}W}$;

$w = \int_0^1 -32e^{-8t}dt = \underline{4(e^{-8}-1)\,J}$.

1.8

(a) $p = vi = 6i^2 = 6(-2e^{-t})^2 = \underline{24e^{-2t}W}$;

$w = \int_0^1 p\,dt = \int_0^1 24e^{-2t}dt = \underline{12(1-e^{-2})\,J}$

(b) $v = 3(2e^{-t}) = 6e^{-t}V$;

$p = (6e^{-t})(-2e^{-t}) = \underline{-12e^{-2t}W}$;

$w = \int_0^1 -12e^{-2t}dt = \underline{6(e^{-2}-1)\,J}$

(c) $v = 2\int_0^t -2e^{-t}dt + 4 = 4e^{-t}V$;

$p = (4e^{-t})(-2e^{-t}) = \underline{-8e^{-2t}W}$;

$w = \int_0^1 -8e^{-2t}dt = \underline{4(e^{-2}-1)\,J}$.

1.9 $w = \int_0^{5(60)} vi\,dt = \int_0^{300}(12)(\frac{t}{6} - 20mA)dt$

$= (300)^2 - (240)(300)$

$= \underline{18\,J}$

1.10 (a) $w = \int_0^{2(3600)} vi\,dt = \int_0^{7200}(12)(3)dt$

$= \underline{259.2\,KJ}$

(b) $q = \int_0^{7200} i\,dt = \int_0^{7200} 3dt = \underline{21.6KC}$

1.11 $q = 21.6KC = \int_0^{10(60)} i\,dt = 600i$

$i = \frac{21.6KC}{600} = \underline{36A}$

1.12 The slope of the v-line is

$m = \frac{12-4}{10(60)-0} = \frac{1}{75}$. v-intercept is

therefore $v = \frac{1}{75}t + 4$

(a) $w = \int_0^{600} vi\,dt = \int_0^{600}(\frac{1}{75}t+4)(3)dt$

$= \underline{14.4\,kJ}$

(b) $q = \int_0^{600} i\,dt = \int_0^{600} 3\,dt = \underline{1800C}$

1.13 $v = 2(3)(4)(-\sin 4t)$ V

$w = \int_0^t (-24\sin 4t)(3\cos 4t)dt$

$= -\frac{72}{8}\sin^2 4t = \underline{-9\sin^2 4t\,J}$

1.14 $q_T = \int_{-\infty}^{\pi/4} i\,dt = \int_0^{\pi/4} 3\cos 4t\,dt$

$= \frac{3}{4}(0) = \underline{0\,C}$

From solution of 1.13,

$v = -24\sin 4t$

$p(\pi/8) = [-24\sin 4(\pi/8)][3\cos 4(\pi/8)]$

$= \underline{0\,W}$

1.15 From the graph,

$q_T = q(10ms) - q(1ms) = 3 - 1 = \underline{2\,mC}$

$$i = \frac{dq}{dt} = \begin{cases} \frac{2-0}{2-0} = 1A & 0 \le t \le 2ms \\ \frac{3-2}{10-2} = \frac{1}{8}A & 2ms \le t \le 10ms \end{cases}$$

$W_T = \int_{10^{-3}}^{10(10^{-3})} vi\,dt$

$= \int_{10^{-3}}^{2(10^{-3})} (6)(1)dt + \int_{2(10^{-3})}^{10(10^{-3})} 6(\frac{1}{8})dt$

$= (6+6)10^{-3} = \underline{12\,mJ}$

1.16 $v = \begin{cases} 10\,V & 0 \le t \le 2s \\ \frac{10}{8}V & 2s \le t \le 10s \end{cases}$

$p(1) = v(1)\,i(1) = (10)(1) = \underline{10\,W}$

$p(4) = (\frac{10}{8})(2 + \frac{2}{8}) = \underline{2.813\,W}$

1.17 $v = \begin{cases} t & 0 \le t \le 2s \\ t/8 + (2 - \frac{2}{8}) & 2s \le t \le 10s \end{cases}$

$p = \underline{\frac{t^2}{12}\,W} \quad 0 \le t \le 2s$

$p = \frac{\left(\frac{t+14}{8}\right)^2}{12} = \underline{\frac{(t+14)^2}{768}}\,W;\; 2s \le t \le 10s$

1.18 (a) $w = \int_0^{2(3600)} vi\,dt = \int_0^{7200} 12\,dt$

$= \underline{86{,}400\,J}$

(b) $q = \int_0^{7200} i\,dt = \int_0^{7200} 2\,dt$

$= \underline{14{,}400\,C}$

1.19 (a) $v = 4\frac{di}{dt} = 4(12\cos 2t)$

$w = \int_{-\infty}^{t} vi\,dt = \int$

$= \int_0^t (48\cos 2t)(6\sin 2t)dt$

$= \underline{72\sin^2 2t} \ge 0$

(b) $v = 2\int_0^t i\,dt = 2\int_0^t 6\sin 2t$

$= -6\cos 2t + 6\,V$

$w = \int_0^t (-6\cos 2t + 6)(6\sin 2t)dt$

$= -9\sin^2 2t - 18\cos 2t + 18$

$= 9(\cos^2 2t - 1 - 2\cos 2t + 2)$

$= 9(\cos^2 2t - 2\cos 2t + 1)$

$= \underline{9(\cos 2t - 1)^2} \ge 0$

Chapter 2

Resistive Circuits

2.1 Ohm's Law

2.1 A 2kΩ resistor is connected to a battery and 6mA flow. What current will flow if the battery is connected to a 600Ω resistor?

2.2 A resistor connected to a 12V source carries a current of 60mA. Find the resistance and the minimum power rating of the resistor. If a 400Ω resistor is inserted in series in the circuit, find the voltage across it.

2.3 A toaster is essentially a resistor that becomes hot when it carries a current. If a toaster is dissipating 400W at a current of 2A, find its voltage and its resistance.

2.4 If a toaster with a resistance of 10Ω is operated at 100V for 8s, find the energy it uses.

2.2 Kirchhoff's Laws

2.5 Find i_1, i_2, v.

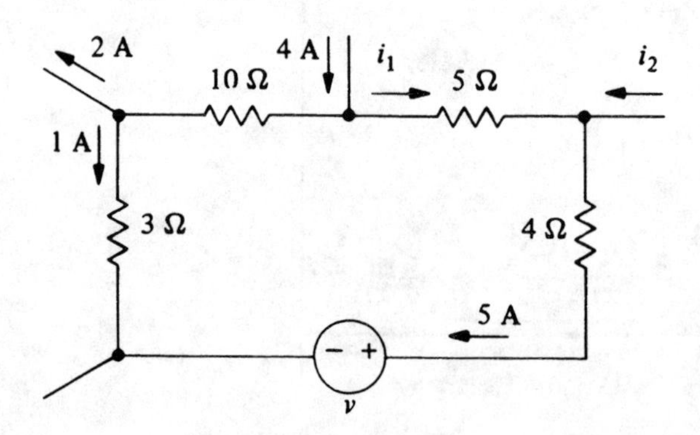

PROBLEM 2.5

2.6 Find v.

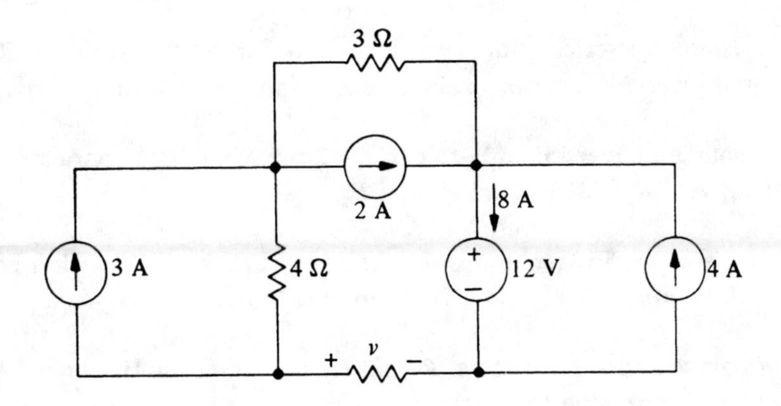

PROBLEM 2.6

2.7 Find i and v_{ab}.

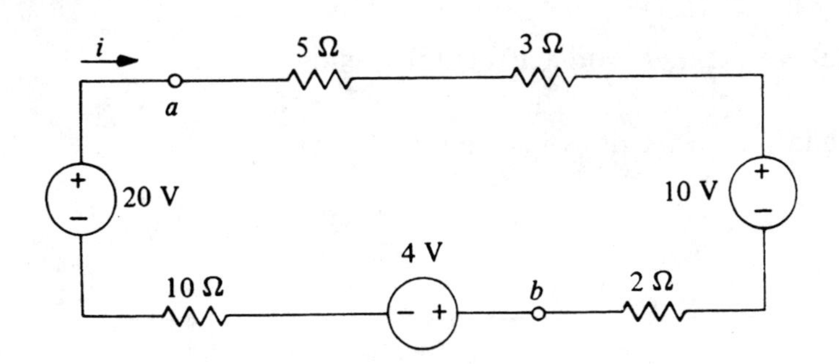

PROBLEM 2.7

2.8 Find i_1, i_2, and v_{ba}.

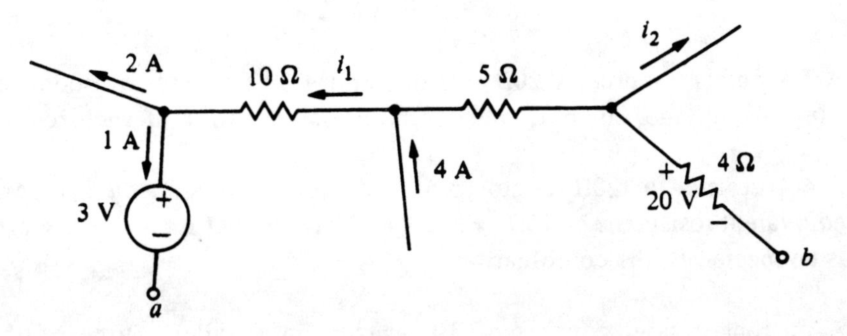

PROBLEM 2.8

2.3 Series Resistance and Voltage Division

2.9 A 10V source in series with several resistors carries a current of 50mA. If a 300Ω resistor is inserted in series in the circuit, find the resulting current.

2.10 A 50V source delivers 500mW to two resistors, R_1 and R_2, connected in series. If the voltage across R_1 is 10V, find R^1 and R_2.

2.11 Design a voltage divider to provide 2, 5, 15, and 35V, all with a common negative terminal, from a 40V source. The source is to deliver 80mW of power.

2.12 A 50V source and two resistors, R_1 and R_2, are connected in series. If $R_2 = 3R_1$, find the voltages across the two resistors.

2.13 A voltage divider is to be constructed with a 60V source and a number of 10kΩ resistors. Find the minimum number of resistors required if the output voltage is (a) 40V and (b) 30V.

2.4 Parallel Resistance and Current Division

2.14 Find i and the power delivered to the 3Ω resistor.

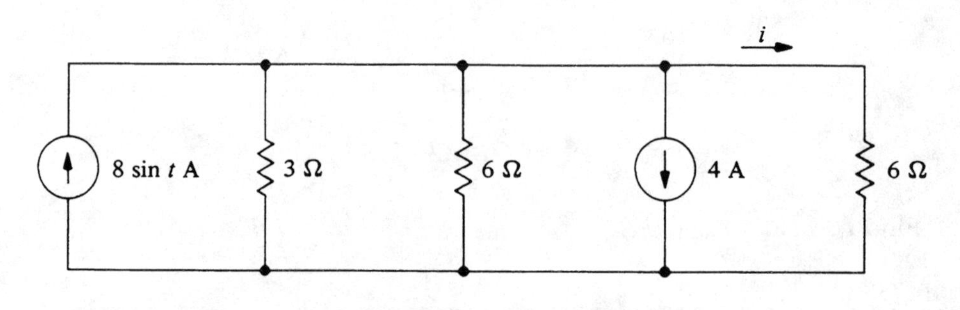

PROBLEM 2.14

2.15 A 5A current source, a 20Ω resistor, and a 30Ω resistor are connected in parallel. Find the voltage, current, and minimum power rating of each resistor.

2.16 A 40Ω resistor, a 120Ω resistor, and a resistor R are connected in parallel to form an equivalent resistance of 15Ω. Find R and the current it carries if a 8A current source is connected to the combination.

2.17 A current divider consists of 10 resistors in parallel. Nine of them have equal resistances of 30kΩ and the tenth is a 10kΩ resistor. Find the equivalent resistance of the divider, and if the total current entering the divider is 20mA, find the current in the tenth resistor.

2.18 A 40Ω resistor, a 60Ω resistor, and a resistor R are connected in parallel to form an equivalent resistance of 8Ω. Find R and the current it carries if a 6A current source is connected to the combination.

2.5 Analysis Examples

2.19 A voltage divider is to be constructed with a 50V source and a number of 10kΩ resistors. If the output voltage is to be 20V, find the minimum number of resistors that can be used. Draw the circuit.

2.20 Find all the possible values of equivalent resistance that can be obtained by someone having three 6Ω resistors.

2.21 Find R_{eq} and the resulting values of i and v if a 50V battery is connected to the open terminals.

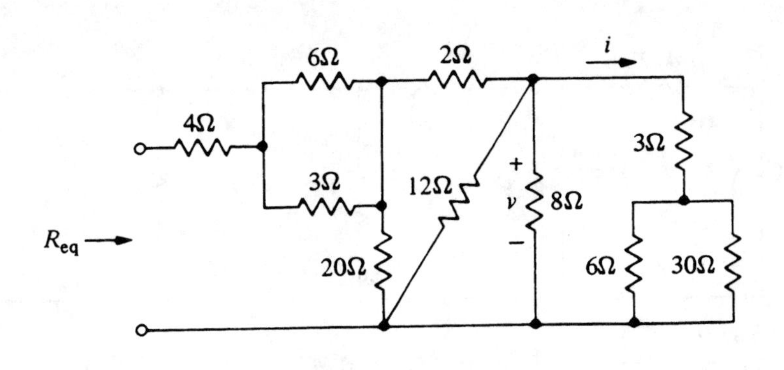

PROBLEM 2.21

2.22 If a current source of 10A is connected to the circuit of Prob. 2.21, find i.

2.23 Find i.

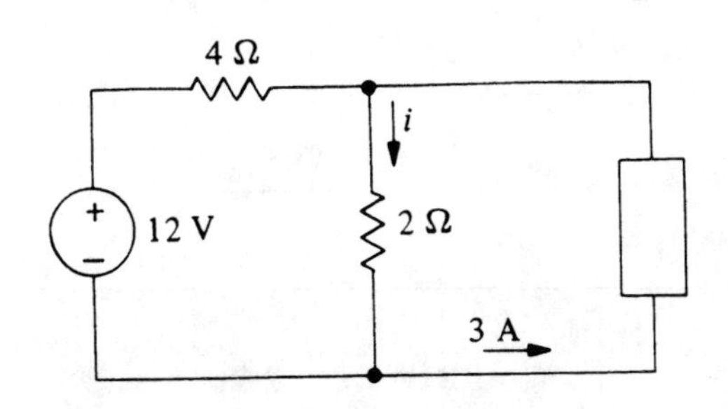

PROBLEM 2.23

2.24 (a) Find the equivalent resistance looking in terminals a-b if terminals c-d are open, and if terminals c-d are shorted together. (b) Find the equivalent resistance looking in terminals c-d if terminals a-b are open, and if a-b are shorted together.

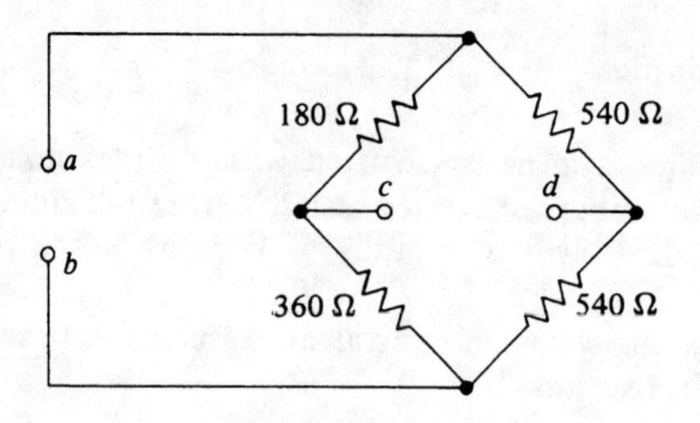

PROBLEM 2.24

2.25 Find v and i.

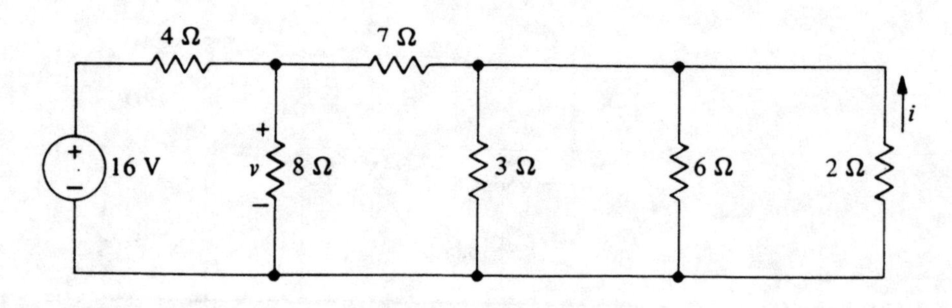

PROBLEM 2.25

2.26 Find i_1 and i_2.

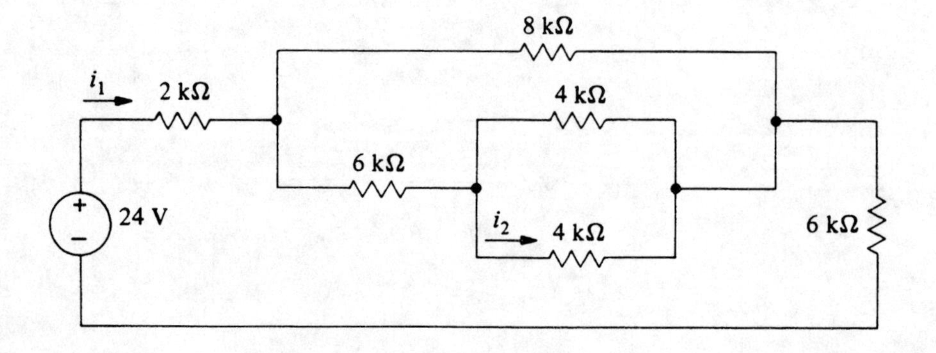

PROBLEM 2.26

2.27 Find i.

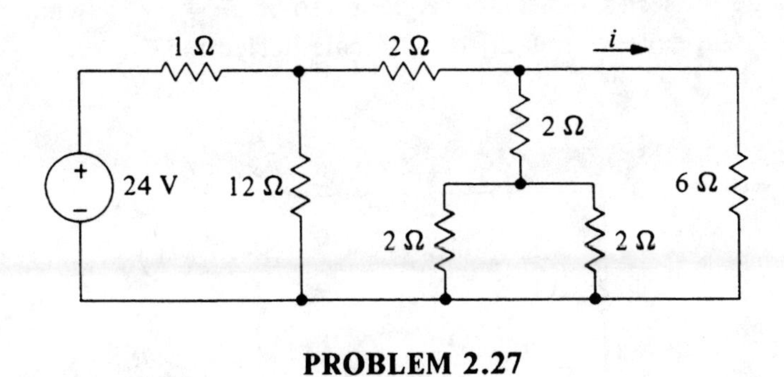

PROBLEM 2.27

2.28 A D'Arsonval meter has a full-scale current of 5mA and a resistance of 9.8Ω. If a 0.2Ω parallel resistor is used in the following figure, what is i_{FS}? What voltage occurs across the meter?

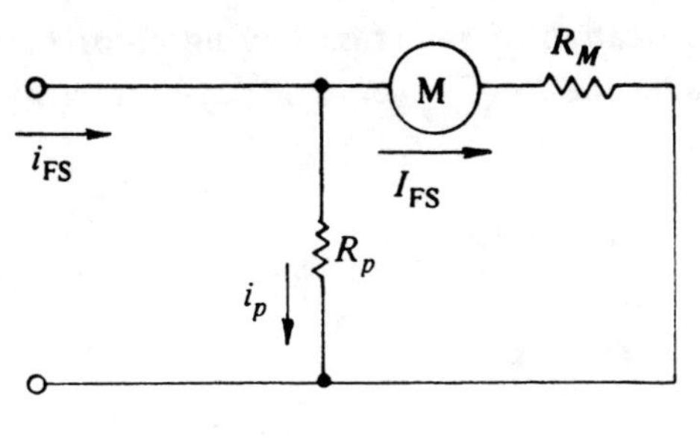

PROBLEM 2.28

2.29 A 36,000 Ω/V voltmeter has a full-scale voltage of 90V. What current flows in the meter when measuring 45V?

2.30 Two 3.3kΩ resistors are connected in series across a 50V source. What voltage will the voltmeter of Prob. 2.29 measure across one of the 3.3kΩ resistors.

2.31 The D'Arsonval meter of Prob. 2.28 is used for the ohmmeter of the following figure. What value of series resistance is required if E = 5V? What value of unknown resistance will cause a one-third full-scale deflection?

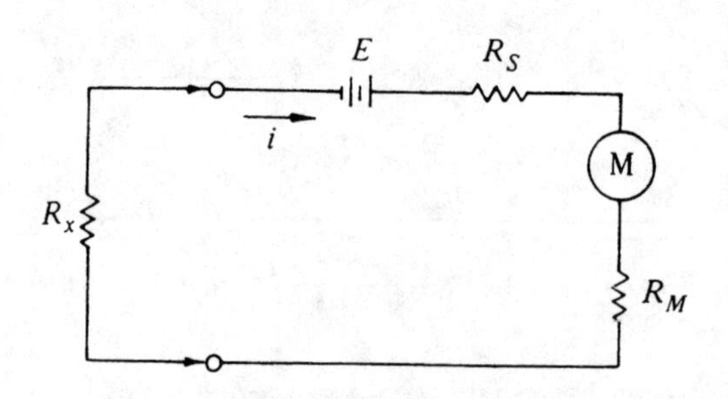

PROBLEM 2.31

2.32 Determine the color codes for resistors having the following resistance ranges: (a) 2.97-3.63MΩ (b) 1.14-1.26kΩ (c) 50.4-61.6Ω.

2.33 Find the resistance range of resistors having color bands of (a) brown, black, red, silver, (b) red, violet, yellow, silver, and (c) blue, gray, silver, gold.

2.1 $v = iR = (2\times10^3\Omega)(6\times10^{-3}A) = 12\,V$;

$i_{600\Omega} = \frac{v}{R} = \frac{12V}{600\Omega} = \underline{20mA}$

2.2 $R = \frac{v}{i} = \frac{12V}{60mA} = \underline{200\Omega}$

$P = vi = (12V)(60mA) = \underline{720mW}$

$i = \frac{v}{R_1+R_2} = \frac{12V}{200+400\Omega} = 20mA$;

$v_{400\Omega} = iR = (20mA)(400\Omega) = \underline{8V}$

2.3 $v = \frac{P}{i} = \frac{400W}{2A} = \underline{200V}$;

$R = \frac{v}{i} = \frac{200V}{2A} = \underline{100\Omega}$

2.4 $P = \frac{v^2}{R} = \frac{(100V)^2}{10\Omega} = 1000W$;

$\Delta W = \int_0^8 p\,dt = \int_0^8 1000\,dt = \underline{8.0\,kJ}$

2.5 By KCL,

$i_3 = 2A + 1A = 3A$;

$i_1 = 4A - i_3 = 4A - 3A = \underline{1A}$;

$i_2 = 5A - i_1 = 5A - 1A = \underline{4A}$;

By KVL, $v = -(5)(4) - (1)(5) + (3)(10) + (1)(3) = \underline{8V}$

2.6 By KCL; $i_{3\Omega} = 8A - 2A - 4A = 2A$ to the right

$i_{4\Omega} = 2A - 3A + 2A = 1A$ up

By KVL; $v = (4)(1) + (3)(2) + 12 = \underline{22V}$

2.7 By KVL;

$(5+3+2+10)i + 10V + 4V - 20V = 0$

$i = \frac{6V}{20\Omega} = \underline{0.3A}$

$v_{ab} = (5+3+2)i + 10V = \underline{13V}$

2.8 By KCL

$i_1 = 2+1 = \underline{3A}$

$i_3 = 4 - i_1 = 1A$

$i_4 = \frac{20V}{4\Omega} = 5A$

2.8 cont.

$i_2 = i_3 - i_4 = 1 - 5 = \underline{-4A}$

By KVL, $v_{ba} - 3 - 10i_1 + 5i_3 + 20 = 0$

$\therefore v_{ba} = 3 + 10(3) - 5(1) - 20 = \underline{8V}$

2.9 R_1 = resistance orginally $= \frac{10V}{50mA} = 200\Omega$

$i = \frac{v}{R_1 + 300\Omega} = \frac{10V}{500\Omega} = \underline{20\,mA}$

2.10 $R_1 + R_2 = \frac{v^2}{P} = \frac{(50V)^2}{500mW} = 5\,K\Omega$

By voltage division;

$10V = \frac{R_1}{R_1+R_2}(50V)$

$R_1 = \frac{10V}{50V}(R_1+R_2) = \frac{(10)(5K\Omega)}{50} = \underline{1K\Omega}$

$R_2 = 5K\Omega - 1K\Omega = \underline{4K\Omega}$

2.11

$i_s = \frac{P_s}{v_s} = \frac{80mW}{40V} = 2mA$

$v_1 = V = R_1 i_s$

$R_1 = \frac{2V}{2mA} = \underline{1K\Omega}$

$v_2 = 5 - 2V = R_2 i_s$

$R_2 = \frac{3V}{2mA} = \underline{1.5K\Omega}$

$v_3 = 15 - 5V = R_3 i_s$

$R_3 = \frac{10V}{2mA} = \underline{5K\Omega}$

$v_4 = 35 - 15V = R_4 i_s$; $R_4 = \frac{20V}{2mA} = \underline{10\,K\Omega}$

$v_5 = 40 - 35V = R_5 i_s$; $R_5 = \frac{5V}{2mA} = \underline{2.5K\Omega}$

2.12 v_i = voltage across R_i, $i = 1,2$

By voltage division,

$v_1 = \frac{R_1}{R_1+R_2}(50) = \frac{50R_1}{R_1+3R_1} = \frac{50}{4} = \underline{12.5V}$

$v_2 = 50 - v_1 = \underline{37.5V}$

2.13 R_1 and R_2 are the voltage divider resistances with output taken across R_2. By voltage division, $\frac{v_{out}}{60} = \frac{R_2}{R_1+R_2}$.

(a) $v_{out} = 40V$, $\therefore 40(R_1+R_2) = 60R_2$ or $4R_1 = 2R_2$. Thus $R_2 = 2R_1$ and <u>three</u> $10K\Omega$ resistors are required using two in series for R_2.

2.13 cont.

(b) $v_{out} = 30V$; $\therefore R_1 + R_2 = 2R_2$ or $R_1 = R_2$. Two $10K\Omega$ resistors are required.

2.14 Equivalent current source has current $8\sin t - 4A$, up. By current division,

$$i = \frac{(1/6)}{(1/3)+(1/6)+(1/6)}(8\sin t - 4) = \underline{2\sin t - 1\text{ A}}$$

$$i_{3\Omega} = \frac{1/3}{1/3 + 1/6 + 1/6}(8\sin t - 4) = 4\sin t - 2$$

$$P_{3\Omega} = (i_{3\Omega})^2(3) = \underline{12(2\sin t - 1)^2\text{ W}}$$

2.15 By current division

$$i_{20\Omega} = \frac{1/20}{1/20 + 1/30}(5A) = \underline{3A}$$

$$i_{30\Omega} = \frac{1/30}{1/20 + 1/30}(5A) = \underline{2A}$$

$$v_{20\Omega} = v_{30\Omega} = (20)(3) = \underline{60V}$$

$$P_{20\Omega} = (60V)(3A) = \underline{180W}$$

$$P_{30\Omega} = (60V)(2A) = \underline{120W}$$

2.16 $G_P = \frac{1}{15} = \frac{1}{40} + \frac{1}{120} + \frac{1}{R}$; $R = \underline{30\Omega}$

By current division,

$$i_R = \frac{1/R}{1/15}(8) = \underline{4A}$$

2.17 R_1 = equivalent resistance of 9 equal R's

$= 30/9 = 10/3\ K\Omega$

R_P = divider resistance

$$= \frac{10/3(10)}{10/3 + 10} = \underline{2.5K\Omega}$$

By current division,

$$i_{10} = \frac{10/3}{10/3 + 10}(20mA) = \underline{5mA},$$ current in the tenth resistor.

2.18 $G_P = \frac{1}{8} = \frac{1}{40} + \frac{1}{60} + \frac{1}{R}$; $R = \underline{12\Omega}$

By current division,

$$i_R = \frac{1/R}{1/8}(6) = \underline{4A}$$

2.19 R_1 and R_2 are the voltage divider resistances with the output taken across R_2. By voltage division, $\frac{R_2}{R_1 + R_2} = \frac{v_{out}}{50}$

$v_{out} = 20V$; $\therefore 50(R_2) = 20(R_1 + R_2)$

Thus $R_2 = 3/2 R_1$, Let $R_2 = 10K\Omega$

Then $R_1 = 15K\Omega = 10K\Omega + \frac{10K\Omega}{2}$

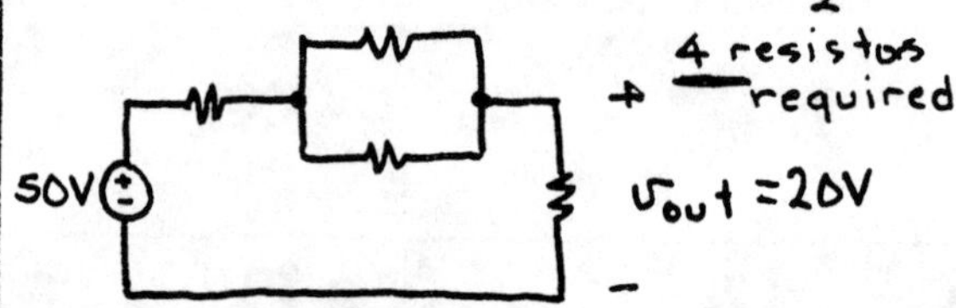

2.20 (1) $\underline{6\Omega}$, one resistor

(2) $\underline{12\Omega}$, (6+6) two resistors in series

(3) $\underline{18\Omega}$, (6+6+6), all three in series.

(4) $\underline{3\Omega}$, (6/2), two resistors in ||.

(5) $\underline{2\Omega}$, (6/3), all three in parallel.

(6) $\underline{9\Omega}$, (6 + 6/2), one in series, with the parallel connection of the other two.

(7) $\underline{4\Omega}$, $\left(\frac{(6+6)(6)}{6+6+6}\right)$, one resistor in parallel with the series connection of the other two.

2.21 Parallel 6Ω & 30Ω is equal to 5Ω
Parallel 12Ω & 8Ω is equal to 4.8Ω
Parallel 6Ω & 3Ω is equal to 2Ω
Equivalent circuit shown:

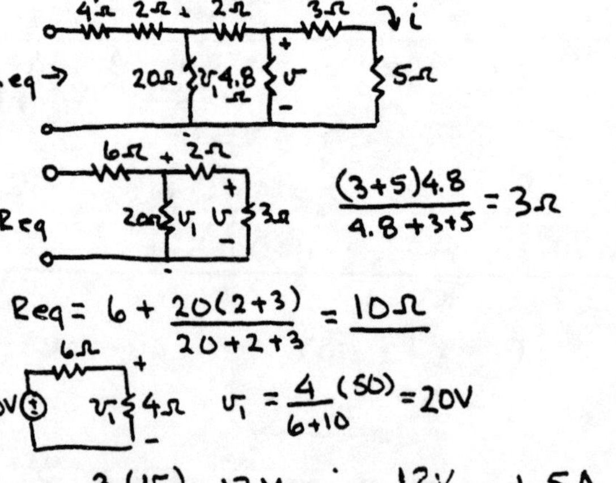

$$\frac{(3+5)4.8}{4.8+3+5} = 3\Omega$$

$$R_{eq} = 6 + \frac{20(2+3)}{20+2+3} = \underline{10\Omega}$$

$$v_1 = \frac{4}{6+10}(50) = 20V$$

$$v = \frac{3}{2+3}(v_1) = \underline{12V};\quad i = \frac{12V}{3+5\Omega} = \underline{1.5A}$$

2.22 $v_1 = (4\Omega)(10A) = 40V$

$v = \frac{3}{5}(40) = 24V$

using Equivalent circuits of Prob. 2.21

$$i = \frac{24V}{8\Omega} = \underline{3A}.$$

2.23 By KCL, the current leaving the positive source terminal is $i-3$. By KVL around the left loop.

$12V - 2i - 4(i-3) = 0$; $i = \underline{4A}$.

2.24 (a) c-d open:

$R_{ab} = \frac{(180+360)(540+540)}{180+360+540+540} = \underline{360\Omega}$

c-d shorted:

$R_{ab} = \frac{(180)(540)}{180+540} + \frac{(360)(540)}{360+540} = \underline{351\Omega}$

(b) a-b open:

$R_{cd} = \frac{(360+540)(180+540)}{360+540+180+540} = \underline{400\Omega}$

a-b shorted:

$R_{cd} = \frac{(180)(360)}{360+180} + \frac{540}{2} = \underline{390\Omega}$

2.25 Replace 3Ω, 6Ω, 2Ω with parallel equivalent:

$R_{eq} = \frac{(7+1)8}{1+7+8} = 4\Omega$

then by voltage division:

$v = \frac{4}{4+4}(16) = \underline{8V}$; $v_1 = \frac{1(8)}{7+1} = 1V$

$i = \frac{-v_1}{2\Omega} = \frac{-1V}{2\Omega} = \underline{-0.5A}$

2.26 R_{eq} = resistance seen by source $= 2+4+6 = 12K\Omega$

$i_1 = \frac{24V}{12K\Omega} = \underline{2mA}$; By current division

$i_3 = i_1/2 = 1mA$, $i_2 = i_3/2 = \underline{0.5mA}$

2.27 combining the two resistors to form equivalent circuit:

$R_{eq} = \frac{(12)\left(2+\frac{(3)(6)}{3+6}\right)}{12+4} = 3\Omega$, By voltage division, $v_2 = \frac{3}{3+1}(24V) = 18V$

$v_1 = \frac{2}{2+2}(v_2) = 9V$, $i = \frac{v_1}{6\Omega} = \underline{1.5A}$

2.28 $i_{FS} = \frac{(R_m+R_p)I_{FS}}{R_p} = \frac{(9.8+0.2)5}{0.2} = \underline{250mA}$

2.29 $i = \frac{v}{v_{FS}(36{,}000\ \Omega/V)} = \frac{45}{(90)(36{,}000)} = 13.9\mu A$

2.30 $R_{eq} = \frac{(3.3K\Omega)(90)(36000)}{3.3K\Omega + 3.24M\Omega} = 3{,}297\Omega$

$i = \frac{50V}{3.3K\Omega + 3297\Omega} = 7.58\ mA$

$v = R_{eq}\,i = (3297)(7.58mA) = \underline{24.99V}$

2.31 $I_{FS} = 5mA$, $R_M = 9.8\Omega$

$(5mA)(9.8+R_s) = 5V$; $R_s = \underline{990.2\Omega}$

$R_x = \frac{5V}{(5mA)/3} - (9.8+990.2) = \underline{2K\Omega}$

2.32 (a) Tolerance $= \frac{3.63-2.97}{2} = 0.33M\Omega$

Nominal resistance $= 2.97+0.33 = 3.3M\Omega$

% tolerance $= \frac{0.33}{3.3} \times 100 = 10\%$

$(10a+b)10^c = 3.3M\Omega$; $c = 5$

$a = b = 3$; Color code is

orange-orange-green-silver

(b) Tolerance $= \frac{1.26-1.14}{2} = 0.06K\Omega$

Nominal resistance $= 1.14+0.06 = 1.2K\Omega$

% tolerance $= \frac{.06}{1.2} \times 100 = 5\%$

$(10a+b)10^c = 1.2K\Omega$; $c = 2$, $a = 1$, $b = 2$ color code is

Brown-Red-Red-Gold.

(c) Tolerance $= \frac{1}{2}(61.6-50.4) = 5.6\Omega$

Nominal Resistance $= 50.4+5.6 = 56\Omega$

% tolerance $= \frac{5.6}{56} \times 100 = 10\%$

$(10a+b)10^c = 56\Omega$; $c = 0$, $a = 5$, $b = 6$: color code is

Green-Blue-Black-silver

2.33 (a) $R = (1\times10+0)\times10^2 \pm 10\%$

$= 1000 \pm 100$

resistance range $= \underline{900-1100\Omega}$

(b) $R = (2\times10+7)\times10^4 \pm 10\%$

$= 270K\Omega \pm 27K\Omega$

resistance range $= \underline{243-297K\Omega}$

(c) $R = (6\times10+8)\times10^{-2} \pm 5\%$

$= 680 \pm 34\ m\Omega$

resistance range $= \underline{646-714m\Omega}$

Chapter 3

Dependent Sources

3.2 Circuits With Dependent Sources

3.1 Find v if $R = 3\Omega$.

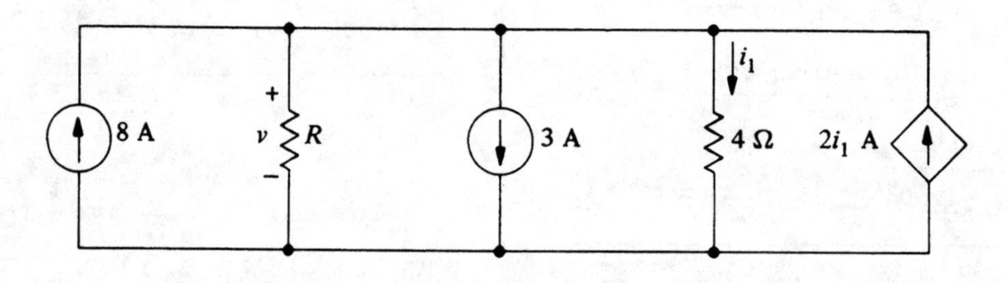

PROBLEM 3.1

3.2 Find v of Prob. 3.1 if $R = 5\Omega$.

3.3 Find the power delivered to the 6Ω resistor.

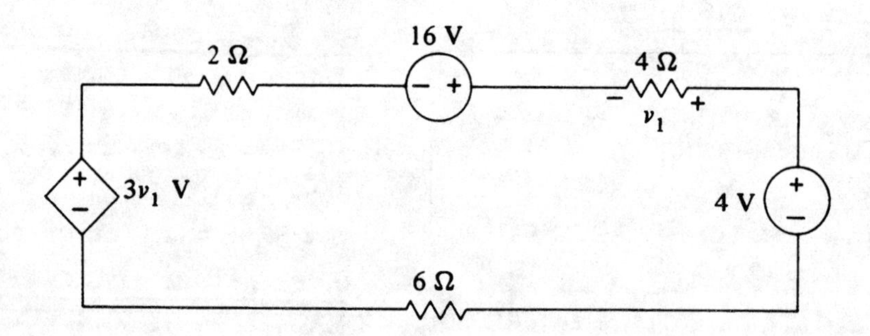

PROBLEM 3.3

3.4 Find i_1 if $R_1 = 1.5\Omega$, $R_2 = 2\Omega$, and $i_g = 3\text{A}$.

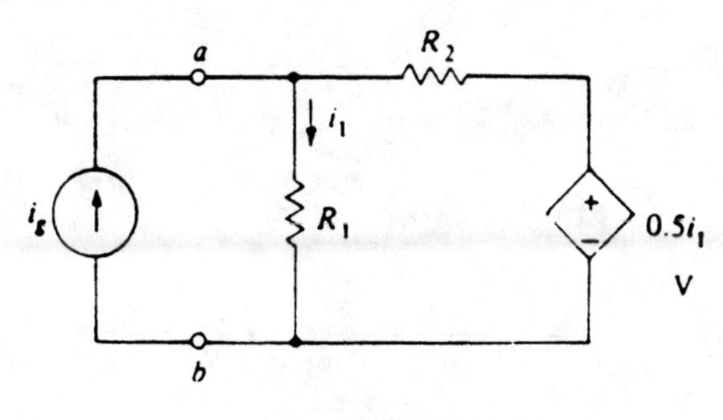

PROBLEM 3.4

3.5 In Prob. 3.4, find the resistance seen by the source looking in terminals a-b.

3.6 In Prob. 3.4, let $R_1 = R_2 = 1/10\ \Omega$. Find the resistance seen by the source.

3.7 Find i if $R = 15\Omega$.

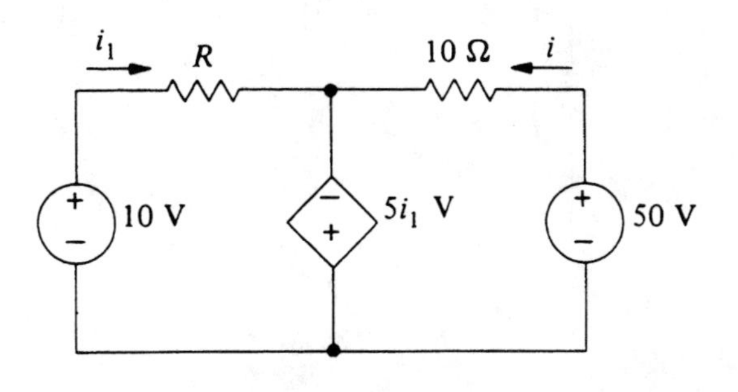

PROBLEM 3.7

3.8 Find R in Prob. 3.7 so that $i = 10\text{A}$.

3.9 Find i and v_1.

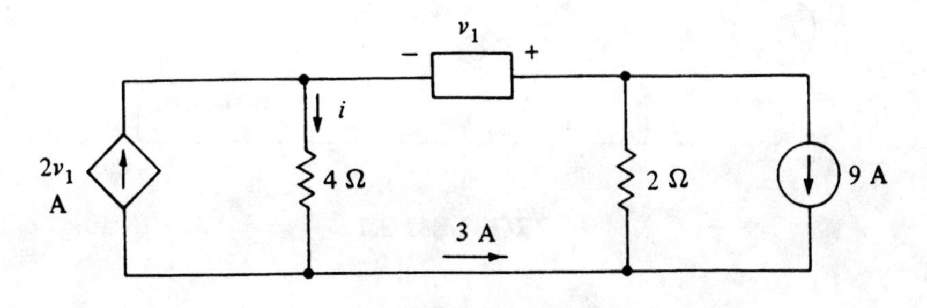

PROBLEM 3.9

3.10 Find i.

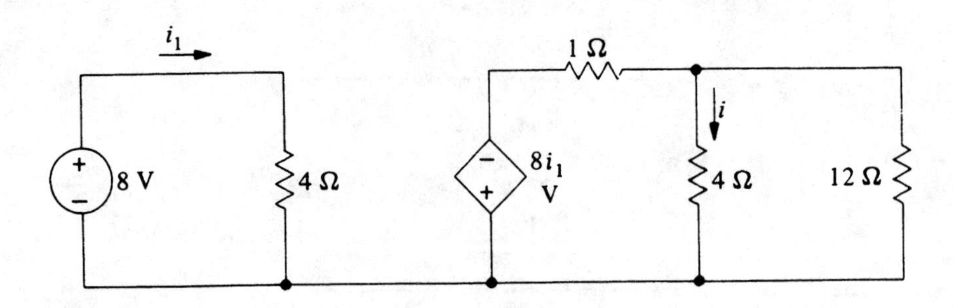

PROBLEM 3.10

3.11 Find (a) the equivalent resistance, in terms of R_1, R_2, and α, seen by the source v_g, and (b) the current i if R_1 = 4kΩ, R_2 = 2kΩ, and v_g = 20V.

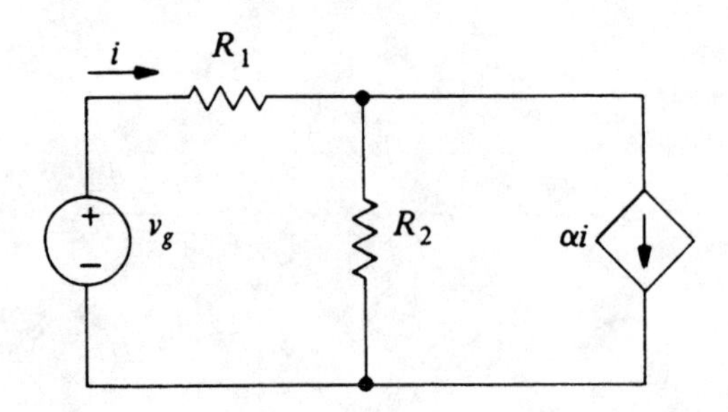

PROBLEM 3.11

3.12 Find v.

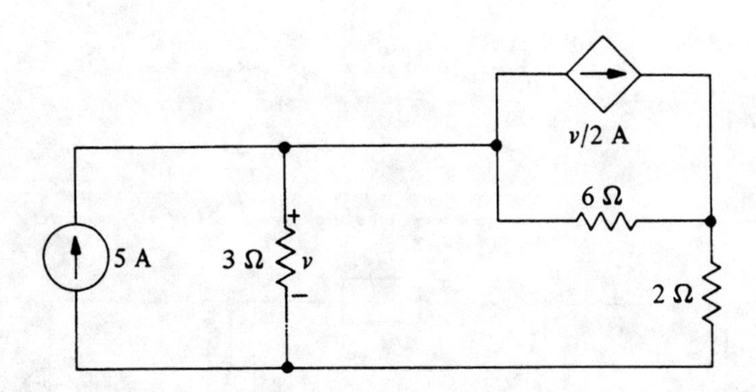

PROBLEM 3.12

3.13 Find v_1 and the power delivered to the 8Ω resistor.

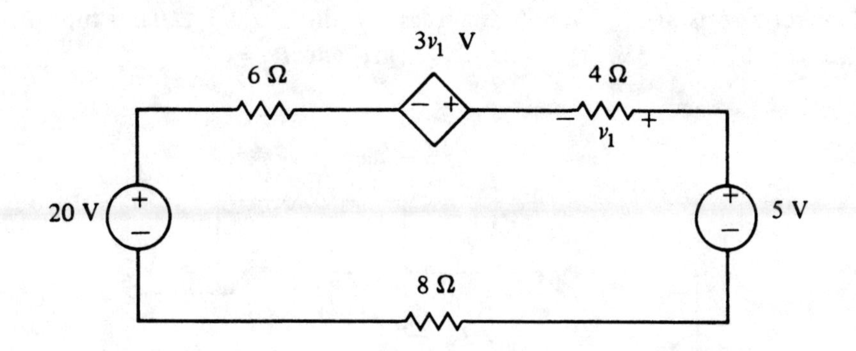

PROBLEM 3.13

3.14 Find v.

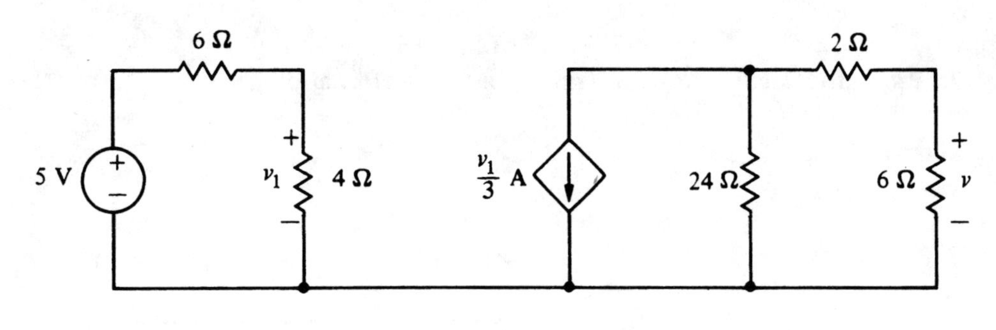

PROBLEM 3.14

3.4 Amplifier Circuits

3.15 Connect a resistance of $R = 5k\Omega$ across terminals a-b and find the resulting current i_{ab} if $v_1 = 1V$, $v_2 = 3V$, $R_0 = 4k\Omega$, $R_1 = 1k\Omega$, and $R_2 = 2k\Omega$.

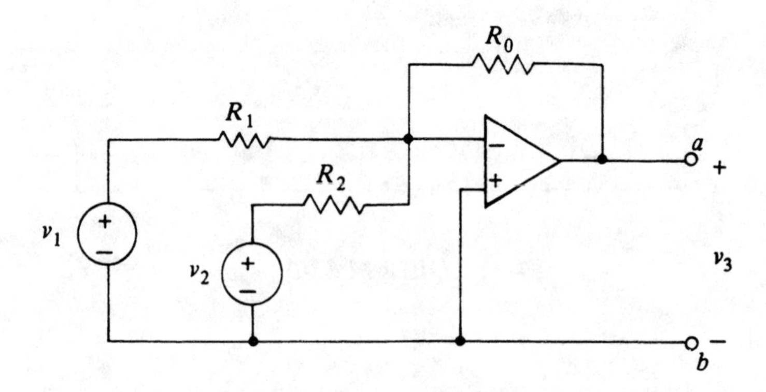

PROBLEM 3.15

3.16 In Prob. 3.15 let $R_1 = R_2 = R_0$ find v_3 in terms of v_1 and v_2.

3.17 Find v.

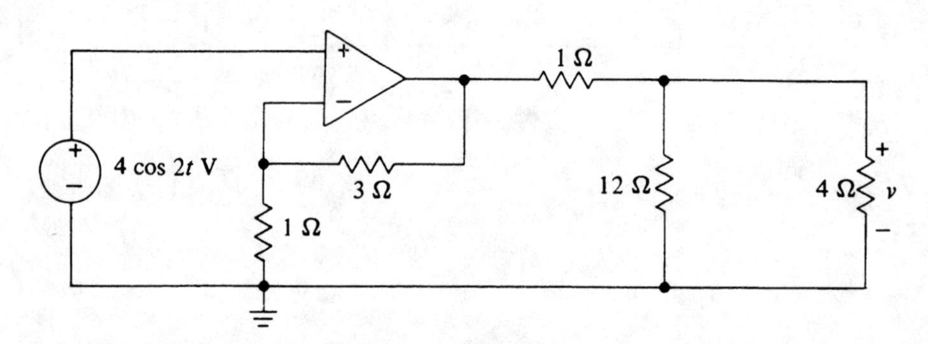

PROBLEM 3.17

3.18 Find i.

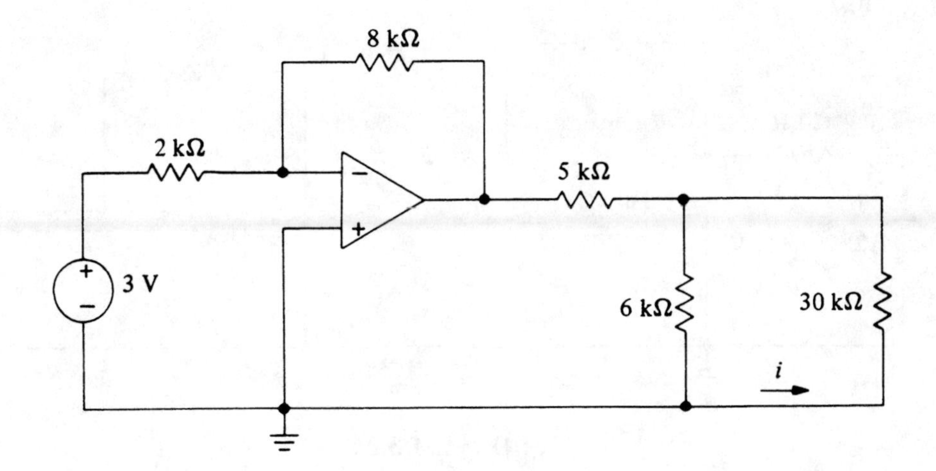

PROBLEM 3.18

3.19 Find i.

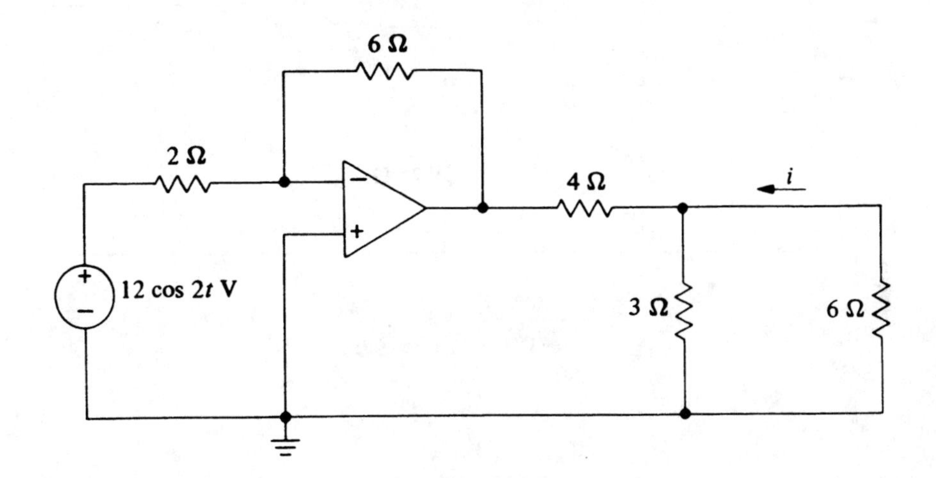

PROBLEM 3.19

3.20 Find v.

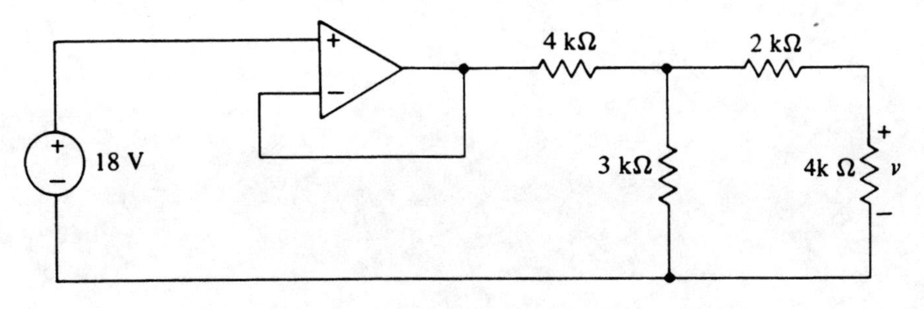

PROBLEM 3.20

3.21 Find i.

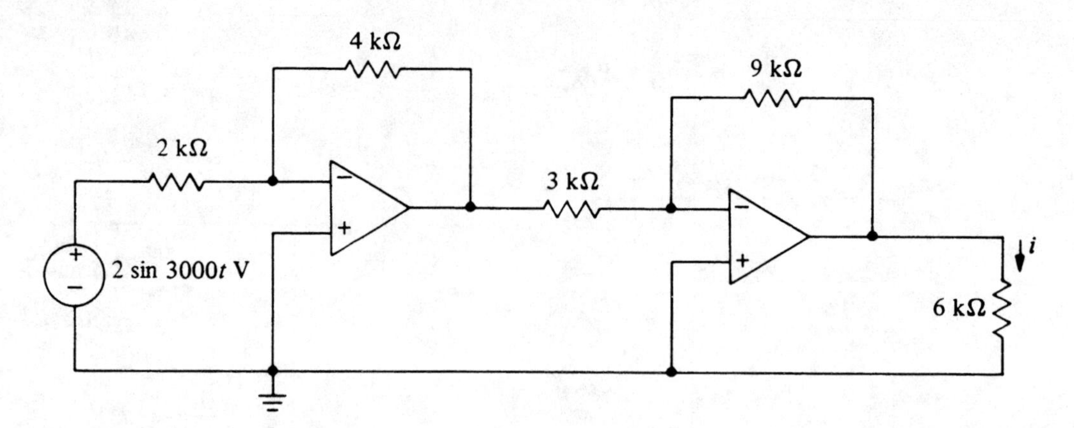

PROBLEM 3.21

3.22 Find v.

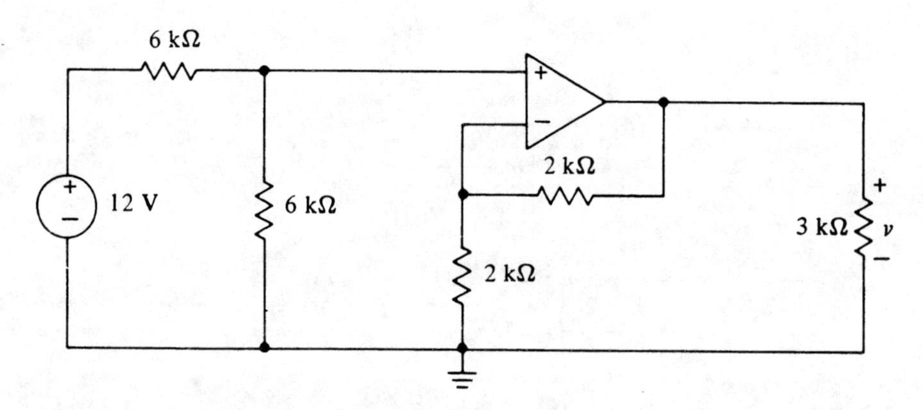

PROBLEM 3.22

3.23 Find v.

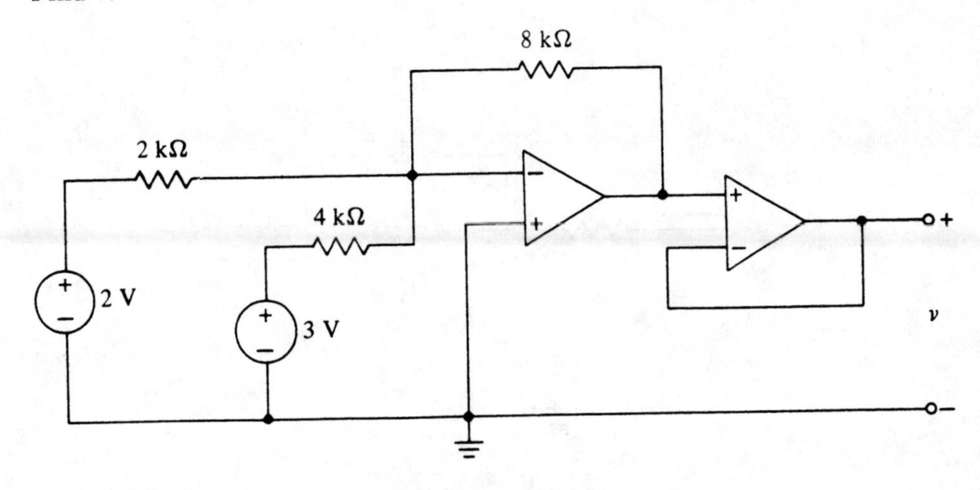

PROBLEM 3.23

3.24 If $v_g = 9V$, find v_1, v_2, and i.

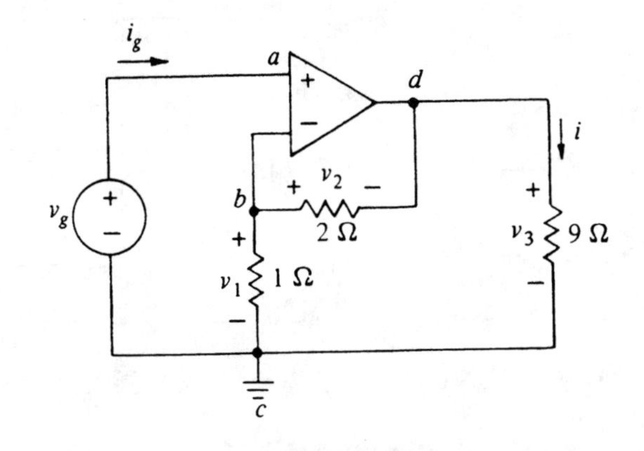

PROBLEM 3.24

3.25 If $v_1 = 2V$, $v_2 = -8V$, and $R_1 = 4k\Omega$, find R_2 and i_1.

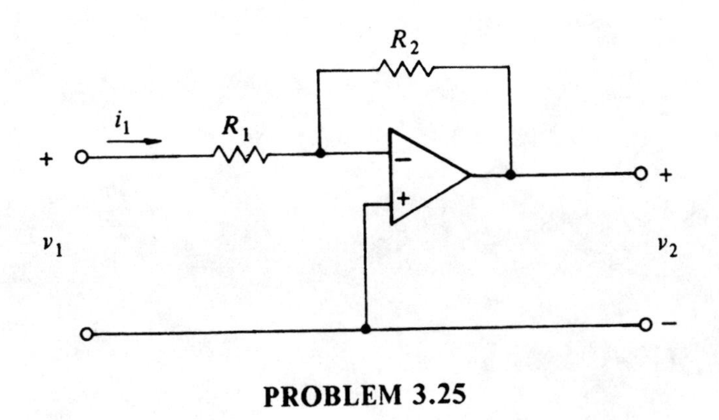

PROBLEM 3.25

3.26 If terminals 3-4 are left open, find i_1 and v_{43}, if $R_1 = 2\text{k}\Omega$, $R_2 = 4\text{k}\Omega$, and $v_1 = 6\text{V}$.

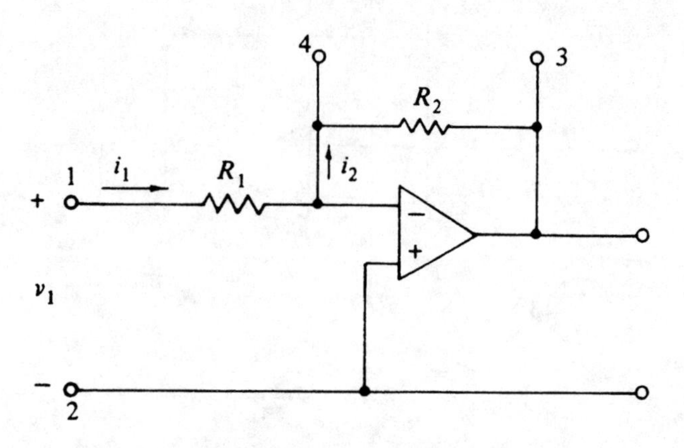

PROBLEM 3.26

3.27 Find v_1 in (a) and v_2 in (b).

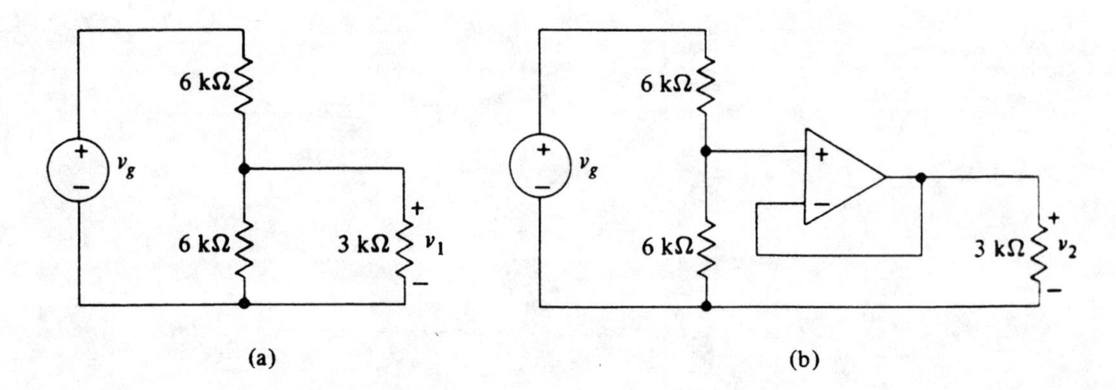

PROBLEM 3.27

3.28 Find v_0.

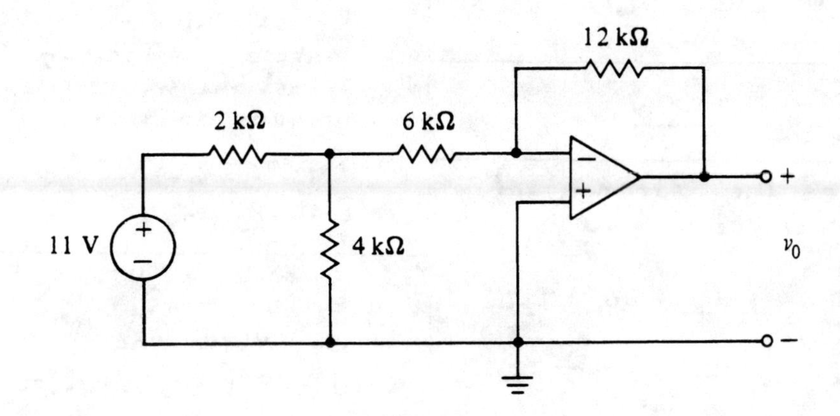

PROBLEM 3.28

3.1 By KCL, $8 - v/R - 3 - i_1 + 2i_1 = 0$,

$i_1 = \frac{v}{4}$, $v = \frac{5}{1/R - 1/4} = \underline{60V}$.

3.2 $v = \frac{5}{1/5 - 1/4} = \underline{-100V}$.

3.3 i leaves the + terminal of the 16-V source :- $i = -\frac{v_1}{4}$,

By KVL,

$16 + v_1 - 4 + 6\frac{v_1}{4} + 3v_1 + 2\frac{v_1}{4} = 0$

$v_1 = \frac{12}{-3/2 - 1/2 - 4} = -2V$, $i = -\frac{2}{4} = -\frac{1}{2}$

$P_{6\Omega} = 6i^2 = \underline{1.5W}$

3.4 By KCL, $i_g - i_1$ is the current to the right in R_2; By KVL around the right loop,

$R_2(i_g - i_1) + 0.5i_1 - R_1 i_1 = 0$

$i_1 = \frac{(2)(3)}{2 + 1.5 - 0.5} = \underline{2A}$

3.5 $v_{a-b} = i_1 R_1 = (2)(1.5) = 3V$

$R_{a-b} = \frac{v_{a-b}}{i_g} = \frac{3V}{3A} = \underline{1\Omega}$

3.6 $i_1 = \frac{(\frac{1}{10})3}{\frac{1}{10} + \frac{1}{10} - 0.5} = -1A$, $v_{a-b} = (-1)(\frac{1}{10}) = -\frac{1}{10}V$

$R_{a-b} = \frac{-1/10 V}{3A} = \underline{-1/30\ \Omega}$.

3.7 By KVL around the left loop,

$10V - Ri_1 + 5i_1 = 0$, $i_1 = 1A$;

By KVL around the right loop,

$50 - 10i + 5i_1 = 0$, $i = \frac{55}{10} = \underline{5.5A}$

3.8 using KVL equation of Prob. 3.7

$i_1 = \frac{50 - 10(10)}{-5} = 10A$;

$R = \frac{10 + 5(10)}{10} = \underline{6\Omega}$.

3.9 By KCL the current going up through the $2\text{-}\Omega$ resistor is 12A.

By KCL, $i = 3 + 2v_1$,

By KVL around the center loop

$v_1 + (12)2 + 4(3 + 2v_1) = 0$

$v_1 = \frac{-36}{1+8} = \underline{-4V}$

$i = 3 - 4(2) = \underline{-5A}$

3.10 By KVL around the left loop

$8 = i_1 4$, $i_1 = 2A$.

By combining the 4Ω & 12Ω to make a 3Ω resistor. The voltage across the 4Ω resistor is, by voltage division.

$v_{4\Omega} = \frac{3}{1+3}(-8i_1) = -12V$

$i = \frac{v_{4\Omega}}{4\Omega} = \underline{-3A}$

3.11 (a) By KCL the current down through R_2 is $i - \alpha i$,

By KVL around the right loop.

$v_g - R_1 i - R_2(i - \alpha i) = 0$

$v_g = i[R_1 + R_2(1-\alpha)]$

R_g = resistance seen by v_g

$= \frac{v_g}{i} = \underline{R_1 + R_2(1-\alpha)\ \Omega}$

(b) using a current divider,

$i = \frac{R_2}{R_2 + R_1 + R_g}(i\alpha)$, the solving for α

$\therefore \alpha = 2$, $i = \frac{v_g}{R_g} = \frac{20\ V}{4 + 2(1-2)k\Omega} = \underline{10mA}$

3.12 Current downward in $3\text{-}\Omega$ resistor is $v/3$. By KCL the current down in $2\text{-}\Omega$ resistor is $5 - v/3$. Current to the right in 6Ω resistor is, by KCL,

$5 - v/3 - v/2 = 5 - \frac{5v}{6}$. By KVL around the loop containing only resistors.

$v - 6(5 - 5v/6) - 2(5 - v/3) = 0$

$\therefore v = \underline{6V}$.

3.13 i leaves + terminal of 20-V source. $\therefore i = -v_1/4$, By KVL

$20 - 6(-\frac{v_1}{4}) + 3v_1 + v_1 - 5 - 8(-\frac{v_1}{4}) = 0$

$\therefore v_1 = \underline{-2V}$, $i = -\frac{(-2)}{4} = \frac{1}{2}A$

$P_{8\Omega} = 8i^2 = \underline{2W}$.

3.14 By voltage division,

$v_1 = \frac{4}{6+4}(15) = 2V$. Let i be the current entering + terminal of v.

Then by current division

$i = \frac{24}{24 + 6 + 2}(-\frac{v_1}{3}) = -\frac{1}{2}A$

$\therefore v = 6(-\frac{1}{2}) = \underline{-3V}$.

3.15 $v_3 = -R_0\left(\frac{v_1}{R_1} + \frac{v_2}{R_2}\right) = 4\left(\frac{1}{1} + \frac{3}{2}\right)$
$= -10\text{V}$

$i_{ab} = \frac{v_3}{R} = \frac{-10\text{V}}{5\text{k}\Omega} = \underline{-2\text{mA}}$

3.16 From equation in Prob. 3.15
$v_3 = \underline{-v_1 - v_2}$

3.17 Let v_2 be VCVS output. Then, since $\frac{R_2}{R_1} = \frac{3}{1}$,
$v_2 = \mu(4\cos 2t) = (1 + \frac{R_2}{R_1})(4\cos 2t)$
$= 16\cos 2t\ \text{V}$
By voltage division,
$v = \frac{4(12)/4+12}{3+1} v_2 = \underline{12\cos 2t\ \text{V}}$

3.18 Let v_2 be the inverter output.
Then $v_2 = -\frac{R_2}{R_1}(3) = \frac{8}{2}(3) = 12\text{V}$
For i in mA, the voltage across the 30-kΩ resistor is $-30i$, + at top, and by voltage division
$-30i = \frac{30(6)/30+6}{5+5} v_2 = \frac{12}{2} = 6\text{V}$
$\therefore i = \underline{-0.2\ \text{mA}}$

3.19 Let v_2 be the inverter output
The $v_2 = -\frac{R_2}{R_1}(12\cos 2t) = \frac{6}{2}(12\cos 2t)$
$= 36\cos 2t\text{V}$
For i, the voltage across the 6Ω resistor is $-6i$, + at top, and by voltage division,
$-6i = \frac{(3)(6)/3+6}{2+4} = \frac{1}{3}(36\cos 2t)$
$\therefore i = \underline{-2\cos 2t\ \text{A}}$.

3.20 Output of voltage follower is 18V. By voltage division
$v_{3k\Omega} = \frac{(3)(2+4)/3+2+4}{4+2}(18) = 6\text{V}$
$v = \frac{4}{2+4} v_{3k\Omega} = \underline{4\text{V}}$.

3.21 v_1 = output voltage of first op amp.
Then $v_1 = -\frac{4}{2}(2\sin 3000t)$
$= -4\sin 3000t\ \text{V}$
v_2 = output voltage of second op amp.
$= -\frac{9}{3}v_1 = 12\sin 3000t\ \text{V}$
$\therefore i = \frac{v_2}{6\text{k}\Omega} = \frac{12\sin 3000t}{6\text{k}\Omega} = \underline{2\sin 3000t\ \text{mA}}$

3.22 current in op amp terminals is zero. By voltage division, the VCVS input voltage is
$v_1 = \frac{6}{6+6}(12) = 6\text{V}$. There fore
$v = (1 + \frac{2}{2})v_1 = \underline{12\text{V}}$

3.23 v = output of the first op amp
$= -8\left(\frac{2}{2} + \frac{3}{4}\right) = \underline{-14\text{V}}$

3.24 Since $V_a = v_b = v_g = v_1 = \underline{9\text{V}}$
$v_3 = (1+\frac{2}{1})v_1 = 27\text{V}$, By KVL,
$V_2 = V_1 - v_3 = \underline{-18\text{V}}$,
$i = \frac{v_3}{9} = \underline{3\text{A}}$

3.25 $v_2 = -\frac{R_2}{R_1}v_1$ or $R_2 = -\frac{v_2 R_1}{v_1} = \frac{8(4\text{k}\Omega)}{2}$
$R_2 = \underline{16\text{k}\Omega}$. $i_1 = \frac{v_1}{R_1} = \frac{2\text{V}}{4\text{k}\Omega} = \underline{0.5\text{mA}}$

3.26 $i_1 = \frac{v_1}{R_1} = \frac{6\text{V}}{2\text{k}\Omega} = \underline{3\text{mA}}$; $i_1 = i_2$,
$v_{43} = i_2 R_2 = (3\text{mA})(4\text{k}\Omega) = \underline{12\text{V}}$.

3.27.(a) By voltage division
$v_1 = \frac{\frac{(6)(3)}{6+3}(v_g)}{\frac{(6)(3)}{6+3} + 6} = \underline{v_g/4}$.

(b) The current into the noninverting terminal is zero. Therefore the two 6-kΩ resistors carry the same current and constitute a voltage divider. Input Voltage for voltage follower is $v_g/2$
$\therefore v_2 = \underline{v_g/2}$.

3.28 Since the op amp input voltage is zero, the 6-kΩ and 4kΩ resistors have same voltages (v_1). By voltage division,
$v_1 = \frac{6(4)/6+4}{2 + 6(4)/6+4}(11) = 6\text{V}$

Since the op amp input current are zero, the 6-kΩ and 12-k have the same current (i_1) to the right. $i_1 = \frac{6\text{V}}{6\text{k}\Omega} = 1\text{mA}$.
KVL around the loop containing v_0
$v_0 + 12(1) + 0 = 0$ or $v_0 = \underline{-12\text{V}}$.

Chapter 4

Analysis Methods

4.1 Nodal Analysis

4.1 Using nodal analysis, find v.

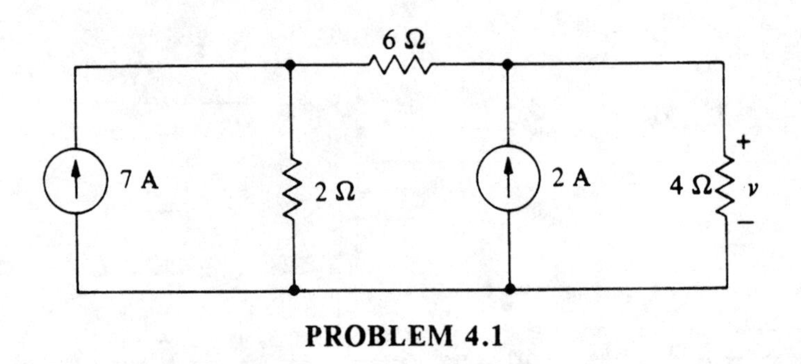

PROBLEM 4.1

4.2 Using nodal analysis, find i.

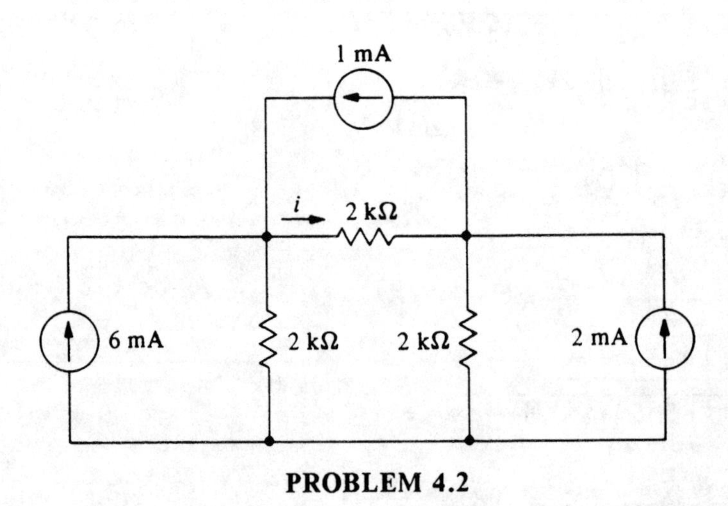

PROBLEM 4.2

4.3 Find the power delivered to the 8Ω using nodal analysis.

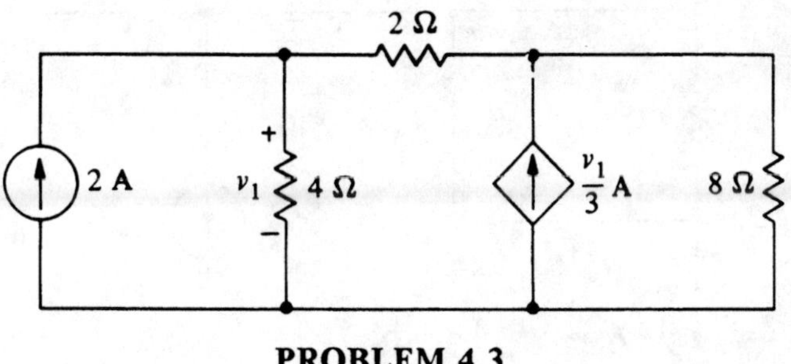

PROBLEM 4.3

4.4 Using nodal analysis, find v_1 and v_2, if R_1 = 2Ω, R_2 = 1Ω, R_3 = 4Ω, i_{g1} = 14A, and i_{g2} = 7A.

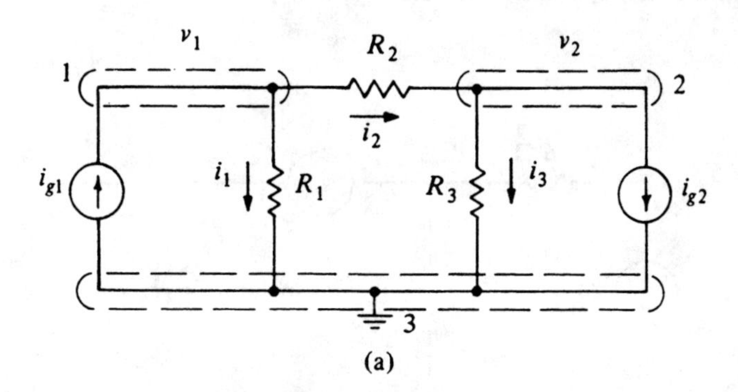

(a)

PROBLEM 4.4

4.5 Using nodal analysis, find v_1, v_2, and i.

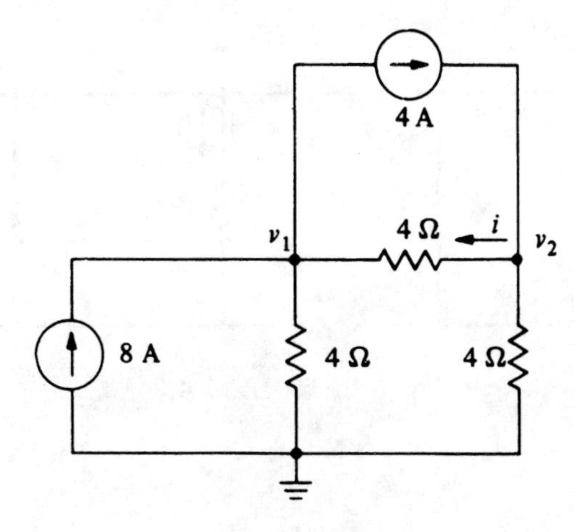

PROBLEM 4.5

4.6 Using nodal analysis, find v_1, v_2, and v_3.

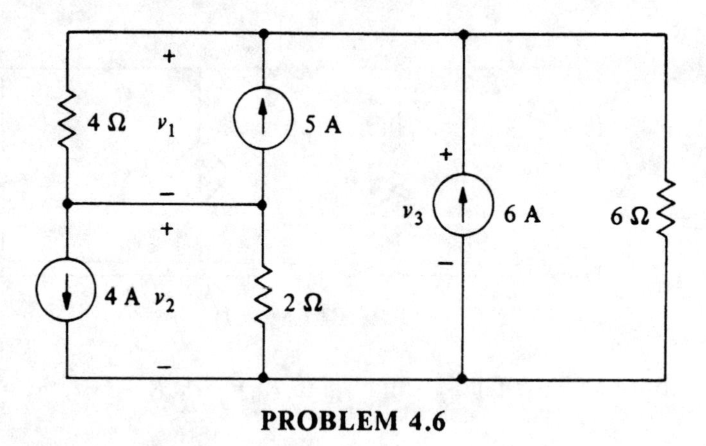

PROBLEM 4.6

4.7 Find v and i using nodal analysis.

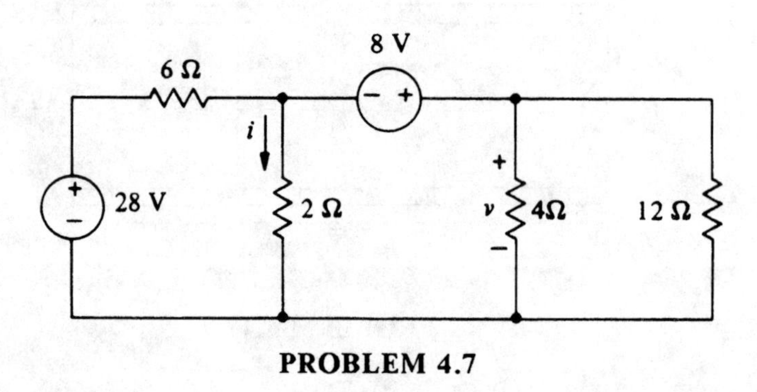

PROBLEM 4.7

4.8 Using nodal analysis, find i_1.

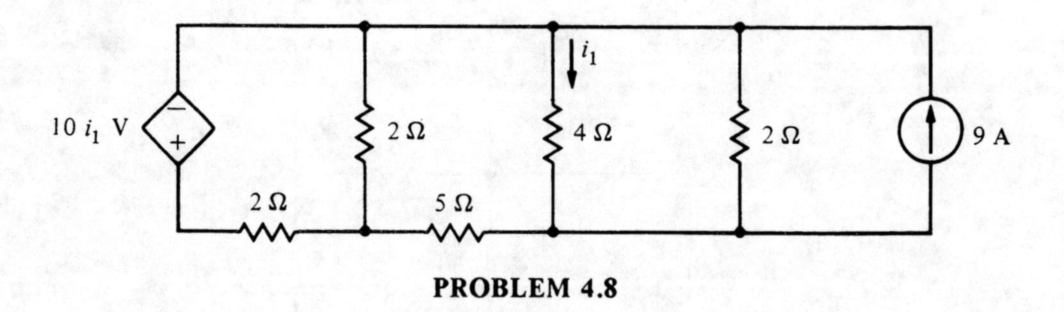

PROBLEM 4.8

4.3 Circuits Containing Voltage Sources

4.9 Find v and i using nodal analysis.

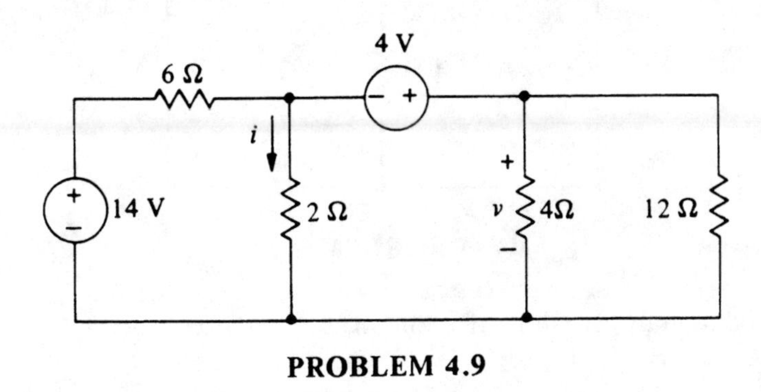

PROBLEM 4.9

4.10 Find v using nodal analysis.

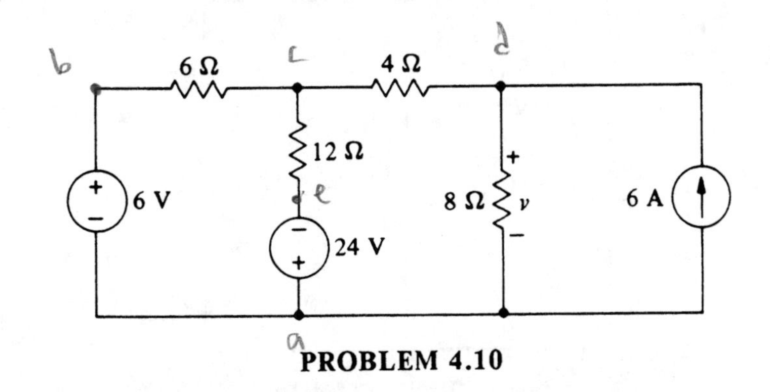

PROBLEM 4.10

4.11 Using nodal analysis, find v.

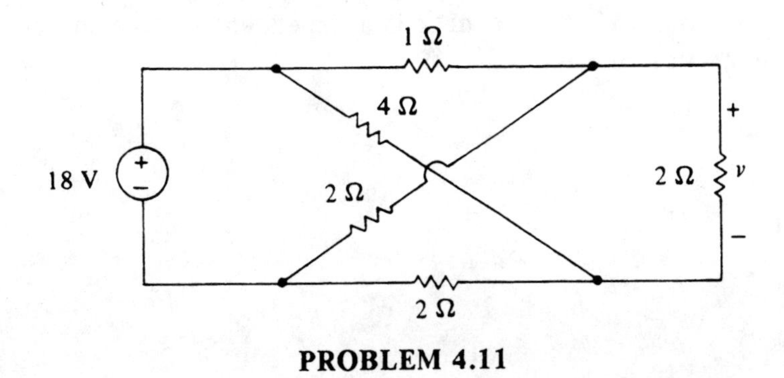

PROBLEM 4.11

4.12 Using nodal analysis, find i_1.

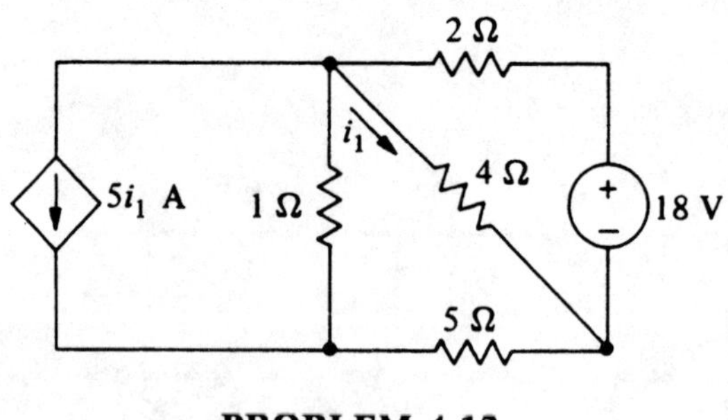

PROBLEM 4.12

4.13 Using nodal analysis, find i if element x is a 4Ω resistor.

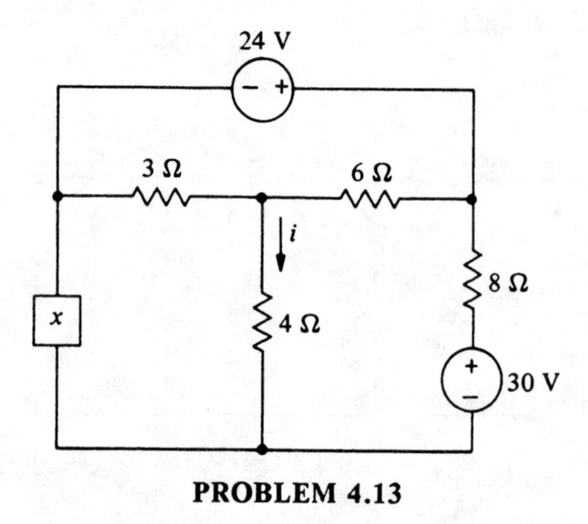

PROBLEM 4.13

4.14 Find i in Prob. 4.13 if element x is a 7A independent current source directed upward.

4.15 Find i in Prob. 4.13 if element x is a dependent voltage source of $5i$ V with the positive terminal at the top.

4.16 Using nodal analysis, find v_2 if $G_1 = 2S$, $G_2 = 0.5S$, $G_3 = 0.25S$, $G_4 = 1S$, $G_5 = 1S$, $\beta = 5$, and $v_g = 4V$.

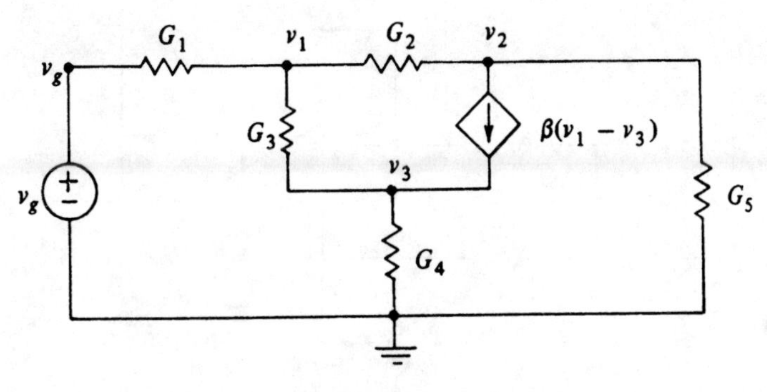

PROBLEM 4.16

4.17 Using nodal analysis, find v_1 for Prob. 4.16.

4.4 Circuits Containing Op Amps

4.18 Find i.

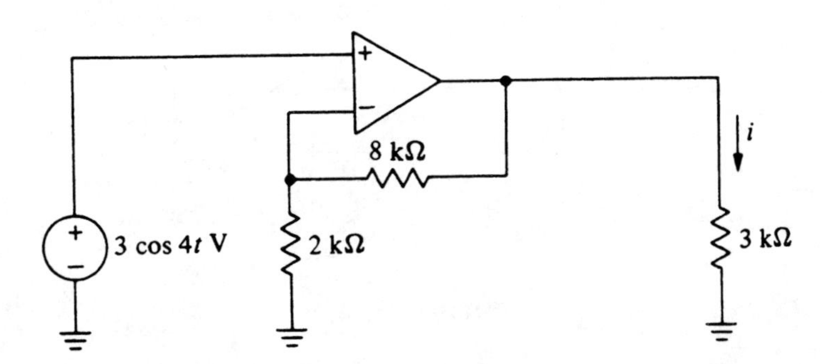

PROBLEM 4.18

4.19 Find v if $v_g = 6\cos(2t)$ V.

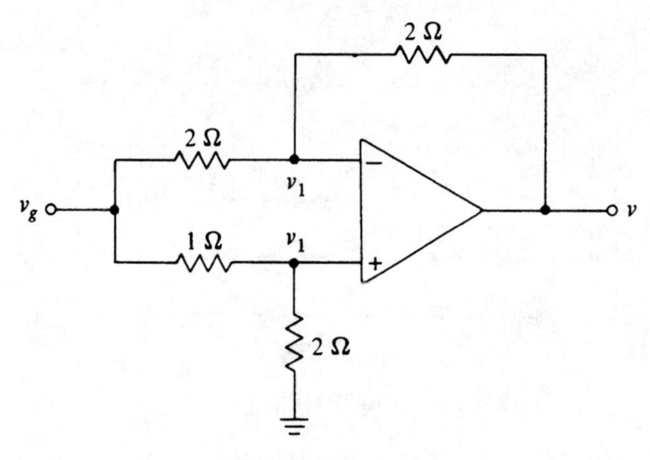

PROBLEM 4.19

4.20 Find v_2 for $G_1 = 0.25$S, $G_2 = 0.5$S, $G_3 = 1$S, $G_4 = 2$S, $G_5 = 0.5$S, $v_1 = 8\cos(1000t)$ V.

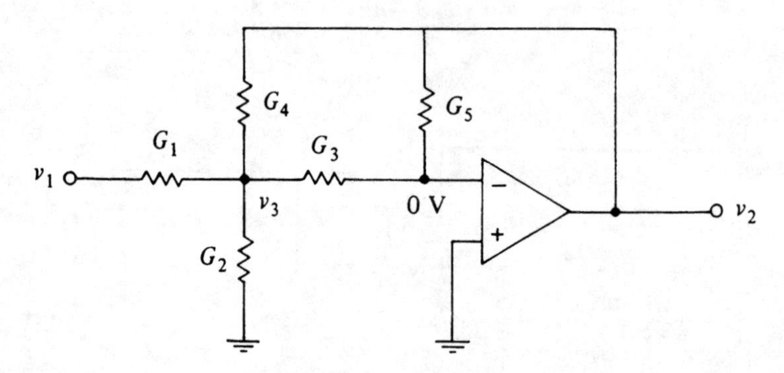

PROBLEM 4.20

4.21 Find v_3 for $v_g = 8$V.

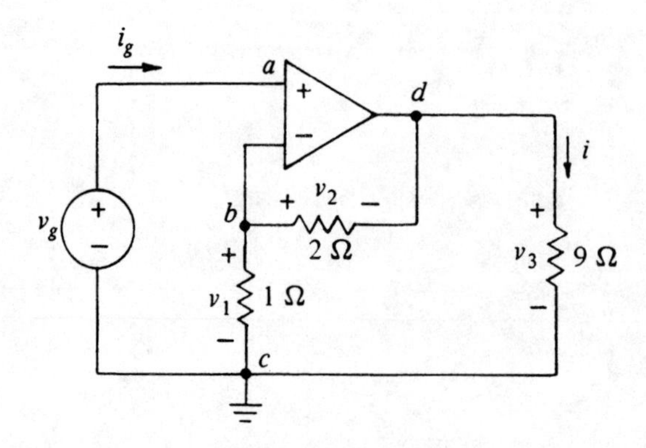

PROBLEM 4.21

4.22 Find the current in the 5kΩ resistor if v_g = 5V.

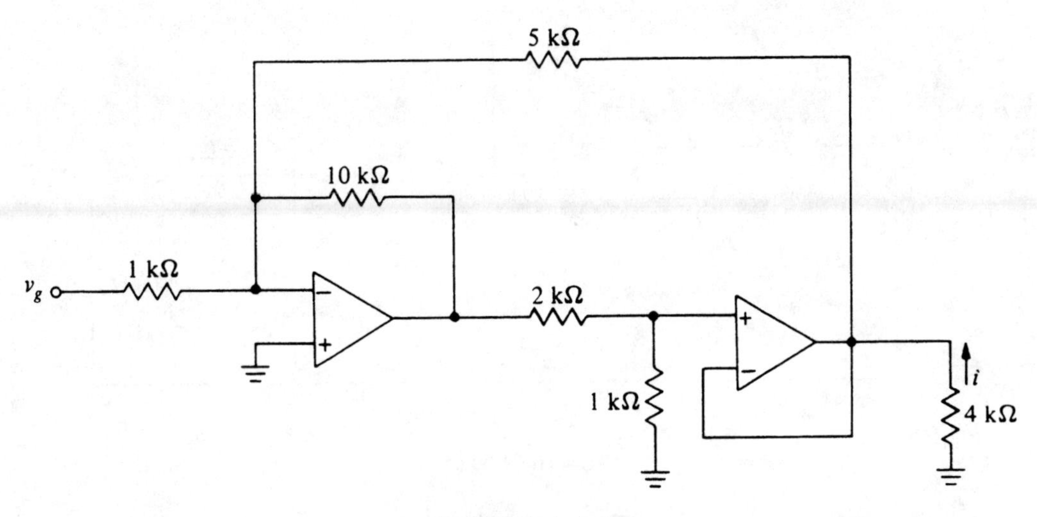

PROBLEM 4.22

4.23 For $R_1 = 1.2k\Omega$, $R_2 = 4.7k\Omega$, $R_3 = 1k\Omega$, $R_4 = 3.3k\Omega$, v_1 = 4V, v_2 = 2cos(377t)V find v_0.

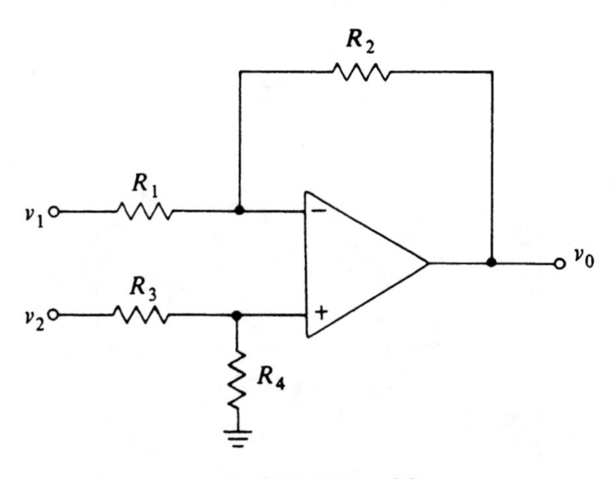

PROBLEM 4.23

4.24 Find i.

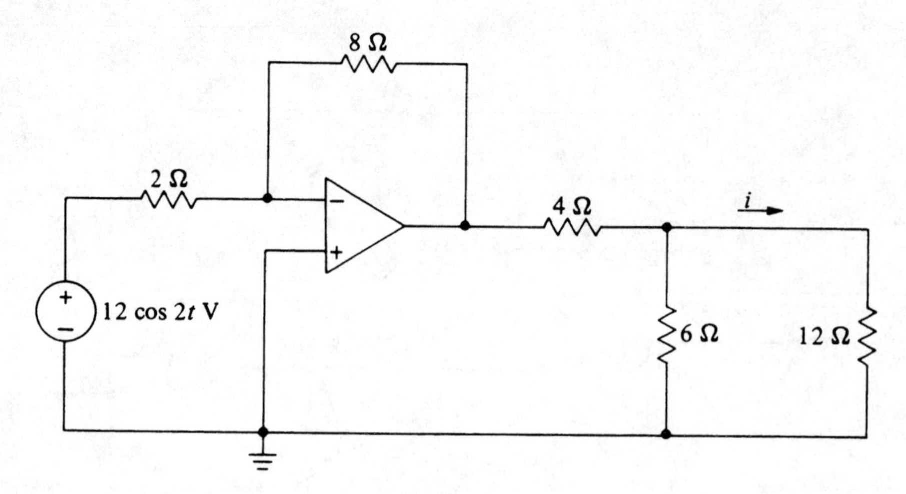

PROBLEM 4.24

4.5 Mesh Analysis

4.25 Using mesh analysis, find i_1 and i_2 if $R_1 = 3\Omega$, $R_2 = 12\Omega$, $R_3 = 6\Omega$, $v_{g1} = 51\text{V}$, and $v_{g2} = 6\text{V}$.

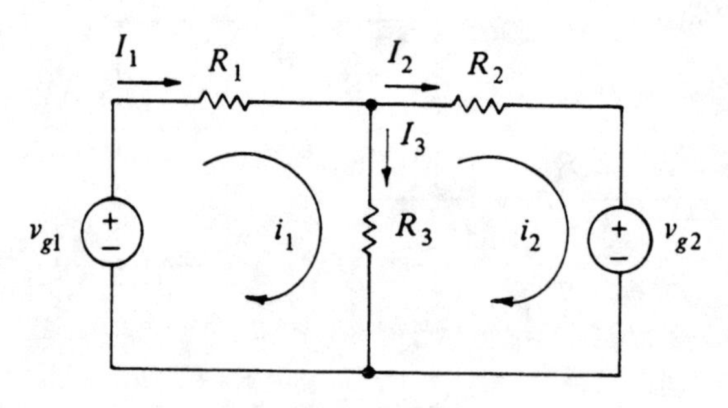

PROBLEM 4.25

4.26 Using mesh analysis, find i_1 and i_2 if element x is a 6Ω resistor.

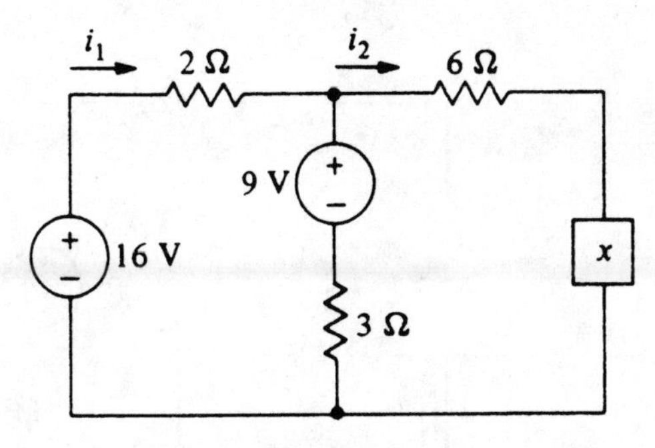

PROBLEM 4.26

4.27 Repeat Prob. 4.26 if element x is a 12V source of with the positive terminal at the bottom.

4.6 Circuits Containing Current Sources

4.28 Solve Prob. 4.2 using mesh analysis.

4.29 Find v_1 using mesh analysis.

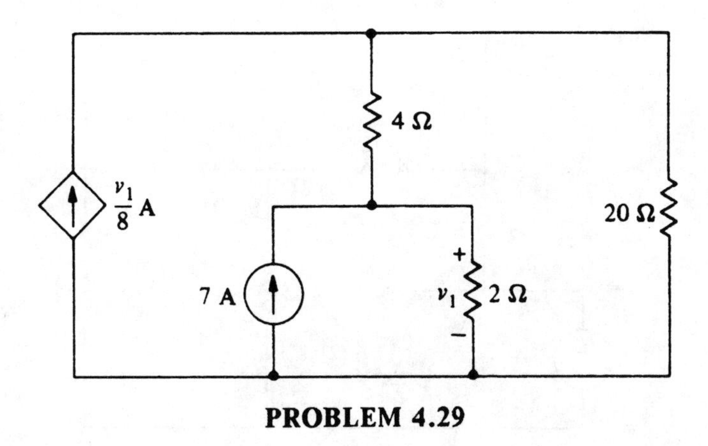

PROBLEM 4.29

4.30 Find the power in the 3kΩ resistor using loop analysis.

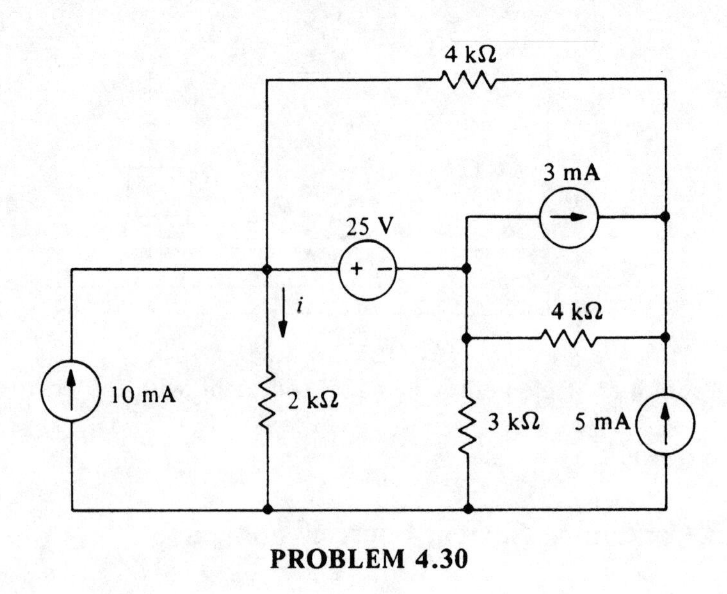

PROBLEM 4.30

4.31 Find the current in the 3Ω resistor using loop analysis.

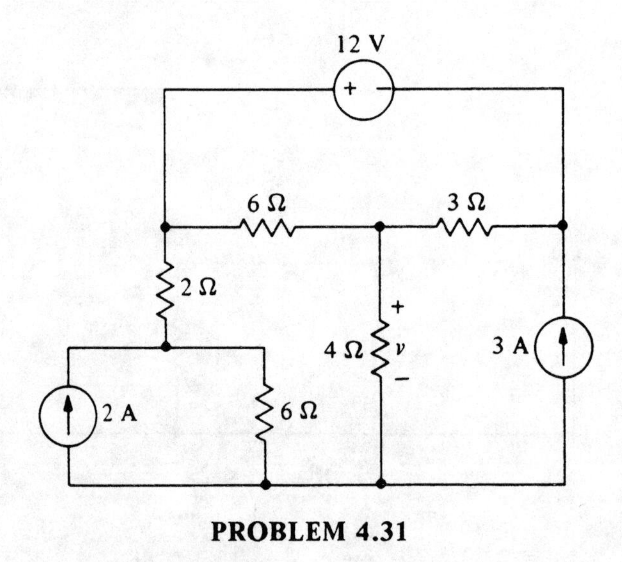

PROBLEM 4.31

4.32 If $i_{g1} = 2\text{A}$, $i_{g2} = 8\text{A}$, $v_{g3} = 24\text{V}$, $R_1 = 1\Omega$, $R_2 = 3\Omega$, $R_3 = 4\Omega$ find the power in R_3.

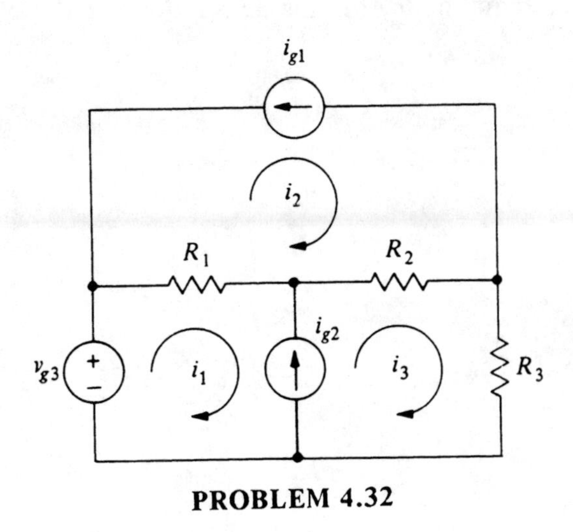

PROBLEM 4.32

4.33 Using mesh analysis, find i.

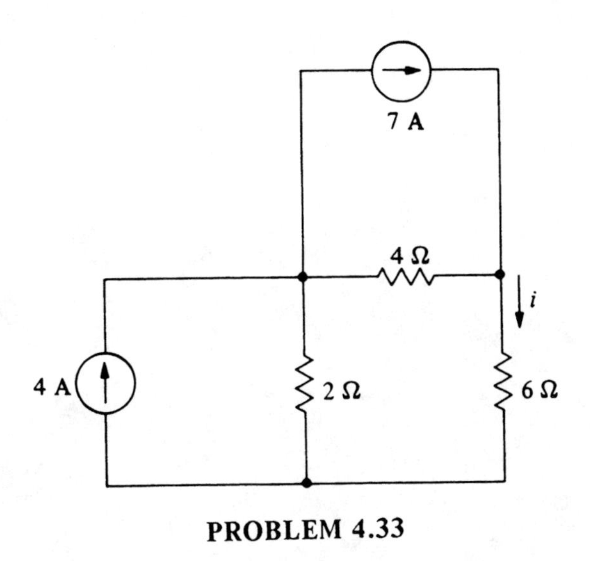

PROBLEM 4.33

4.7 Duality

4.34 Draw the dual of the circuit in Prob. 4.1.

4.35 Draw the dual of the circuit in Prob. 4.2.

4.8 Computer-Aided Circuit Analysis Using SPICE

4.36 Create a SPICE circuit file to find the dc solution for v of Prob. 4.11.

4.1 With v_1 & v as node voltages, at terminals of 6-Ω resistor, KCL,

$$\left(\frac{1}{2}+\frac{1}{6}\right)v_1 - \frac{1}{6}v = 7 \text{ or } 4v_1 - v = 42$$

$$\left(-\frac{1}{6}\right)v_1 + \left(\frac{1}{6}+\frac{1}{4}\right)v = 2 \text{ or } -2v_1 + 5v = 24$$

Adding, $9v = 90$ and $v = \underline{10V}$.

4.2 Let the node voltages be v_1 and v_2 at the terminals of the 2-kΩ resistor. The node equations are:

$$\left(\frac{1}{2}+\frac{1}{2}\right)v_1 - \frac{1}{2}v_2 = 6+1 \text{ or } 2v_1 - v_2 = 14$$

$$-\frac{1}{2}v_1 + \left(\frac{1}{2}+\frac{1}{2}\right)v_2 = 2-1 \text{ or } -v_1 + 2v_2 = 2$$

Adding, $3v_2 = 18$ and $v_2 = 6V$,

$v_1 = \frac{14+v_2}{2} = 10V$, $i = \frac{v_1 - v_2}{2k\Omega} = \underline{2\,mA}$.

4.3 Let v_2 be the voltage across the 2-Ω resistor. The node equations are:

$$\left(\frac{1}{4}+\frac{1}{2}\right)v_1 - \frac{1}{2}v_2 = 2 \text{ or } 3v_1 - 2v_2 = 8$$

$$\left(-\frac{1}{2}-\frac{1}{3}\right)v_1 + \left(\frac{1}{2}+\frac{1}{8}\right)v_2 = 0 \text{ or } -4v_1 + 3v_2 = 0$$

$$v_2 = \begin{vmatrix} 3 & 8 \\ -4 & 0 \end{vmatrix} \Big/ \begin{vmatrix} 3 & -2 \\ -4 & 3 \end{vmatrix} = \frac{(4)(8)}{9-8} = 32V$$

$P_{8\Omega} = \frac{v_2^2}{8} = \underline{128\,W}$.

4.4 $\left(\frac{1}{2}+1\right)v_1 - v_2 = 14$ or $3v_1 - 2v_2 = 28$

$$-v_1 + \left(1+\frac{1}{4}\right)v_2 = -7 \text{ or } -4v_1 + 5v_2 = -28$$

$$v_1 = \begin{vmatrix} 28 & -2 \\ -28 & 5 \end{vmatrix} \Big/ \begin{vmatrix} 3 & -2 \\ -4 & 5 \end{vmatrix} = \frac{140-56}{15-8} = \underline{12V}$$

$$v_2 = \begin{vmatrix} 3 & 28 \\ -4 & -28 \end{vmatrix} \Big/ 7 = \frac{-84+112}{7} = \underline{4V}.$$

4.5 $\left(\frac{1}{4}+\frac{1}{4}\right)v_1 - \frac{1}{4}v_2 = 8-4$ or $2v_1 - v_2 = 16$

$$\left(-\frac{1}{4}\right)v_1 + \left(\frac{1}{4}+\frac{1}{4}\right)v_2 = 4 \text{ or } -v_1 + 2v_2 = 16$$

adding, $3v_1 = 48$ and $v_1 = \underline{16V}$

$v_2 = \frac{16+v_1}{2} = \underline{16V}$ $\quad i = \frac{v_2 - v_1}{4\Omega} = \underline{0\,A}$

4.6 $\left(\frac{1}{4}+\frac{1}{6}\right)v_3 - \frac{1}{4}v_2 = 6+5$ or $-3v_2 + 5v_3 = 132$

$$\left(\frac{1}{2}+\frac{1}{4}\right)v_2 - \frac{1}{4}v_3 = -5-4 \text{ or } 3v_2 - v_3 = -36$$

adding, $4v_3 = 96$ and $v_3 = \underline{24V}$,

$v_2 = \frac{-36+v_3}{3} = \underline{-4V}$,

$v_1 = v_3 - v_2 = \underline{28V}$.

4.7 The node voltages are 28, $v-8$, and v. KCL for the supernode containing the 8-V source yeilds

$$\frac{v-8-28}{6} + \frac{v-8}{2} + \frac{v}{4} + \frac{v}{12} = 0 \text{ or}$$

$12v = 120$ ∴ $v = \underline{10V}$ and

$i = (v-8)/2 = \underline{1A}$.

4.8 With v_1 and v_2 as node voltages, at terminals of 5-Ω resistor. The node equations are:

$$v_1\left(\frac{1}{4}+\frac{1}{2}+\frac{1}{5}\right) - v_2\frac{1}{5} = -9$$

$$v_1\left(-\frac{1}{5}\right) + v_2\left(\frac{1}{5}+\frac{1}{2}+\frac{1}{2}\right) - \frac{10i_1}{2} = 0$$

since $i_1 = \frac{-v_1}{4}$, the node equations become,

$19v_1 - 4v_2 = -180$, $21v_1 + 24v_2 = 0$

Adding six times the first equation

$135v_1 = -1{,}080$ ∴ $v_1 = -8V$

$i_1 = -\frac{(-8V)}{4} = \underline{2\,A}$.

4.9 The node voltages are 14, $v-4$, and v. KCL for the supernode containing the 4V source yeilds

$$\frac{v-4-14}{6} + \frac{v-4}{2} + \frac{v}{4} + \frac{v}{12} = 0 \text{ or}$$

$12v = 60$ ∴ $v = \underline{5V}$ and

$i = (v-4)/2 = \underline{0.5\,A}$

4.10 With v and v_1 as node voltages, at the terminals of the 4-Ω. The node equations are:

$$v\left(\frac{1}{8}+\frac{1}{4}\right) - v_1\frac{1}{4} = 6 \text{ or } 3v - 2v_1 = 48$$

$$-\frac{v}{4} + v_1\left(\frac{1}{12}+\frac{1}{6}+\frac{1}{4}\right) = \frac{6}{6} - \frac{24}{12}$$

$-3v + 6v_1 = -12$, Adding equation

$4v_1 = 36V$, $v = \frac{48+2v_1}{3} = \underline{22V}$

4.11 With v_1 and v_2 as node voltages at terminals of 2-Ω resistor. The node equations are

$$v_1\left(\frac{1}{2}+\frac{1}{2}+1\right) - \frac{v_2}{2} = \frac{18}{1}$$

$$-\frac{v_1}{2} + v_2\left(\frac{1}{4}+\frac{1}{2}+\frac{1}{2}\right) = \frac{18}{4} \quad \text{or}$$

$4v_1 - v_2 = 36$, $-2v_1 + 5v_2 = 18$

Adding twice the second equation

$9v_2 = 72$, or $v_2 = 8V$, $v_1 = \frac{36+v_2}{4} = 11V$

$v = v_1 - v_2 = \underline{3V}$.

4.12 with v_1 and v_2 as node voltages at terminals of 1-Ω resistor and $i_1 = v_1/4$. The node equations are

$v_1(1+\frac{1}{4}+\frac{1}{2}+\frac{5}{4}) - v_2 = \frac{18}{2}$ or $12v_1 - 4v_2 = 36$

$-v_1(1+\frac{5}{4}) + v_2(\frac{1}{5}+1) = 0$ or $-45v_1 + 24v_2 = 0$

Adding 6 times the first equation

$27v_1 = 216$ or $v_1 = 8\text{V}$ ∴ $i_1 = 8/4 = \underline{2\text{ A}}$.

4.13 The node voltages are v_1, v_2 and $v_2 - 24$ with $i = v_1/4$. The node equations are:

$$v_1\left(\frac{1}{3}+\frac{1}{4}+\frac{1}{6}\right) - \frac{v_2}{6} - \frac{(v_2-24)}{3} = 0$$

$$\frac{v_2-30}{8} + \frac{v_2}{6} - \frac{v_1}{6} + \frac{v_2-24}{x} + \frac{v_2-24}{3} - \frac{v_1}{3} = 0$$

or $9v_1 - 6v_2 = 96$

$-12v_1 + 21v_2 = 426$

$$v_1 = \begin{vmatrix} 96 & -6 \\ 426 & 21 \end{vmatrix} \Big/ \begin{vmatrix} 9 & -6 \\ -12 & 21 \end{vmatrix} = \frac{4572}{117} = 39.08\text{V}$$

$$\therefore i = \frac{39.08}{4} = \underline{9.77\text{A}}$$

4.14 Node equations are using node voltages of Prob. 4.13.

$9v_1 - 6v_2 = 96$,

$$\frac{v_2-30}{8} + \frac{v_2-v_1}{6} + \frac{v_2-24-v_1}{3} = 7 \text{ or}$$

$-12v_1 + 15v_2 = 289$

$$v_1 = \begin{vmatrix} 96 & -6 \\ 289 & 15 \end{vmatrix} \Big/ \begin{vmatrix} 9 & -6 \\ -12 & 15 \end{vmatrix} = \frac{3174}{63} = 50.38\text{V}$$

$$\therefore i = \frac{50.38\text{V}}{4\,\Omega} = \underline{12.60\text{ A}}$$

4.15 The node voltages are 30, $5i$, $5i+24$ an v_1 of Prob. 4.13. $i = v_1/4$. The node equation is:

$$\frac{v_1-5i}{3} + \frac{v_1}{4} + \frac{v_1-(5i+24)}{6} = 0$$

$3v_1 = 96$, $v_1 = 32\text{V}$

$$i = \frac{32}{4} = \underline{8\text{ A}}.$$

4.16 The node equation are

$v_1(2+0.5+0.25) - v_2(0.5) - v_3(0.25) = 4(2)$

or $11v_1 - 2v_2 - v_3 = 32$

$-v_1(0.5) + v_2(0.5+1) + 5(v_1 - v_3) = 0$

or $9v_1 + 3v_2 - 10v_3 = 0$

4.16 cont.

$-v_1(0.25) - 5(v_1 - v_3) + v_3(0.25+1) = 0$

or $-21v_1 + 25v_3 = 0$

$$v_2 = \frac{\begin{vmatrix} 11 & 32 & -1 \\ 9 & 0 & -10 \\ -21 & 0 & 25 \end{vmatrix}}{\begin{vmatrix} 11 & -2 & -1 \\ 9 & 3 & -10 \\ -21 & 0 & 25 \end{vmatrix}} = \underline{-0.606\text{ V}}$$

4.17 using equation from Prob. 4.16

$$v_1 = \frac{\begin{vmatrix} 32 & -2 & -1 \\ 0 & 3 & -10 \\ 0 & 0 & 25 \end{vmatrix}}{\begin{vmatrix} 11 & -2 & -1 \\ 9 & 3 & -10 \\ -21 & 0 & 25 \end{vmatrix}} = \underline{3.03\text{V}}$$

4.18 v_o = node voltage at top of 3-Ω resistor. Voltages between input terminals of op-amp is zero. Node analysis at (+) terminal

$$\frac{3\cos 4t}{2\text{k}\Omega} + \frac{3\cos 4t - v_o}{8\text{k}\Omega} = 0$$

$v_o = 15\cos 4t\text{ V}$, $i = v_o/3\text{k}\Omega$

$i = \underline{5\cos 4t\text{ mA}}$

4.19 Using Node analysis

$v_1(\frac{1}{2}+1) = 6\cos 2t$ or $v_1 = 4\cos 2t$

$v_1(\frac{1}{2}+\frac{1}{2}) - \frac{v}{2} = \frac{6}{2}\cos 2t$ or

$2v_1 - v = 6\cos 2t$, $v = \underline{2\cos 2t\text{ V}}$.

4.20 using node analysis at v_3 & 0V.

$-v_1(\frac{1}{4}) - v_2(2) + v_3(\frac{1}{4}+\frac{1}{2}+1+2) = 0$ or

$-8v_2 + 15v_3 = v_1 = 8\cos 1000t$,

$-v_2(\frac{1}{2}) - v_3(1) + 0 = 0$ or $-v_2 - 2v_3 = 0$

Adding 15/2 times the second

$-15.5v_2 = 8\cos 1000t$ ∴

$v_2 = \underline{0.516\cos 1000t\text{ V}}$.

4.21 By KVL, $v_1 = v_g = 8\text{V}$, and

$v_1(1+\frac{1}{2}) - v_3(\frac{1}{2}) = 0$ or

$v_3 = 3v_1 = 3(8) = \underline{24\text{V}}$.

4.22 Let v_1 be the output of the first op amp and v_2 be the output of the second. The voltage at the noninverting input of the voltage follower is also v_2. KCL at inverting input of first op amp yeilds (currents in mA)

$\frac{v_g}{1}+\frac{v_1}{10}+\frac{v_2}{5}$ or $v_1+2v_2=-(10)(5)$

KCL at noninverting input of voltage follower yields.

$v_2(1+\frac{1}{2})-\frac{v_1}{2}=0$ or $-v_1+3v_2=0$

$\therefore\ 5v_2=-50$ or $v_2=-10\text{V}$

$i_{5k\Omega}=\frac{v_2}{5}=\underline{-2\text{mA}}$ to the left.

4.23 Let v_3 be the node voltage of the op amp input terminals (the input voltage is zero.) KCL at these inputs yeilds

noninverting: $v_3\left(\frac{1}{1}+\frac{1}{3.3}\right)-\frac{1}{1}v_2=0$

inverting: $v_3\left(\frac{1}{1.2}+\frac{1}{4.7}\right)-\frac{1}{1.2}v_1-\frac{1}{4.7}v_0=0$

$v_3=\frac{3.3}{1+3.3}v_2=1.535\cos 377t\ \text{V}$

$v_0=-\frac{4.7}{1.2}v_1+\frac{1.2+4.7}{(1.2)}v_3$

$=\underline{-15.67+7.547\cos 377t\ \text{V}}$.

4.24 Let v_0 be the op amp output voltage. Then

$v_0=-\frac{8}{2}(12\cos 2t)=-48\cos 2t\ \text{V}$

By voltage division

$12i=\frac{(12)(6)/(12+6)}{4+4}v_0,\quad i=\frac{v_0}{(2)(12)}$

$\therefore\ \underline{i=-2\cos 2t\ \text{A}}$

4.25 Apply KVL to each mesh loop,

$i_1(3+6)-6i_2=51$ or $9i_1-6i_2=51$

$-i_1 6+i_2(6+12)=-6$ or $-6i_1+18i_2=-6$

Adding 3 times the first equation

$21i_1=147$ or $i_1=\underline{7\text{A}}$.

$i_2=\frac{-6+6i_1}{18}=\underline{2\text{A}}$.

4.26 The mesh equations are

$i_1(2+3)-3i_2=16-9$ or $5i_1-3i_2=7$

$-3i_1+i_2(6+6+3)=9$ or $-3i_1+15i_2=9$

Adding 5 times the first

$22i_1=44$ or $i_1=\underline{2\text{A}}$.

$i_2=\frac{9+3i_1}{15}=\underline{1\text{A}}$.

4.27 The second mesh equation is

$-3i_1+i_2(6+3)=9+12$ or $-3i_1+9i_2=21$

Adding 3 times the first of Prob 4.26

$12i_1=42$ or $i_1=\underline{3.5\text{A}}$.

$i_2=\frac{21+3i_1}{9}=\underline{3.5\text{A}}$.

4.28 Let i_1 be the mesh current through the 2-Ω resistor and the other mesh currents be the current sources in those loops.

$i_1(2+2+2)+1(2)+2(2)-6(2)=0$

$6i_1=6$ or $i_1=1\text{mA}$

$i=i_1+1\text{mA}=\underline{2\text{mA}}$.

4.29 For the mesh with only resistors, KVL yeilds.

$(2+4+20)i-7(2)+20\left(\frac{v_1}{8}\right)=0$

since $v_1=2(7-i)$

or $i=\frac{14-v_1}{2}$ then $\frac{26}{2}(14-v_1)+\frac{20}{8}v_1=14$

or $v_1=\underline{16\text{V}}$.

4.30

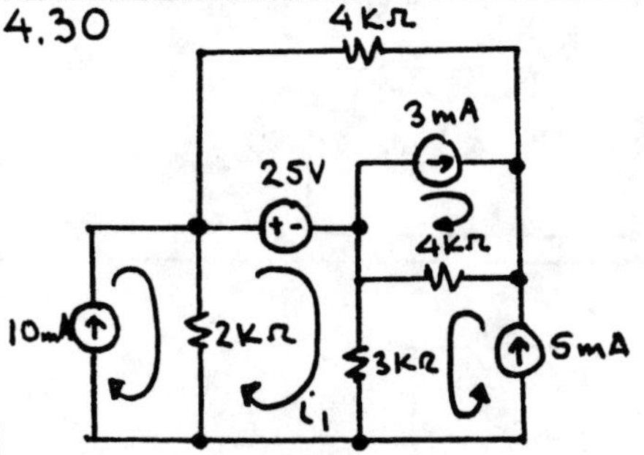

i_1 in mA, KVL for the mesh of i_1

$2(i_1-10)+3(i_1+5)=-25$ then $i_1=-4\text{mA}$

$i_{3k\Omega}=i_1+5\text{mA}=1\text{mA}$,

$P_{3k\Omega}=i_{3k\Omega}^2R=(3k\Omega)(1\text{mA})^2=\underline{3\text{mW}}$

4.31

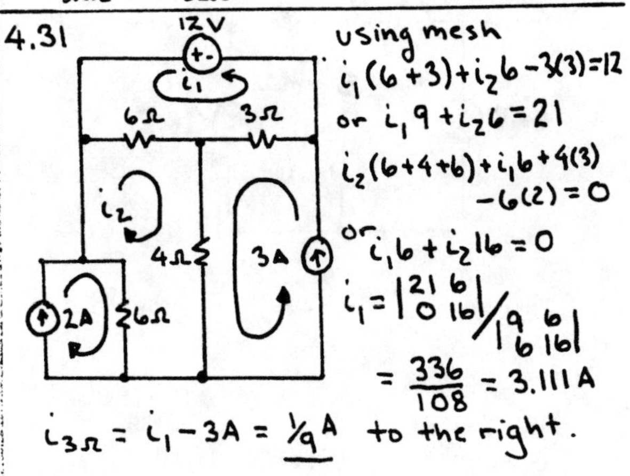

using mesh

$i_1(6+3)+i_2 6-3(3)=12$

or $i_1 9+i_2 6=21$

$i_2(6+4+6)+i_1 6+4(3)-6(2)=0$

or $i_1 6+i_2 16=0$

$i_1=\begin{vmatrix}21&6\\0&16\end{vmatrix}\Big/\begin{vmatrix}9&6\\6&16\end{vmatrix}$

$=\frac{336}{108}=3.111\text{A}$

$i_{3\Omega}=i_1-3\text{A}=\underline{1/9\ \text{A}}$ to the right.

4.32

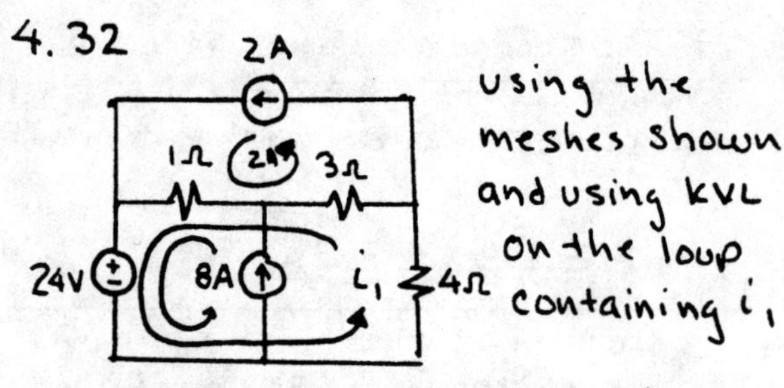

using the meshes shown and using KVL on the loop containing i_1

$i_1(4+3+1)+8(1)-2(3+1)=-24$ or

$i_1 8=-24$ or $i_1=-3\text{ A}$

$P_{R_3}=R_3 i_1^2=(4)(-3)^2=\underline{36\text{W}}$.

4.33 Let i be the mesh current in the all resistor loop then by KVL in that loop.

$i(6+4+2)-4(2)-7(4)=0$ or

$12i=36 \therefore i=\underline{3\text{A}}$.

4.34

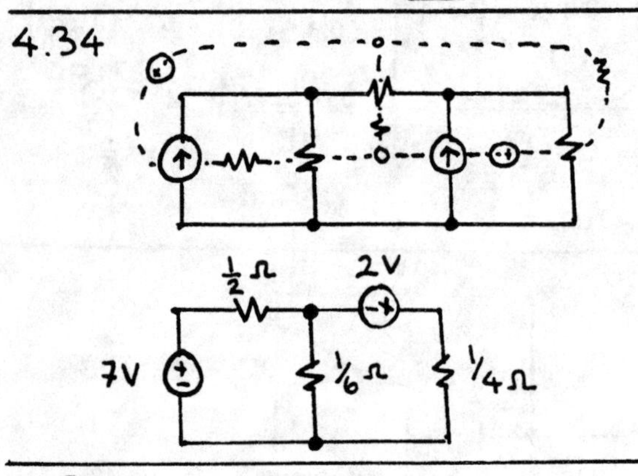

4.35

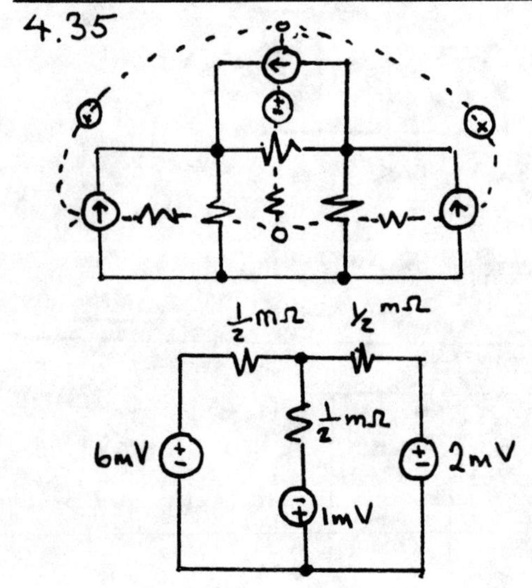

4.36

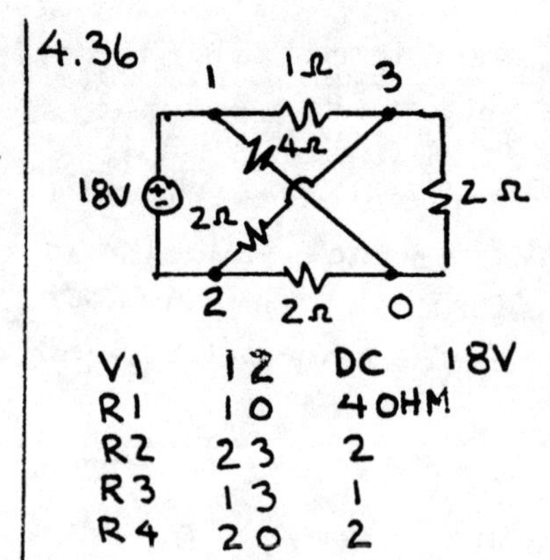

V1	1 2	DC	18V
R1	1 0	4OHM	
R2	2 3	2	
R3	1 3	1	
R4	2 0	2	

Chapter 5

Network Theorems

5.1 Linear Circuits

5.1 Find v_1, v_2, v_3 with (a) the source values as shown, (b) the source values divided by 2, and (c) the source values multiplied by 2.

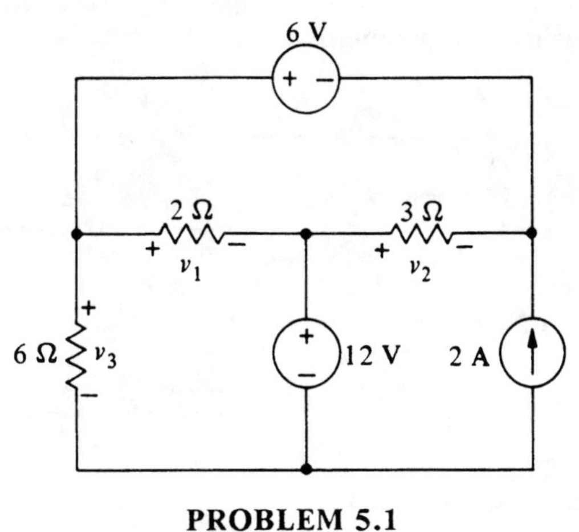

PROBLEM 5.1

5.2 Find v using the principle of proportionality.

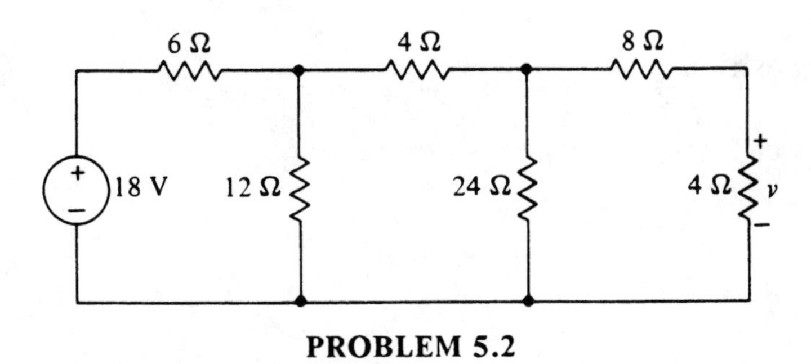

PROBLEM 5.2

5.3 A circuit is made of a voltage source v_g, a 2Ω resistor, and a nonlinear resistor in series. The nonlinear resistor is described by $v = 3 + 3i$ where v is the voltage across the resistor and i, which is constrained to be nonnegative, is the current flowing into the positive terminal. Find i if (a) $v_g = 8$V and (b) $v_g = 16$V.

5.4 Find i and v using proportionality.

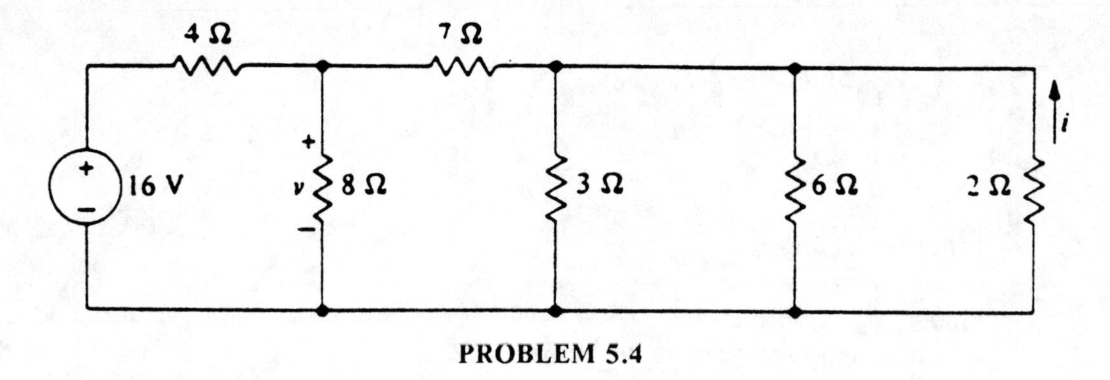

PROBLEM 5.4

5.5 Find i_1 and i_2 using proportionality.

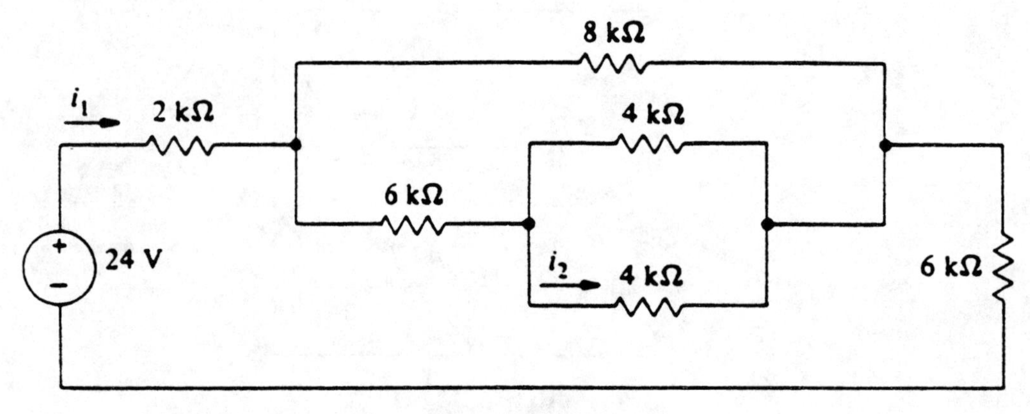

PROBLEM 5.5

5.6 Find i of Prob. 4.18 using proportionality.

5.2 Superposition

5.7 Find v_1 of Prob. 4.6 using superposition.

5.8 Show that $v_0 = \dfrac{R_4(R_1+R_2)}{R_1(R_3+R_4)}v_2 - \dfrac{R_2}{R_1}v_1$ using superposition.

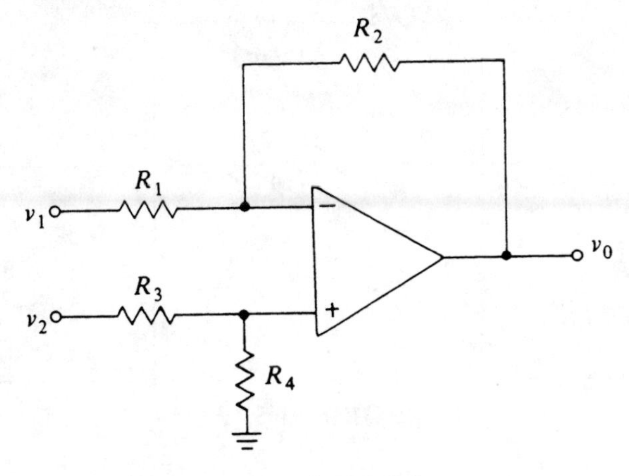

PROBLEM 5.8

5.9 If $R = 8\Omega$ find the voltage across R by superposition.

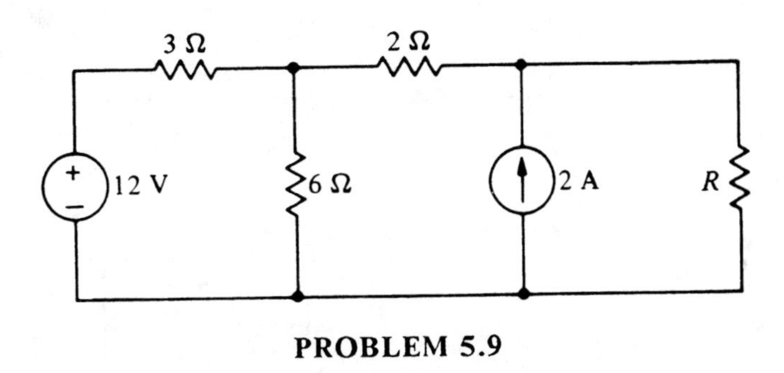

PROBLEM 5.9

5.10 Find i of Prob. 4.33 by superposition.

5.11 Find v using superposition.

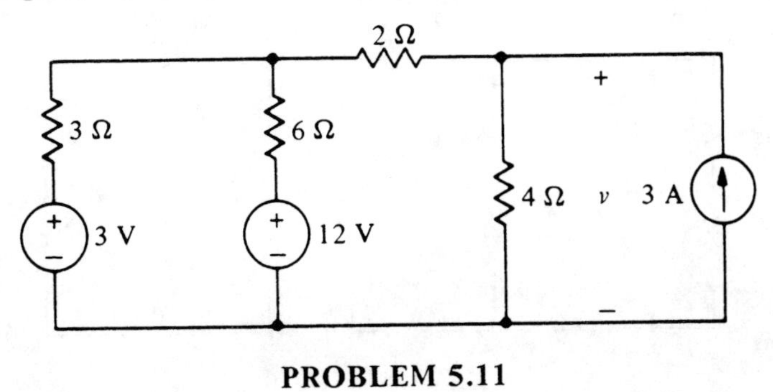

PROBLEM 5.11

5.12 Find v using superposition if $R = 12\Omega$.

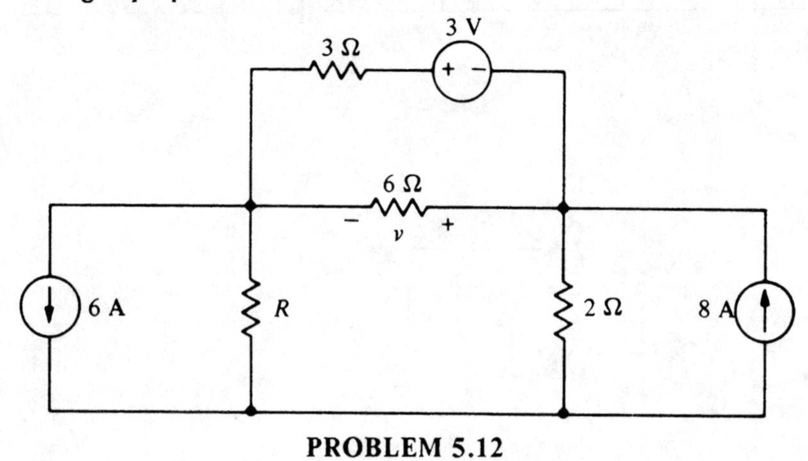

PROBLEM 5.12

5.13 Using superposition, find the power delivered to the 2Ω resistor.

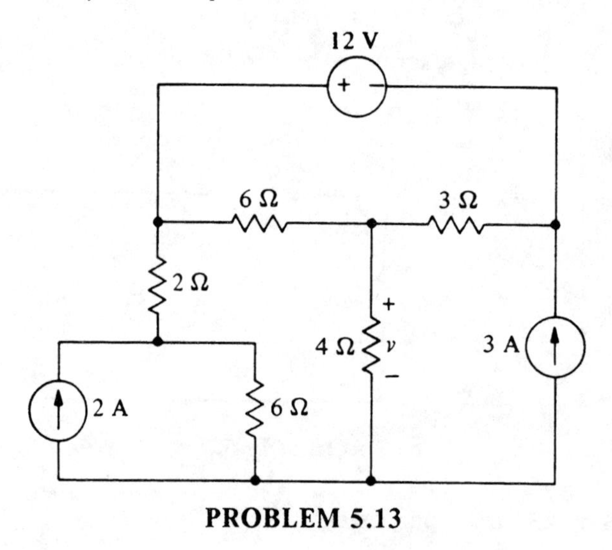

PROBLEM 5.13

5.14 Find v using superposition.

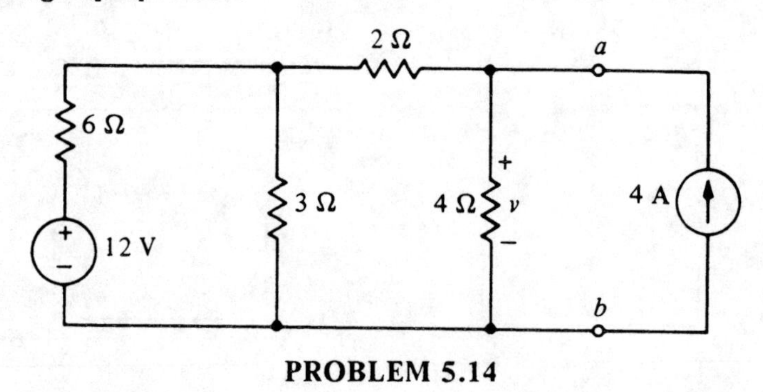

PROBLEM 5.14

5.15 Find i using superposition.

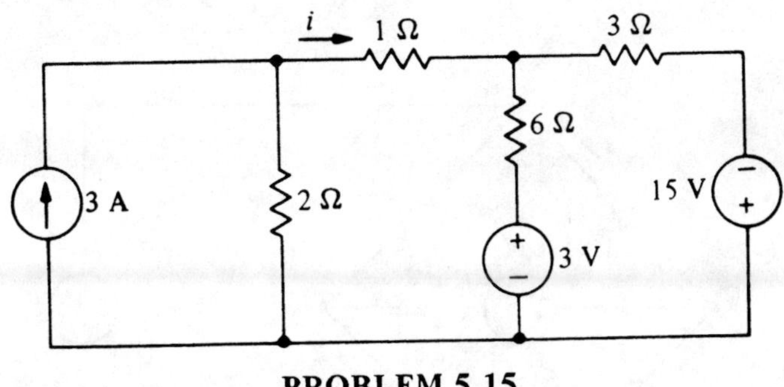

PROBLEM 5.15

5.16 Find v using superposition.

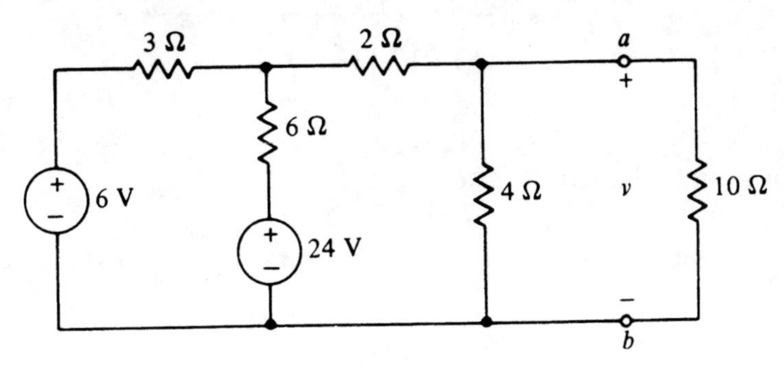

PROBLEM 5.16

5.17 Find v using superposition.

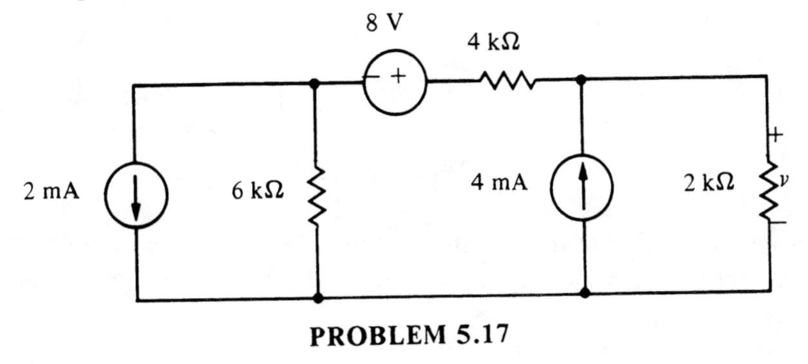

PROBLEM 5.17

5.18 Find i using superposition.

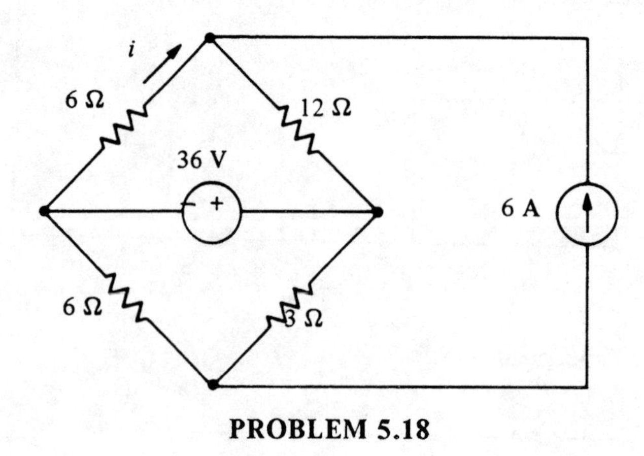

PROBLEM 5.18

5.3 Thevenin's and Norton's Theorems

5.19 Replace the network to the left of terminals a-b by its Thevenin equivalent and find i.

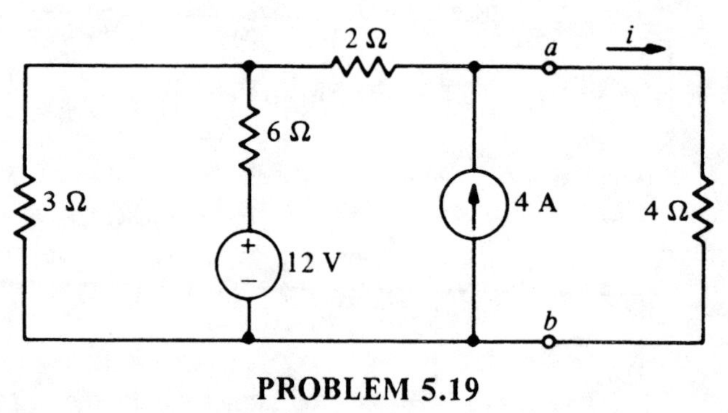

PROBLEM 5.19

5.20 Replace everything in the circuit of Prob. 5.19 except the 2Ω resistor by its Thevenin equivalent circuit and use the result to find the power delivered to the 2Ω resistor.

5.21 Repeat Prob. 5.20 using the Norton equivalent rather than the Thevenin equivalent circuit.

5.22 Replace everything in Prob. 5.2 except the 4Ω resistor with its Thevenin equivalent circuit and find v.

5.23 Find the Norton equivalent seen by the 4Ω resistor at v in Prob. 5.2 and find v.

5.24 Find the Norton equivalent seen by the 4Ω resistor in Prob. 5.11.

5.25 In Prob. 5.14, replace the network to the left of terminals a-b by its Thevenin equivalent and use the result to find v.

5.26 In Prob. 5.15, find the Thevenin equivalent of everything except the 1Ω resistor and use the result to find i.

5.27 Find v by replacing everything in the circuit except the 4Ω resistor by its Thevenin equivalent.

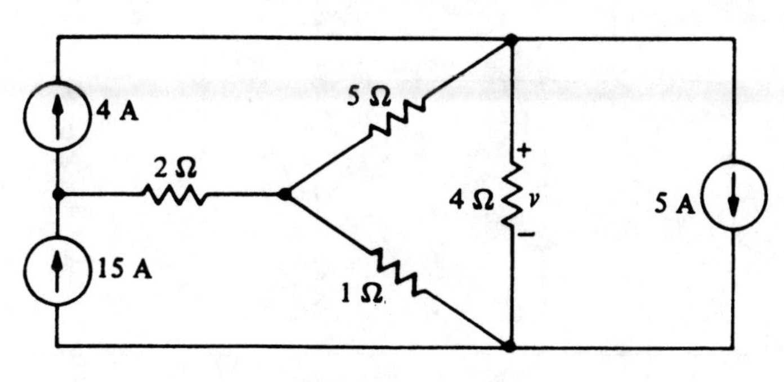

PROBLEM 5.27

5.4 Practical Sources

5.28 By successive source transformations reduce the circuit of Prob. 5.16 to the left of terminals a-b to its Thevenin equivalent and use the result to find v.

5.29 Convert all the sources in Prob. 4.6 to voltage sources and find v_3.

5.30 By source transformation, replace everything except the 2kΩ resistor by a practical current source and find v.

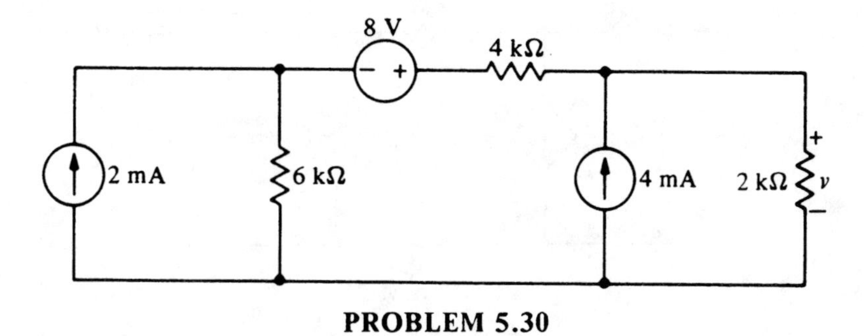

PROBLEM 5.30

5.31 Find i by using source transformations to obtain an equivalent circuit insofar as the $1 k\Omega$ resistor is concerned, containing only one source and one resistor, in addition to the $1 k\Omega$ resistor.

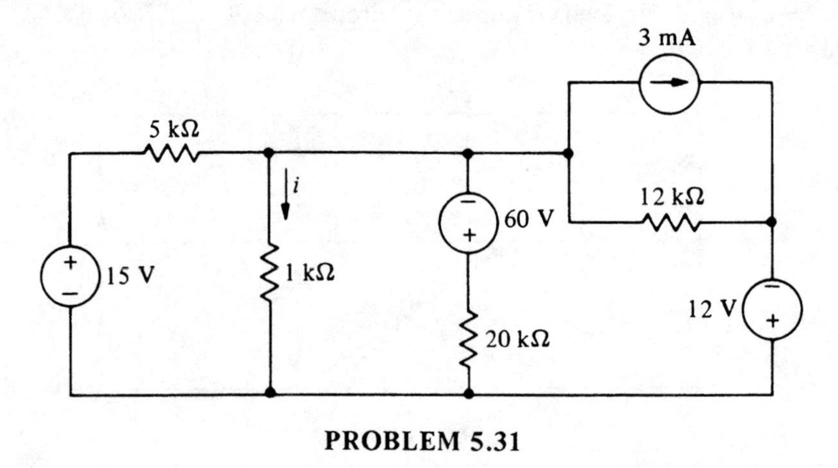

PROBLEM 5.31

5.5 Maximum Power Transfer

5.32 Find R for maximum power transfer to R in Prob. 5.12.

5.33 Find the value of R that will draw the maximum power from the rest of the circuit.

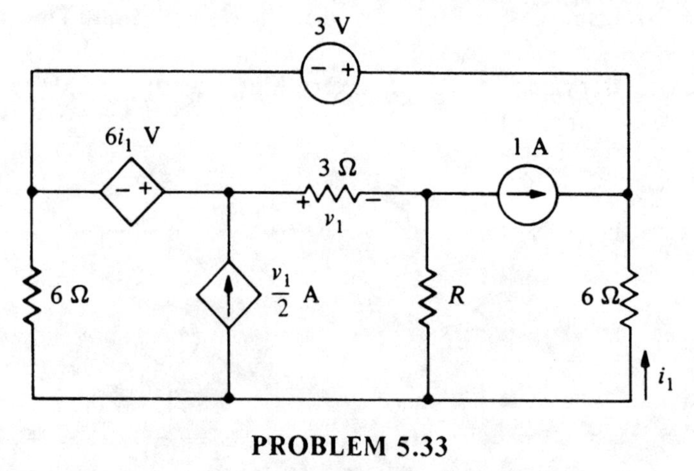

PROBLEM 5.33

5.34 Find the maximum power in R of Prob. 5.33.

5.35 Find the value of a resistor R that will give maximum power transfer when placed across the open terminals.

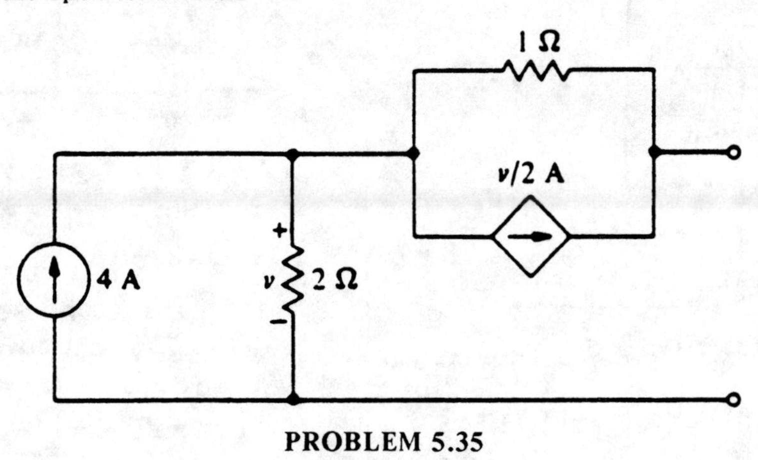

PROBLEM 5.35

5.1

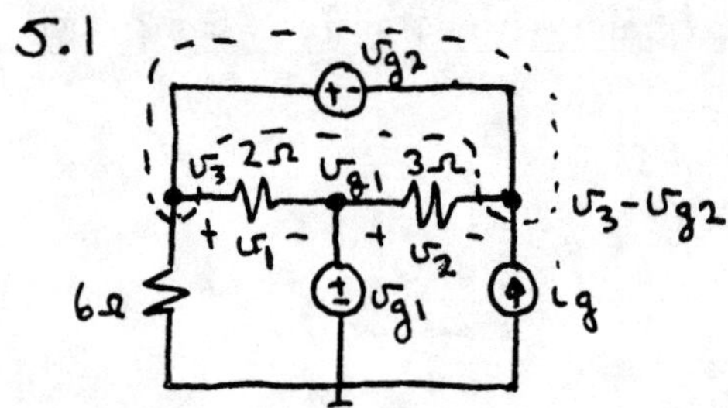

KVL for supernode:

$$\frac{v_3}{6} + \frac{v_3 - v_{g1}}{2} + \frac{v_3 - v_{g2} - v_{g1}}{3} = i_g$$

$$\left(\frac{1}{6} + \frac{1}{2} + \frac{1}{3}\right)v_3 = v_3 = \frac{5}{6}v_{g1} + \frac{1}{3}v_{g2} + i_g$$

$v_1 = v_3 - v_{g1}$, $v_2 = v_{g1} - v_3 + v_{g2}$

(a) $v_{g1} = 12V$, $v_{g2} = 6V$, $i_g = 2A$,

$v_3 = \frac{5}{6}(12) + \frac{6}{3} + 2 = \underline{14V}$,

$v_1 = 14 - 12 = \underline{2V}$, $v_2 = 12 - 14 + 6 = \underline{4V}$.

(b) $v_{g1} = 6V$, $v_{g2} = 3V$, $i_g = 1A$,

$v_3 = \frac{5}{6}(6) + \frac{3}{3} + 1 = \underline{7V}$

$v_1 = 7 - 6 = \underline{1V}$, $v_2 = 6 - 7 + 3 = \underline{2V}$.

(c) $v_{g1} = 24V$, $v_{g2} = 12V$, $i_g = 4A$.

$v_3 = \frac{5}{6}(24) + \frac{12}{3} + 4 = \underline{28V}$,

$v_1 = 28 - 24 = \underline{4V}$, $v_2 = 24 - 28 + 12 = \underline{8V}$

5.2

assume $i = 1A$. Then

$v_1 = (8+4)i = 12V$, $i_1 = \frac{v_1}{24} = \frac{1}{2}A$,

$i_2 = i_1 + i = \frac{3}{2}A$, $v_2 = 4i_2 = 6V$,

$v_3 = v_2 + v_1 = 18V$, $i_3 = \frac{v_3}{12} = \frac{3}{2}A$,

$i_4 = i_3 + i_2 = 3A$, $v_g = v_3 + i_4(6) = 36V$

since v_g should be 18V instead all voltages and currents are multiplied by $\frac{18}{36} = \frac{1}{2}$

$\therefore\ i = \frac{1}{2}A$, $v = 4i = \underline{2V}$.

5.3 KVL gives

(a) $3 + 3i + 2i = 8V$; $i = \frac{5}{5} = \underline{1A}$.

(b) $3 + 3i + 2i = 16V$, $i = \frac{13}{5} = \underline{2.6A}$

5.4

Assume $v_1 = 1V$ then

$i_2 = v_1\left(\frac{1}{3} + \frac{1}{6} + \frac{1}{2}\right) = 1A$, $v_2 = i_2 7 = 7V$

$v = v_1 + v_2 = 8V$, $i_3 = \frac{v}{8} = 1A$

$i_4 = i_2 + i = 2A$, $v_g = v + i_4 4 = 16V$

since v_g is suppose to be 16V therefore all currents and voltages are correct and

$\therefore\ i = -\frac{v_1}{2} = \underline{-\frac{1}{2}A}$, $v = \underline{8V}$.

5.5

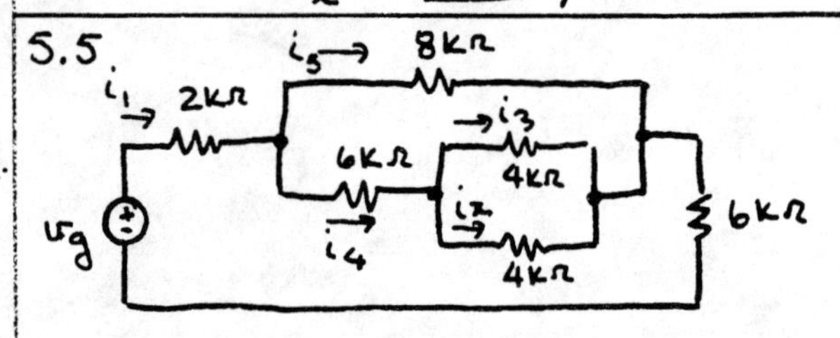

Assume $i_2 = 1mA$ then

$v_2 = i_2 4 = v_3 = 4V$, $i_3 = \frac{v_3}{4} = 1mA$,

$i_4 = i_3 + i_2 = 2mA$, $v_4 = i_4 6 = 12V$,

$v_5 = v_4 + v_3 = 16V$, $i_5 = \frac{v_5}{8} = 2mA$

$i_1 = i_5 + i_4 = 4mA$, $v_g = v_5 + i_1(2+6) = 48V$

Since v_g should be 24V, then all voltages and currents are multiplied by $\frac{24}{48} = \frac{1}{2}$,

$\therefore\ i_1 = \frac{1}{2}(4mA) = \underline{2mA}$, $\underline{i_2 = \frac{1}{2}mA}$.

5.6 Assume $i = 1mA$, Let v_0 be output voltage of o and v_g be the negative input of amp, then.

$v_0 = i(3) = 3V$, $v_g = \frac{2}{2+8}v_0 = 0.6V$

since v_g is suppose to be $3\cos 4t$ then the voltages and currents are multiplied by $\frac{3\cos 4t}{0.6} = 5\cos 4t$

$\therefore\ i = \underline{5\cos 4t\ \text{-mA}}$.

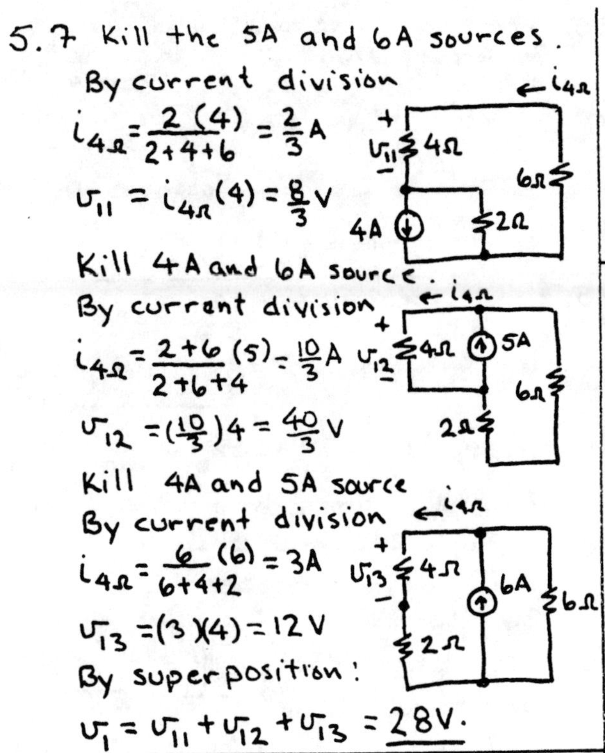

5.7 Kill the 5A and 6A sources.

By current division

$$i_{4\Omega} = \frac{2\,(4)}{2+4+6} = \frac{2}{3}\text{A}$$

$$v_{11} = i_{4\Omega}(4) = \frac{8}{3}\text{V}$$

Kill 4A and 6A source.

By current division,

$$i_{4\Omega} = \frac{2+6}{2+6+4}(5) = \frac{10}{3}\text{A}$$

$$v_{12} = \left(\frac{10}{3}\right)4 = \frac{40}{3}\text{V}$$

Kill 4A and 5A source

By current division

$$i_{4\Omega} = \frac{6}{6+4+2}(6) = 3\text{A}$$

$$v_{13} = (3)(4) = 12\text{ V}$$

By superposition:

$$v_1 = v_{11} + v_{12} + v_{13} = \underline{28\text{V}}.$$

5.8 Set $v_2 = 0$, then

$$v_0 = -\frac{R_2}{R_1}v_1\,;$$ Set $v_1 = 0$, then

$$v_0 = \left(1+\frac{R_2}{R_1}\right)\left(\frac{R_4}{R_3+R_4}\right)v_2 = \frac{R_4(R_1+R_4)}{R_1(R_3+R_4)}v_2$$

then by superposition,

$$v_0 = \underline{\frac{R_4(R_1+R_4)}{R_1(R_3+R_4)}v_2 - \frac{R_2}{R_1}v_1}$$

5.9 Kill the 12 V source,

By current division

$$i = \frac{2+\dfrac{(3)(6)}{3+6}}{2+2+R}(2) = \frac{2}{3}\text{A}, \quad v_1 = iR = \frac{16}{3}\text{V}.$$

Kill the 2A source

By voltage division:

$$v = \frac{\dfrac{(2+R)6}{2+R+6}}{3+3.75}(12) = \frac{20}{3}\text{V}$$

$$v_2 = \frac{R}{2+R}(v) = \frac{8}{10}\left(\frac{20}{3}\right) = \frac{16}{3}\text{V}$$

By superposition:

$$v_R = v_1 + v_2 = \frac{16}{3} + \frac{16}{3} = \underline{32/3\text{V}}$$

5.10 Kill the 7A source

$$i_1 = \frac{2}{2+6+4}(4) = \frac{2}{3}\text{A}$$

5.10 cont. Kill the 2A source

$$i_2 = \frac{4}{2+6+4}(7) = \frac{7}{3}\text{A}$$

By superposition

i = current through 4Ω resistor

$$= i_1 + i_2 = \frac{2}{3} + \frac{7}{3} = \underline{3\text{ A}}.$$

5.11 With only the 3V source active

By voltage division

$$v_4 = \frac{\dfrac{(6)(2+4)}{6+2+4}}{3+3}(3) = \frac{3}{2}\text{V}$$

$$v_1 = \frac{4}{2+4}(v_4) = 1\text{V}.$$

with only 12V source active

$$v_4 = \frac{\dfrac{3(2+4)}{3+2+4}}{6+2}(12) = 3\text{V}$$

$$v_2 = \frac{4}{2+4}(v_4) = 2\text{V}.$$

with only 3A source active

By current division:

$$i_3 = \frac{\dfrac{(6)(3)}{6+3}+2}{4+2+2}(3) = \frac{3}{2}\text{A}$$

$v_3 = i_3(4\Omega) = 6\text{V}$, By superposition,

$$v = v_1 + v_2 + v_3 = 1+2+6 = \underline{9\text{V}}.$$

5.12 with only 6A source active,

By current division,

$$i = \frac{R}{\dfrac{6(3)}{6+3}+2+R}(6) = 4.5\text{A}$$

$$i_1 = \frac{3}{3+6}(i) = 1.5\text{A}$$

$$v_1 = i_1(6\Omega) = 9\text{V}.$$

with only 3V source active,

By voltage division,

$$v_2 = -\frac{\dfrac{(R+2)(6)}{R+2+6}}{3+4.2}(3) = -1.75\text{V}$$

with only 8A source active,

By current division,

$$i = \frac{2}{\dfrac{6(3)}{6+3}+R+2}(8) = 1\text{A}$$

$$v_3 = 2i = 2\text{ A}$$

By superposition,

$$v = v_1 + v_2 + v_3 = 9 - 1.75 + 2 = \underline{9.25\text{V}}$$

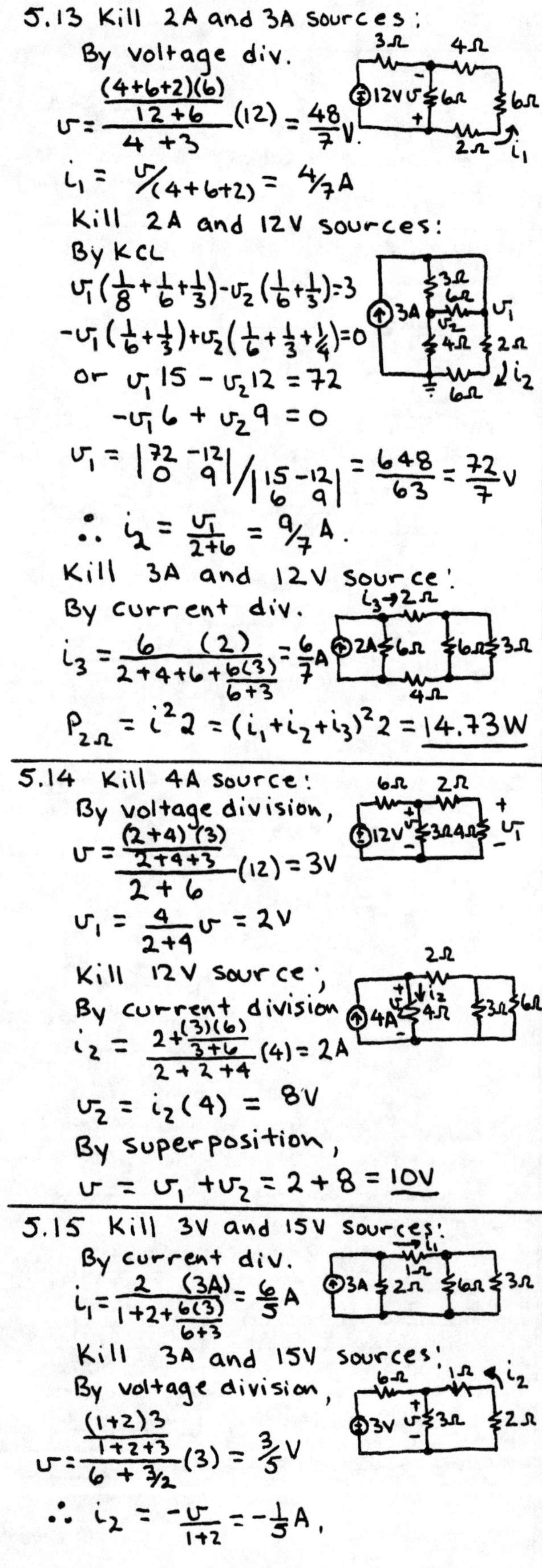

5.13 Kill 2A and 3A sources:

By voltage div.

$$v = \frac{\frac{(4+6+2)(6)}{12+6}}{4+3}(12) = \frac{48}{7}\text{V}.$$

$$i_1 = v/(4+6+2) = 4/7\text{ A}$$

Kill 2A and 12V sources:

By KCL

$$v_1\left(\tfrac{1}{8}+\tfrac{1}{6}+\tfrac{1}{3}\right) - v_2\left(\tfrac{1}{6}+\tfrac{1}{3}\right) = 3$$

$$-v_1\left(\tfrac{1}{6}+\tfrac{1}{3}\right) + v_2\left(\tfrac{1}{6}+\tfrac{1}{3}+\tfrac{1}{4}\right) = 0$$

or $v_1 15 - v_2 12 = 72$

$-v_1 6 + v_2 9 = 0$

$$v_1 = \begin{vmatrix}72 & -12\\ 0 & 9\end{vmatrix} \Big/ \begin{vmatrix}15 & -12\\ 6 & 9\end{vmatrix} = \frac{648}{63} = \frac{72}{7}\text{V}$$

$$\therefore\ i_2 = \frac{v_1}{2+6} = 9/7\text{ A}.$$

Kill 3A and 12V source:

By current div.

$$i_3 = \frac{6}{2+4+6+\frac{6(3)}{6+3}}(2) = \frac{6}{7}\text{A}$$

$$P_{2\Omega} = i^2 2 = (i_1+i_2+i_3)^2 2 = \underline{14.73\text{W}}$$

5.14 Kill 4A source:

By voltage division,

$$v = \frac{\frac{(2+4)(3)}{2+4+3}}{2+6}(12) = 3\text{V}$$

$$v_1 = \frac{4}{2+4}v = 2\text{V}$$

Kill 12V source;

By current division

$$i_2 = \frac{2+\frac{(3)(6)}{3+6}}{2+2+4}(4) = 2\text{A}$$

$$v_2 = i_2(4) = 8\text{V}$$

By superposition,

$$v = v_1 + v_2 = 2 + 8 = \underline{10\text{V}}$$

5.15 Kill 3V and 15V sources:

By current div.

$$i_1 = \frac{2}{1+2+\frac{6(3)}{6+3}}(3\text{A}) = \frac{6}{5}\text{A}$$

Kill 3A and 15V sources:

By voltage division,

$$v = \frac{\frac{(1+2)3}{1+2+3}}{6+3/2}(3) = \frac{3}{5}\text{V}$$

$$\therefore\ i_2 = \frac{-v}{1+2} = -\frac{1}{5}\text{A}.$$

5.15 cont. Kill 3A and 3V sources,

By voltage division,

$$v = \frac{\frac{(1+2)6}{1+2+6}}{3+2}(15) = 6\text{V}$$

$i_3 = \frac{6}{1+2} = 2\text{A}$. By superposition,

$$i = i_1 + i_2 + i_3 = \frac{6}{5} - \frac{1}{5} + 2 = \underline{3\text{A}}.$$

5.16 Kill the 24V source:

By KCL,

$$v\left(\tfrac{1}{3}+\tfrac{1}{6}+\tfrac{1}{2}\right) - v_1\left(\tfrac{1}{2}\right) = \tfrac{6}{3}$$

or $2v - v_1 = 4$

$-v\left(\tfrac{1}{2}\right) + v_1\left(\tfrac{1}{2}+\tfrac{1}{4}+\tfrac{1}{10}\right) = 0$ or $-10v + 17v_1 = 0$

Adding 5 times the first equations

$12v_1 = 20$ or $v_1 = \frac{10}{6}$ V

Kill the 6V source:

By KCL

$$v\left(\tfrac{1}{6}+\tfrac{1}{3}+\tfrac{1}{2}\right) - v_2\left(\tfrac{1}{2}\right) = \tfrac{24}{6}$$

or $2v - v_2 = 8$

$-v\left(\tfrac{1}{2}\right) + v_2\left(\tfrac{1}{2}+\tfrac{1}{4}+\tfrac{1}{10}\right) = 0$ or $-10v + 17v_2 = 0$

Adding 5 times the first equation

$12v_2 = 40$ or $v_2 = \frac{20}{6}$V.

By superposition,

$$v = v_1 + v_2 = \frac{10}{6} + \frac{20}{6} = \underline{5\text{V}}$$

5.17 with only the 2mA source active,

By current division and ohm's law

$$v_1 = -\left[\frac{6(2)}{6+4+2}\text{mA}\right](2\text{k}\Omega) = -2\text{V}$$

with only 8-V source active,

Voltage division

$$v_2 = \frac{2}{2+4+6}(8) = \frac{4}{3}\text{V}$$

with only the 4-mA source active,

current division

$$v_3 = \left[\frac{(4+6)(4)}{4+6+2}\text{mA}\right](2\text{k}\Omega) = \frac{20}{3}\text{V}$$

By superposition,

$$v = v_1 + v_2 + v_3 = -2 + \frac{4}{3} + \frac{20}{3} = \underline{6\text{V}}.$$

5.18 with the 6-A source dead

$$i_1 = \frac{-36}{6+12} = -2\text{A}$$

with the 36V source dead,

current division gives

$$i_2 = \frac{12}{6+12}(-6) = -4\text{A}$$

By superposition,

$$i = i_1 + i_2 = -2 - 4 = \underline{-6\text{A}}.$$

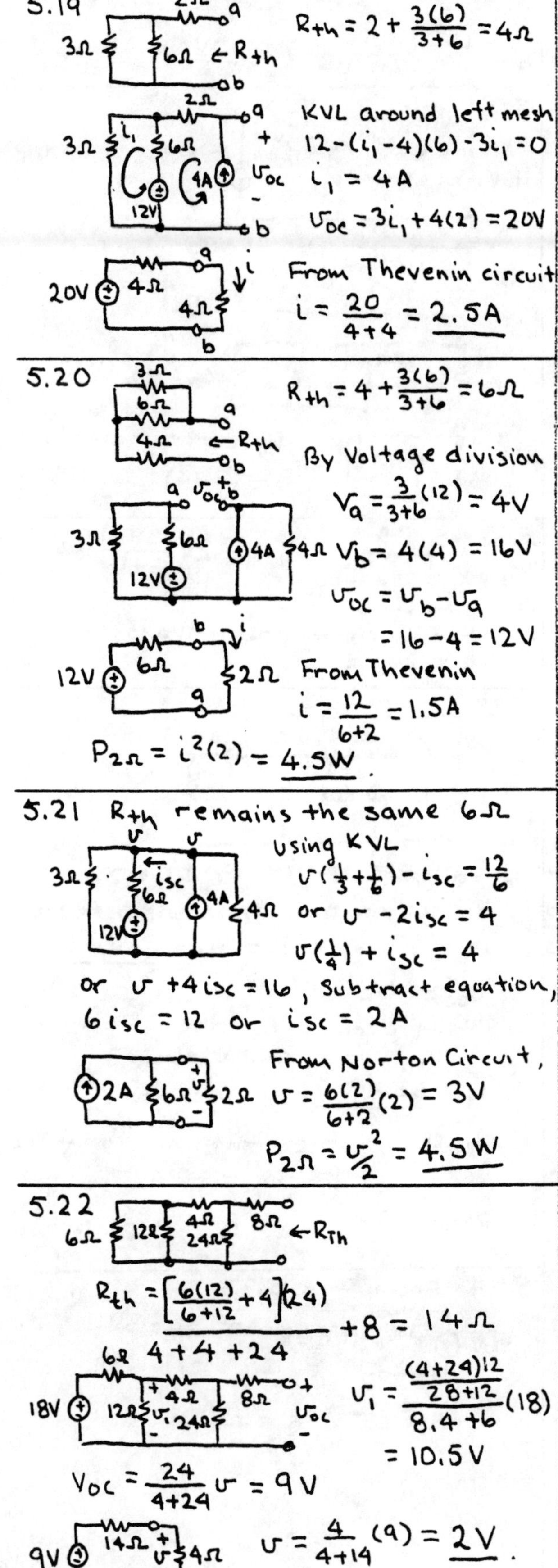

5.19 $R_{th} = 2 + \frac{3(6)}{3+6} = 4\Omega$

KVL around left mesh

$12 - (i_1 - 4)(6) - 3i_1 = 0$

$i_1 = 4A$

$v_{oc} = 3i_1 + 4(2) = 20V$

From Thevenin circuit

$i = \frac{20}{4+4} = 2.5A$

5.20 $R_{th} = 4 + \frac{3(6)}{3+6} = 6\Omega$

By Voltage division

$V_a = \frac{3}{3+6}(12) = 4V$

$V_b = 4(4) = 16V$

$v_{oc} = v_b - v_a = 16 - 4 = 12V$

From Thevenin

$i = \frac{12}{6+2} = 1.5A$

$P_{2\Omega} = i^2(2) = 4.5W$

5.21 R_{th} remains the same 6Ω

using KVL

$v(\frac{1}{3}+\frac{1}{6}) - i_{sc} = \frac{12}{6}$ or $v - 2i_{sc} = 4$

$v(\frac{1}{4}) + i_{sc} = 4$

or $v + 4i_{sc} = 16$, Subtract equation,

$6i_{sc} = 12$ or $i_{sc} = 2A$

From Norton Circuit,

$v = \frac{6(2)}{6+2}(2) = 3V$

$P_{2\Omega} = \frac{v^2}{2} = 4.5W$

5.22 $R_{th} = \frac{\left[\frac{6(12)}{6+12}+4\right](24)}{4+4+24} + 8 = 14\Omega$

$v_1 = \frac{\frac{(4+24)12}{28+12}}{8.4+6}(18) = 10.5V$

$V_{oc} = \frac{24}{4+24}v = 9V$

$v = \frac{4}{4+14}(9) = 2V$.

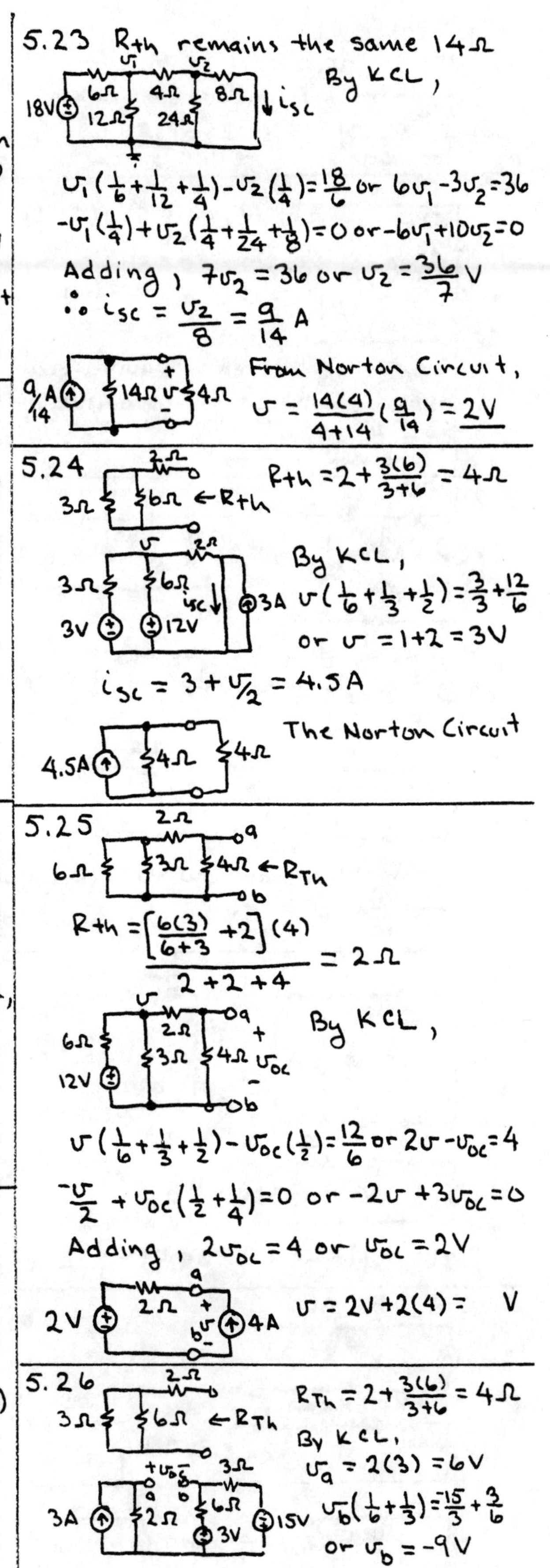

5.23 R_{th} remains the same 14Ω

By KCL,

$v_1(\frac{1}{6}+\frac{1}{12}+\frac{1}{4}) - v_2(\frac{1}{4}) = \frac{18}{6}$ or $6v_1 - 3v_2 = 36$

$-v_1(\frac{1}{4}) + v_2(\frac{1}{4}+\frac{1}{24}+\frac{1}{8}) = 0$ or $-6v_1 + 10v_2 = 0$

Adding, $7v_2 = 36$ or $v_2 = \frac{36}{7}V$

$\therefore i_{sc} = \frac{v_2}{8} = \frac{9}{14}A$

From Norton Circuit,

$v = \frac{14(4)}{4+14}(\frac{9}{14}) = 2V$

5.24 $R_{th} = 2 + \frac{3(6)}{3+6} = 4\Omega$

By KCL, $v(\frac{1}{6}+\frac{1}{3}+\frac{1}{2}) = \frac{3}{3} + \frac{12}{6}$

or $v = 1 + 2 = 3V$

$i_{sc} = 3 + v/2 = 4.5A$

The Norton Circuit

5.25 $R_{th} = \frac{\left[\frac{6(3)}{6+3}+2\right](4)}{2+2+4} = 2\Omega$

By KCL,

$v(\frac{1}{6}+\frac{1}{3}+\frac{1}{2}) - v_{oc}(\frac{1}{2}) = \frac{12}{6}$ or $2v - v_{oc} = 4$

$-\frac{v}{2} + v_{oc}(\frac{1}{2}+\frac{1}{4}) = 0$ or $-2v + 3v_{oc} = 0$

Adding, $2v_{oc} = 4$ or $v_{oc} = 2V$

$v = 2V + 2(4) = \quad V$

5.26 $R_{Th} = 2 + \frac{3(6)}{3+6} = 4\Omega$

By KCL,

$v_a = 2(3) = 6V$

$v_b(\frac{1}{6}+\frac{1}{3}) = \frac{-15}{3} + \frac{3}{6}$

or $v_b = -9V$

5.26 cont.

$v_{oc} = v_a - v_b = 6 + 9 = 15\text{V}$

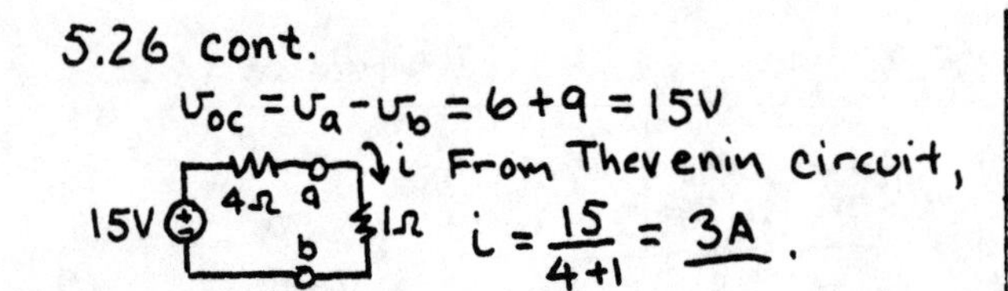

From Thevenin circuit,

$i = \frac{15}{4+1} = \underline{3\text{A}}$.

5.27

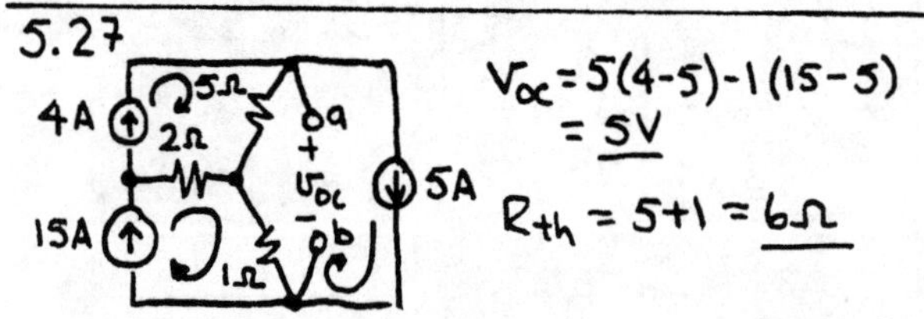

$V_{oc} = 5(4-5) - 1(15-5)$
$= \underline{5\text{V}}$

$R_{th} = 5 + 1 = \underline{6\Omega}$

From the thevenin equivalent circuit, voltage division gives

$v = \frac{4}{4+6}(5) = \underline{2\text{ V}}$.

5.28

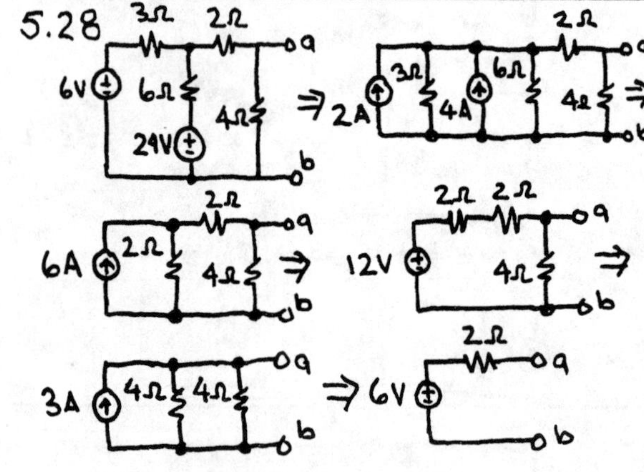

From Thevenin and voltage div.

$v = \frac{10}{10+2}(6) = \underline{5\text{V}}$.

5.29

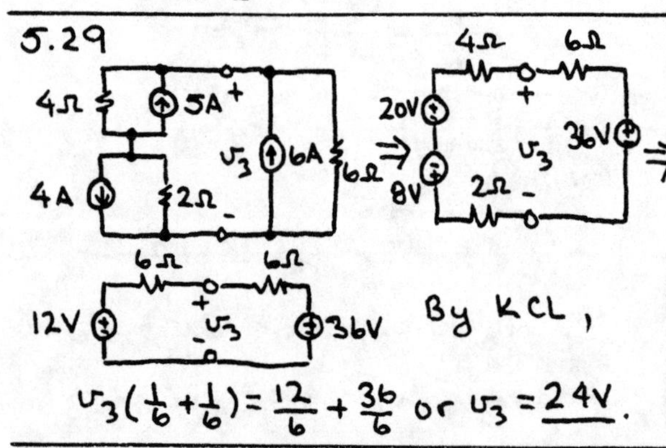

By KCL,

$v_3\left(\frac{1}{6} + \frac{1}{6}\right) = \frac{12}{6} + \frac{36}{6}$ or $v_3 = \underline{24\text{V}}$.

5.30

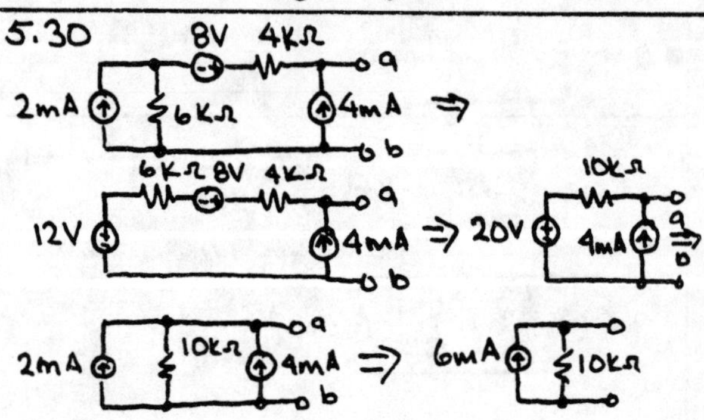

5.30 cont. From Norton Circuit

$v = \frac{(10)(2)}{2+10}(6) = \underline{10\text{V}}$.

5.31

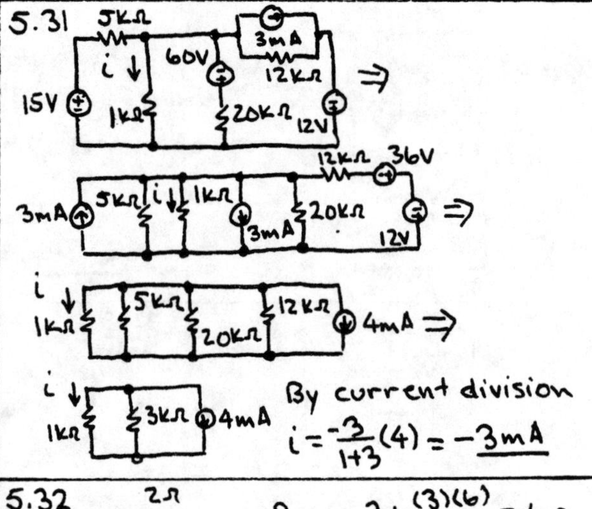

By current division

$i = \frac{-3}{1+3}(4) = \underline{-3\text{mA}}$

5.32

2Ω, 3Ω, 6Ω ← R_{Th}

$R_{Th} = 2 + \frac{(3)(6)}{3+6} = 4\Omega$

For maximum power transfer

$R = R_{Th} = \underline{4\Omega}$.

5.33

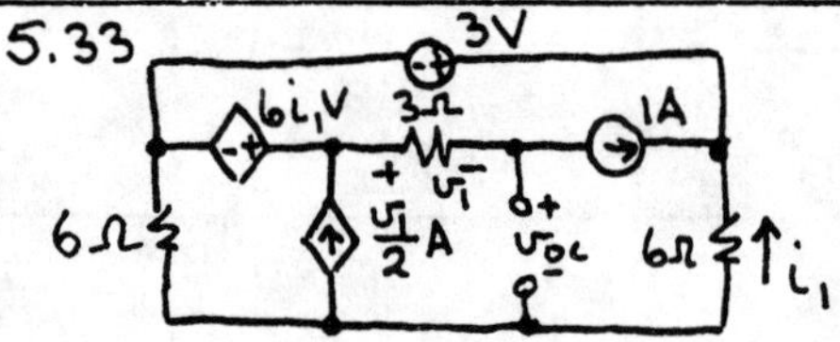

By Supernode of V_{oc}, $v_{oc} + v_1$, $v_{oc} + v_1 - 6i_1$ and $v_{oc} + v_1 - 6i_1 + 3$ then

$v_1 = (1)(3) = 3\text{V}$, $i_1 = \frac{6i_1 - v_{oc} - v_1 - 3}{6}$ or

$v_{oc} = -6\text{V}$.

Short across V_{oc}, then supernode is v_1, $v_1 - 6i_1$ and $v_1 - 6i_1 + 3$,

$i_1 = \frac{6i - v_1 - 3}{6}$ or $v_1 = -3\text{V}$.

By KCL, i_{sc} = current down through V_{oc}

$= \frac{v_1}{3} - 1 = -2\text{A}$.

$R_{th} = \frac{v_{oc}}{i_{sc}} = \frac{-6}{-2} = 3\Omega \quad \therefore R = R_{th} = \underline{3\Omega}$

5.34 From Prob. 5.33

−6V, 3Ω, + v −, $3\Omega = R$

$v = \frac{6}{2} = 3\text{V}$

$P_R = \frac{v^2}{R} = \frac{(3)^2}{3} = \underline{3\text{W}}$.

5.35

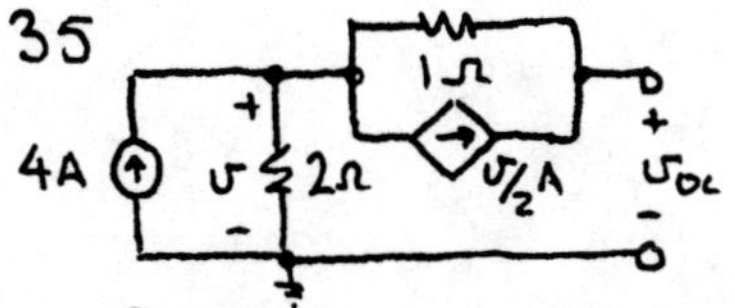

By KCL,

$v\left(\frac{1}{2}+1\right)-\frac{v_{oc}}{1}+\frac{v}{2}=4$ or

$2v-v_{oc}=4$

$v_{oc}-v=\frac{v}{2}$ or $-3v+2v_{oc}=0$

Adding $2/3$ times the second equation, $\frac{1}{3}v_{oc}=4$ or $v_{oc}=12$ V

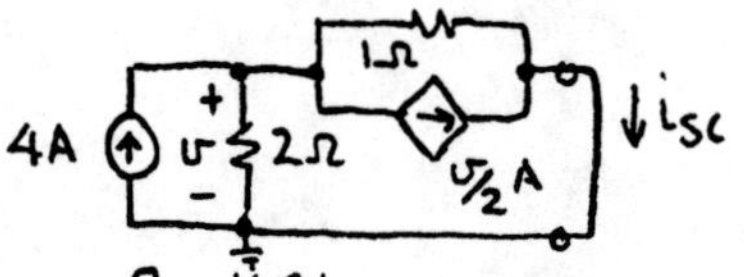

By KCL,

$v\left(\frac{1}{2}+1\right)+\frac{v}{2}=4$ or $v=2$V

$i_{sc}=\frac{v}{1}+\frac{v}{2}=2+1=3$A

$R_{th}=\frac{v_{oc}}{i_{sc}}=\frac{12}{3}=4\Omega$

$R=R_{th}=\underline{4\Omega}$.

Chapter 6

Independent Equations

6.1 Graph of a Network

6.1 Show the graph is planar.

PROBLEM 6.1

6.2 Trees and Cotrees

6.2 Find the number of tree branches and cotrees in the graph.

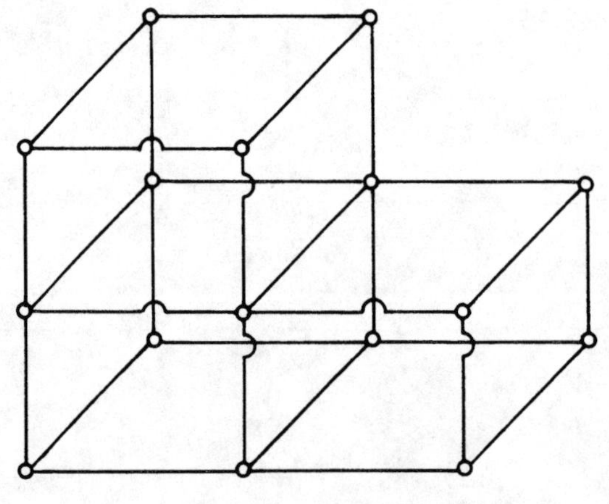

PROBLEM 6.2

6.3 Find the number of tree branches and cotrees in Prob. 4.1.

6.4 Find the number of tree branches and cotrees in Prob. 4.2.

6.5 Find the number of tree branches and cotrees in Prob. 4.3.

6.6 Find the number of tree branches and cotrees in Prob. 4.9.

6.7 Find the number of tree branches and cotrees in Prob. 4.29.

6.8 Find the number of tree branches and cotrees in Prob. 4.33.

6.9 Find the cotrees for tree (1,2,4).

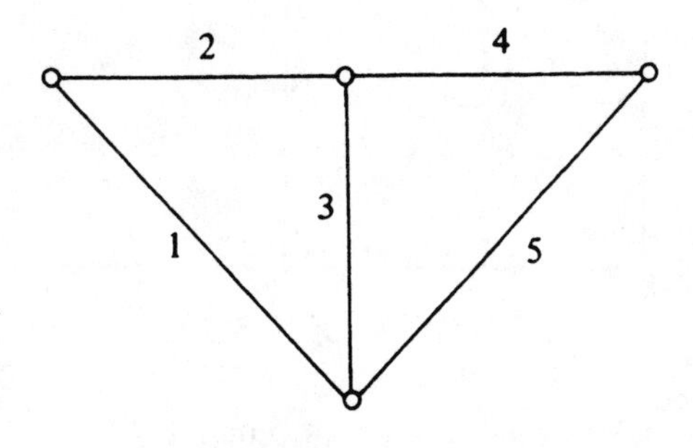

PROBLEM 6.9

6.10 In Prob. 6.9, let i_2 and i_4 be cotree currents in elements 2 and 4, directed to the right. Let i_1, i_3, and i_5 be tree currents directed downward. Find the cotree currents in terms of the tree currents.

6.3 Independent Voltage Equations

6.11 In the graph of the circuit of Prob. 4.9, select the tree of the voltage sources and the 4Ω resistor and use the method of this section to find v.

6.12 Select the tree of the voltage source, and the 3Ω and 6Ω resistors, and use the method of this section to find i.

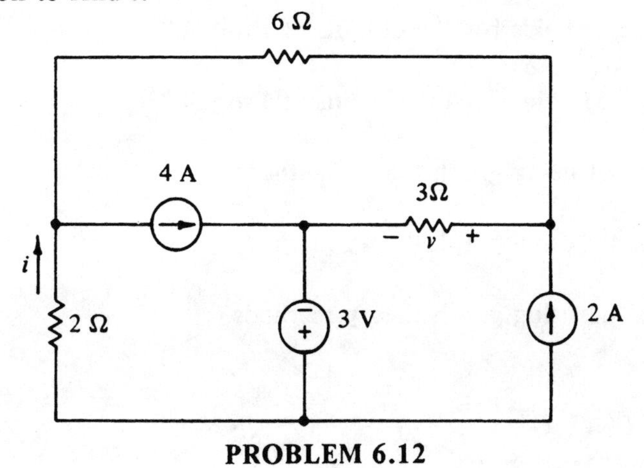

PROBLEM 6.12

6.13 Select a tree and use an appropriate graph theory method to find i.

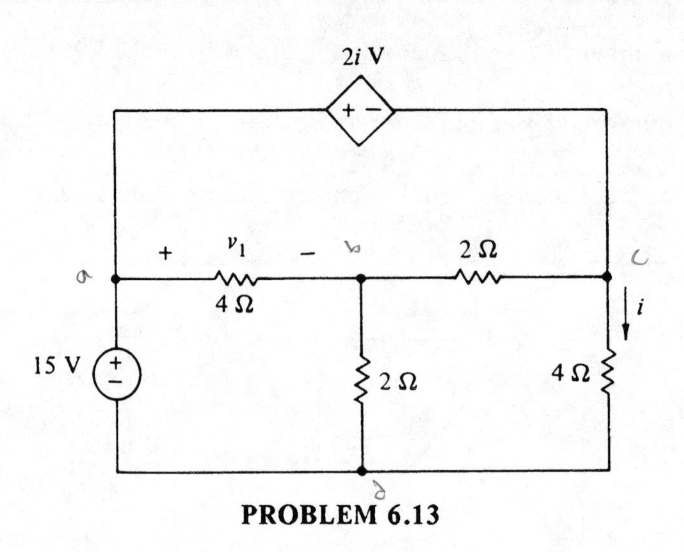

PROBLEM 6.13

6.14 Solve Prob. 4.10 using graph theory methods.

6.15 Solve Prob. 4.11 using graph theory methods.

6.4 Independent Current Equations

6.16 Show that (6.3) holds for the circuits of Prob. 4.1.

6.17 Show that (6.3) holds for the circuits of Prob. 4.2.

6.18 Show that (6.3) holds for the circuits of Prob. 4.3.

6.19 Show that (6.3) holds for the circuits of Prob. 4.9.

6.20 Show that (6.3) holds for the circuits of Prob. 4.29.

6.21 Show that (6.3) holds for the circuits of Prob. 4.33.

6.22 Solve Prob. 4.1 using graph theory methods.

6.23 Solve Prob. 4.2 using graph theory methods.

6.24 Solve Prob. 4.3 using graph theory methods.

6.1 Identify the nodes and branches of the graph as in (a) then redraw as in (b).

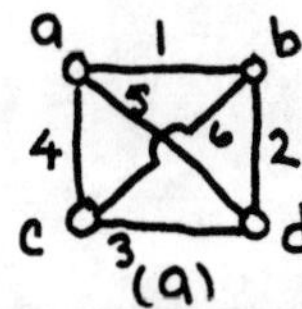

(a)

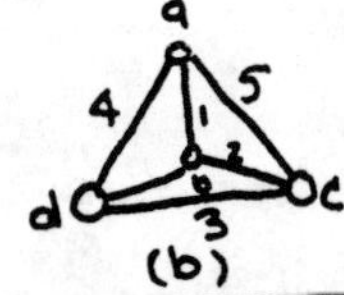

(b)

6.2 The number of tree branches is $N-1 = 16-1 = \underline{15}$. The remaining branches, $B-(N-1) = 28-15 = \underline{13}$, are Cotrees.

6.3 The number of tree branches is $N-1 = 3-1 = \underline{2}$. The remaining branches, $B-(N-1) = 5-2 = \underline{3}$.

6.4 Branches, $N-1 = 3-1 = \underline{2}$.
Cotrees, $B-(N-1) = 6-2 = \underline{4}$.

6.5 Branches, $N-1 = 3-1 = \underline{2}$.
Cotrees, $B-(N-1) = 5-2 = \underline{3}$.

6.6 Branches, $N-1 = 4-1 = \underline{3}$.
Cotrees, $B-(N-1) = 6-3 = \underline{3}$.

6.7 Branches, $N-1 = 3-1 = \underline{2}$.
Cotrees, $B-(N-1) = 5-2 = \underline{3}$.

6.8 Branches, $N-1 = 3-1 = \underline{2}$.
Cotrees, $B-(N-1) = 5-2 = \underline{3}$.

6.9 The branches not in the tree are cotrees $\underline{(3,5)}$.

6.10 By KCL,
$i_2 = -i_1, \; i_4 = i_5$.

6.11

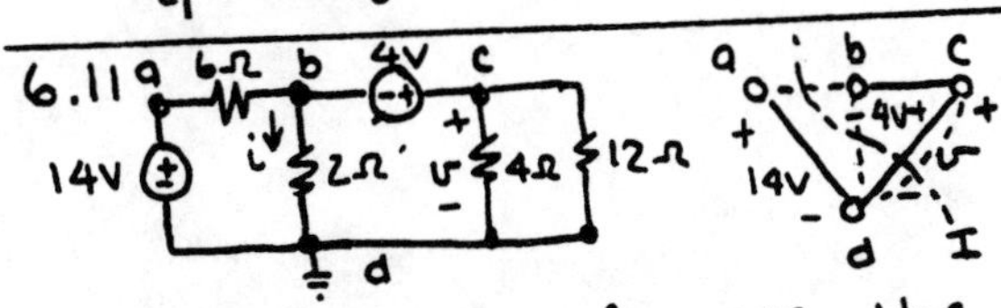

with d as reference, the node voltages are, $v_a = 14V$, $v_b = v-4$, $v_c = v$,
KCL across I yields,

$$\frac{v-4-14}{6} + \frac{v-4}{2} + \frac{v}{4} + \frac{v}{12} = 0 \text{ or}$$

$$v(2+6+3+1) = 8+28+24$$

$v = \underline{5V}$.

6.12

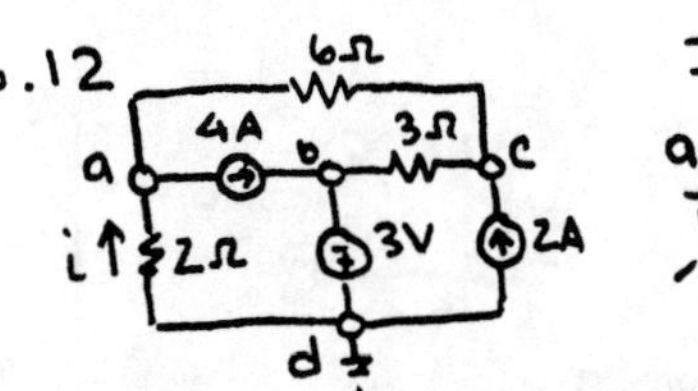

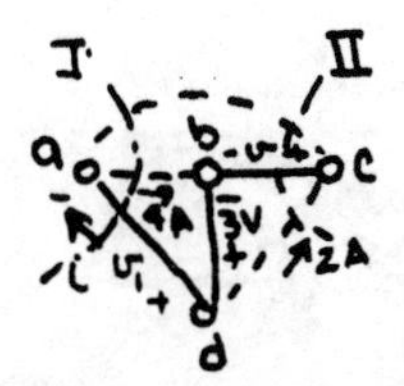

with d as reference, the node voltages are $v_a = -v_1$, $v_b = -3$, $v_c = -3+v$; KCL across I and II

$$\frac{-v_1+3-v}{6} + 4 - \frac{v_1}{2} = 0 \text{ or } 4v_1 + v = 27$$

$$\frac{-3+v+v_1}{6} + \frac{v}{3} - 2 = 0 \text{ or } v_1 + 3v = 15$$

subtract, 3 times the first equation.

$11 v_1 = 66$ or $v_1 = 6V$

$\therefore i = \frac{v_1}{2} = \frac{6}{2} = \underline{3A}$.

6.13

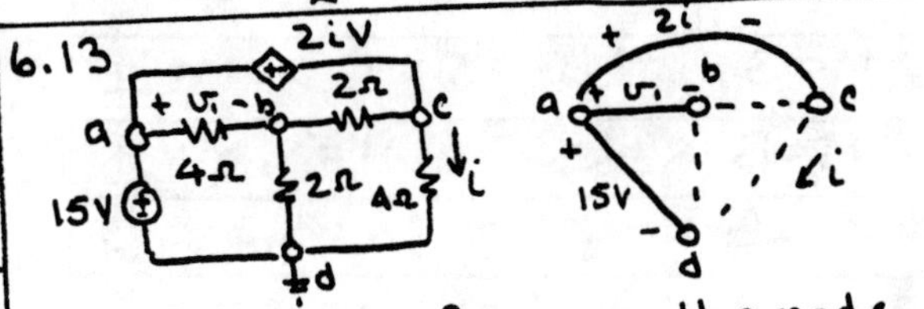

with d as reference, the node voltages are $v_a = 15$, $v_b = 15 - v_1$,
$v_c = 15 - 2i$,
$i = \frac{v_c}{4}$, then $v_c = 15 - \frac{v_c}{2}$ or $v_c = 10V$.

$\therefore i = \frac{10}{4} = \underline{2.5A}$.

6.14

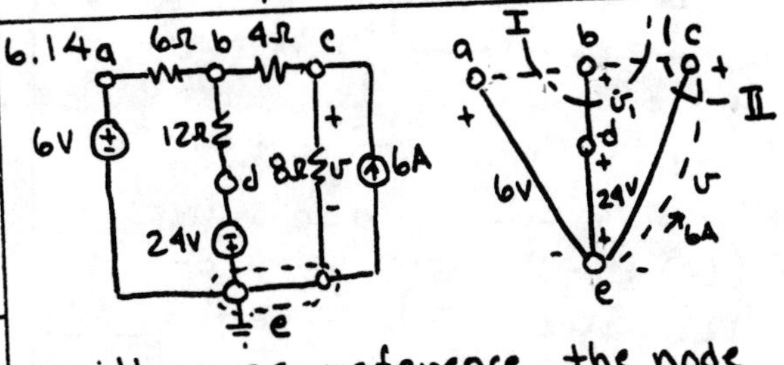

with e as reference, the node voltages are $v_a = 6V$, $v_b = v_1 - 24$,
$v_c = v$, $v_d = -24V$.
KCL across I and II yields.

$$\frac{v_1-24-6}{6} + \frac{v_1-24+24}{12} + \frac{v_1-24-v}{4} = 0 \text{ or}$$

$6v_1 - 3v = 132$ or $2v_1 - v = 44$

$$\frac{v-v_1+24}{4} + \frac{v}{8} - 6 = 0 \text{ or } -2v_1 + 3v = 0$$

Adding, $2v = 44$ or $v = \underline{22V}$.

6.15

With d as reference, node voltages are, $v_a = 18 + v_1$, $v_b = v$, $v_c = v_1$; By KCL around I and II,

$$\frac{18+v_1-v}{1} + \frac{18+v_1}{2} + \frac{v_1}{2} + \frac{v_1-v}{2} = 0 \text{ or}$$

$$3v_1 - 2v = -30$$

$$\frac{v-18-v_1}{1} + \frac{v-v_1}{2} + \frac{v}{2} = 0 \text{ or}$$

$$-3v_1 + 4v = 36$$

Adding, $2v = 6$ or $v = \underline{3V}$.

6.16 $M = B - N + 1$, $B = 5$, $N = 3$, $M = \underline{3}$.

6.17 $M = B - N + 1$, $B = 6$, $N = 3$, $M = \underline{4}$.

6.18 $M = B - N + 1$, $B = 5$, $N = 3$, $M = \underline{3}$

6.19 $M = B - N + 1$, $B = 6$, $N = 4$, $M = \underline{3}$

6.20 $M = B - N + 1$, $B = 5$, $N = 3$, $M = \underline{3}$

6.21 $M = B - N + 1$, $B = 5$, $N = 3$, $M = \underline{3}$

6.22

By KCL,

$i_{bc} = i$, $i_{ca} = i - 9$, $i_{ab} = i - 2$;

By KVL around bcab gives

$$i(4) + (i-9)2 + (i-2)6 = 0 \text{ or}$$

$$12i = 30 \text{ or } i = 10/4;\ v = 4i = \underline{10V}.$$

6.23

By KCL, (currents in mA)

$i_{ab} = i$, $i_{bc} = i + 1$, $i_{ca} = i - 7$;

By KVL around abca yeilds

$$i(2) + (i+1)2 + (i-7)2 = 0 \text{ or}$$

$$6i = 12 \text{ or } i = \underline{2\,mA}.$$

6.24

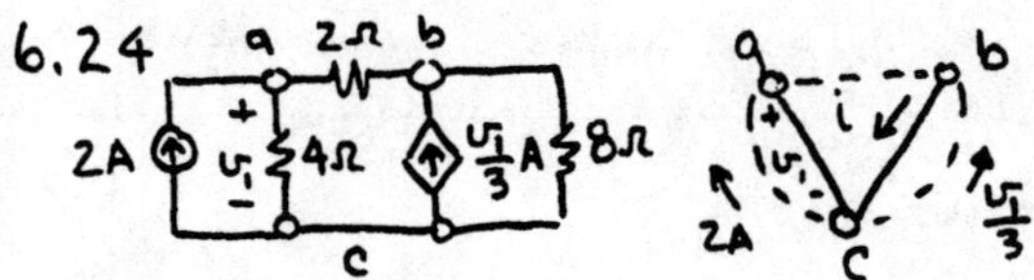

By KCL,

$i_{bc} = i$, $i_{ca} = i - \frac{v_1}{3} - 2$, $i_{ab} = i - \frac{v_1}{3}$;

since $v_1 = -i_{ca}4 = (2 + \frac{v_1}{3} - i)4$ or

$$v_1 = 12i - 24.$$

then $i_{ca} = 6 - 3i$, $i_{ab} = 8 - 3i$;

By KVL around abca gives

$$i8 + (6-3i)4 + (8-3i)2 = 0$$

$$10i = 40 \text{ or } i = 4A$$

$\therefore P_{8\Omega} = i^2 8 = \underline{128W}$

Chapter 7

Energy Storage Elements

7.1 Capacitors

7.1 A constant current of $10\mu A$ is charging a $5\mu F$ capacitor. If the capacitor was previously uncharged, find the charge and voltage on it after 20s.

7.2 A $10\mu F$ capacitor has a voltage $v(t) = f(t)$ V, as shown. Find the current i at $t = -7, -3, 1, 3,$ and 7ms.

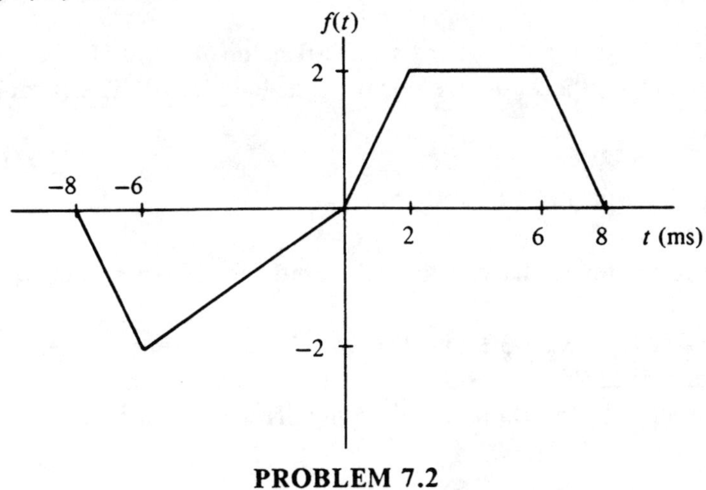

PROBLEM 7.2

7.3 If $f(t)$ in Prob. 7.2 is the current in mA of a $1\mu F$ capacitor, find its voltage at $t = -6, 0, 2,$ and 8ms.

7.4 A $0.1\mu F$ capacitor has a current of $3\cos(2000t)$mA. Find its voltage $v(t)$ if $v(0) = -15$V.

7.5 Find the charge residing on each plate of a $2\mu F$ capacitor that is charged to 50V. If the same charge resides on a $1\mu F$ capacitor, what is the voltage?

7.6 Determine the voltage required to store $100\mu C$ on a $2\mu F$ capacitor. What time will be required for a constant current of 50mA to deliver this charge?

7.7 The voltage across a $10\mu F$ capacitor is as shown. Find the current i at t = 5ms, and the power p at t = 20ms.

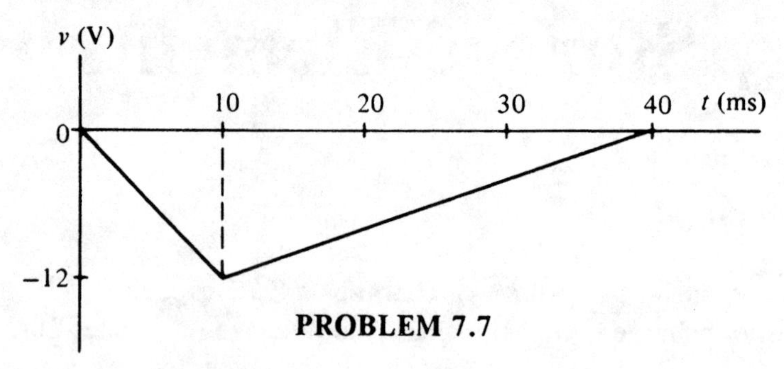

PROBLEM 7.7

7.8 How long will it take for a constant 25mA current to deliver a charge of $100\mu C$ to a $1\mu F$ capacitor? What will then be the capacitor voltage?

7.9 Find the voltage for $t > 0$ across a 2Ω resistor and a 0.5F capacitor in series if the current entering their positive terminals is $4\cos(2t)$ A. The capacitor is uncharged at $t = 0$.

7.2 Energy Storage in Capacitors

7.10 A $10\mu F$ capacitor is charged to 20V. Find the charge and energy.

7.11 Let $C = \frac{1}{3}$F, $R_1 = R_2 = 3\Omega$, and V = 9V. If the current in R_2 at $t = 0^-$ is 1A directed downward, find at $t = 0^-$ and at $t = 0^+$ (a) the charge on the capacitor, (b) the current in R_1 directed to the right, and (c) the current in C directed downward.

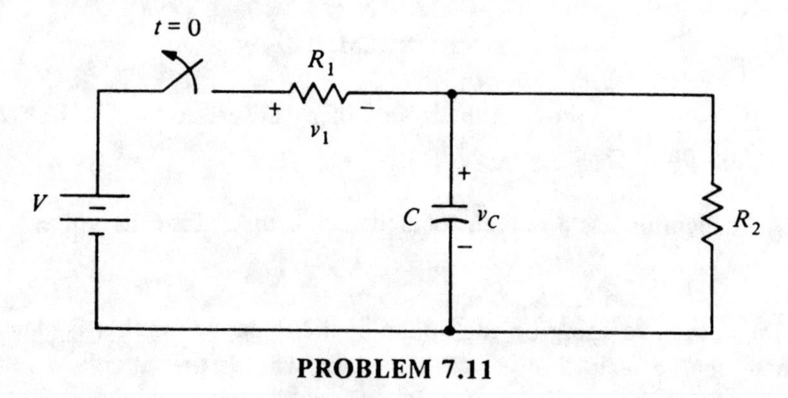

PROBLEM 7.11

7.12 The current entering the positive terminal of a 0.25F capacitor is $i = 6\cos(2t)$A. If the initial capacitor voltage (at $t = 0$) is 4V, find the energy stored in the capacitor at $t = 3\pi/4$ s.

7.13 The current in a 0.5F capacitor is $i = 6t$ A and the voltage at $t = 0$ is $v(0)$ = 2V. Find the energy stored in the capacitor at t = 1s.

7.14 Find the work required to charge a $0.1\mu F$ capacitor to 200V.

7.15 A voltage of $4e^{-t}$V appears across a parallel combination of a 1Ω resistor and a 0.25F capacitor. Find the power absorbed by the combination.

7.16 Find $i_1(0^-)$, $i_1(0^+)$, $i_2(0^-)$, and $i_2(0^+)$ if the switch is opened at $t = 0$, and $v_1(0^-) = 18$V and $v_2(0^-) = 6$V.

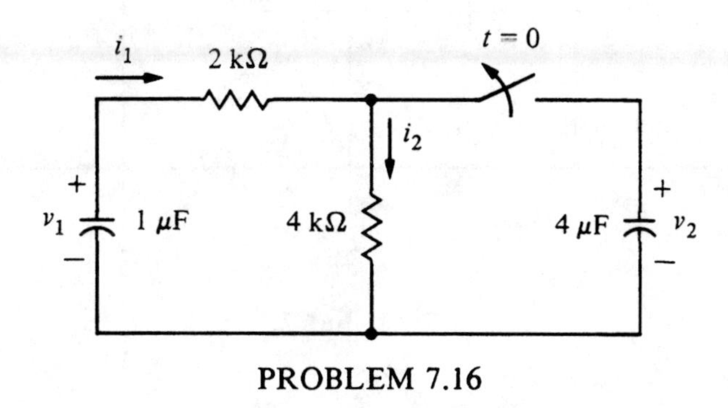

PROBLEM 7.16

7.3 Series and Parallel Capacitors

7.17 Find the equivalent capacitance and the initial voltage.

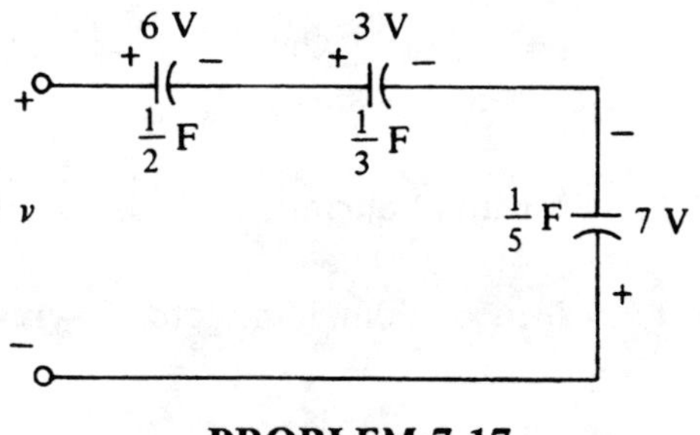

PROBLEM 7.17

7.18 Find the equivalent capacitance.

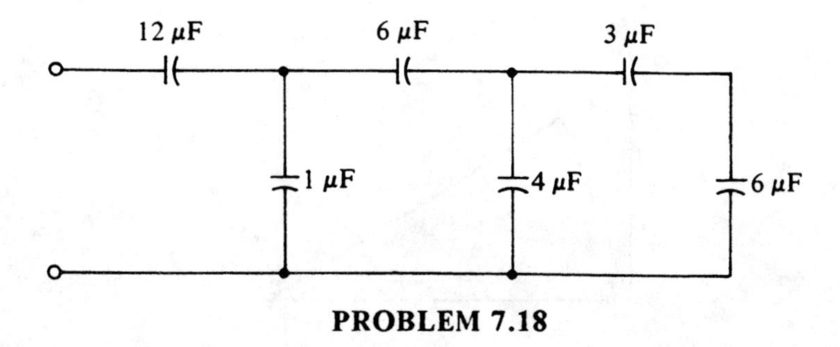

PROBLEM 7.18

7.19 The capacitances shown are all in μF. Find C_{eq} at terminals a-b.

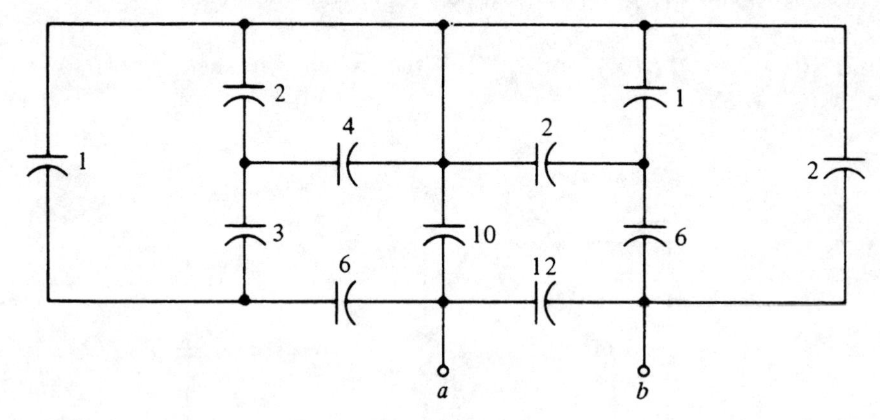

PROBLEM 7.19

7.4 Inductors

7.20 If a 600mH inductor has $i(t) = f(t)$ mA, where $f(t)$ is given in Prob. 7.2, find its voltage at t = -3, 1, 3 and 7ms.

7.21 If a 1mH inductor has $v(t) = f(t)$ mV, where $f(t)$ is given in Prob. 7.2 find its current at -6, 0 and 8ms, if the current at -8ms is 0.

7.22 A 100mH inductor has a terminal current of 100sin(100t)mA. Find $v(t)$.

7.23 Find the current for $t > 0$ in a 100mH inductor having a terminal voltage of $10e^{-100t}$V if $i(0)$ = 0A.

7.24 Find the flux linkage of a 50mH inductor at 10ms, 40ms and 80ms if the current is $f(t)$ mA, as shown.

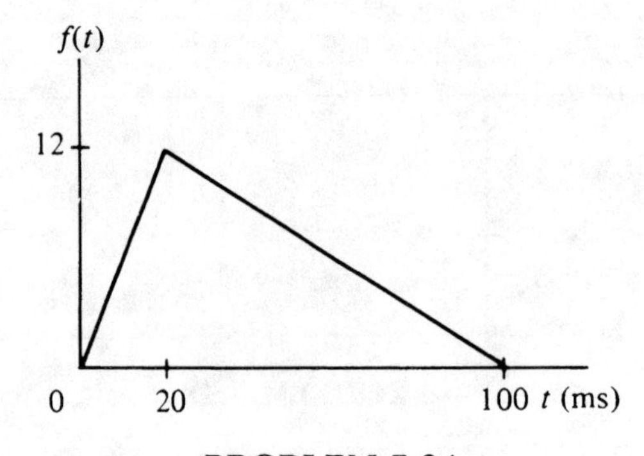

PROBLEM 7.24

7.25 Find the voltage across a 10H inductor a t = 10ms, 40ms, and 80ms if its current is $f(t)$ mA, given in Prob. 7.24.

7.26 Find the current i through a 120H inductor at t = 20ms, 60ms and 100ms if its voltage is $f(t)$ V, given in Prob. 7.24 and $i(0)$ = 2mA.

7.27 Find the terminal voltage of a 100mH inductor if its current is (a) 5A, (b) $100t$ A, (c) $10\sin(200t)$A, (d) $10e^{-50t}$A.

7.28 Find the current $i(t)$ of a 10mH inductor with a voltage of $2\sin(1000t)$V, if $i(0) = 0$.

7.5 Energy Storage in Inductors

7.29 A 5mH inductor has a current of 200mA. Find the flux linkage and the energy.

7.30 Let I = 3A, R_1 = 6Ω, R_2 = 12Ω, L = 2H, and $i_1(0^-)$ = 2A. If the switch is open at $t = 0^-$, find $i_L(0^-)$, $i_L(0^+)$, $i_1(0^+)$, and $di_L(0^+)/dt$.

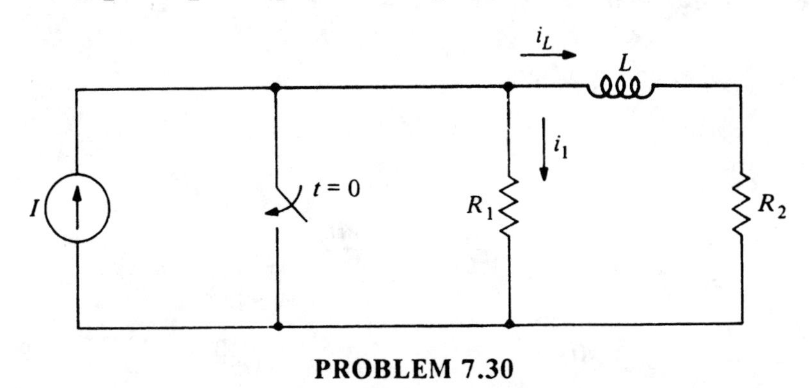

PROBLEM 7.30

7.31 The voltage of a 0.5H inductor is $v = 6t$ V and the initial current is $i(0)$ = 2A. Find the energy stored in the inductor at t = 1s.

7.32 The voltage of a 0.25H inductor is $v = 6\cos(2t)$V and the current at $t = 0$ is 0. Find the energy stored in the inductor at $3\pi/4$ s and the power delivered to the inductor at $\pi/8$ s.

7.33 Find the work required to establish a current of 60mA in a 100mH inductor.

7.34 If $v(0^-) = 9\text{V}$, $i(0^-) = 1\text{A}$, and the switch is opened at $t = 0$, find $i(0^+)$ and $di(0^+)/dt$.

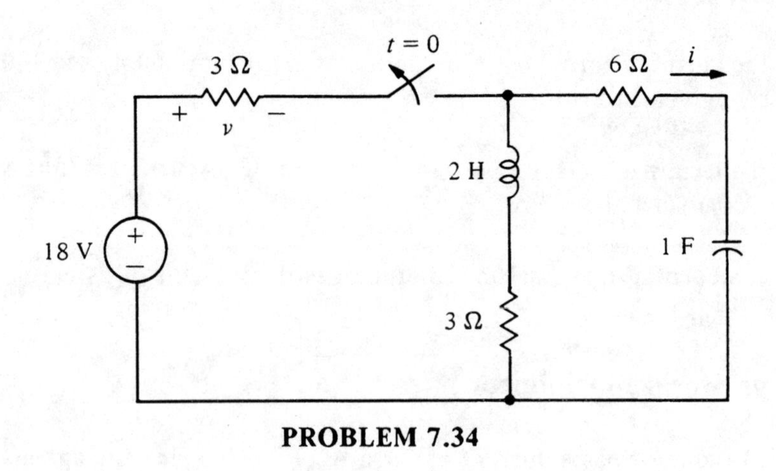

PROBLEM 7.34

7.6 Series and Parallel Inductors

7.35 Find the equivalent inductance.

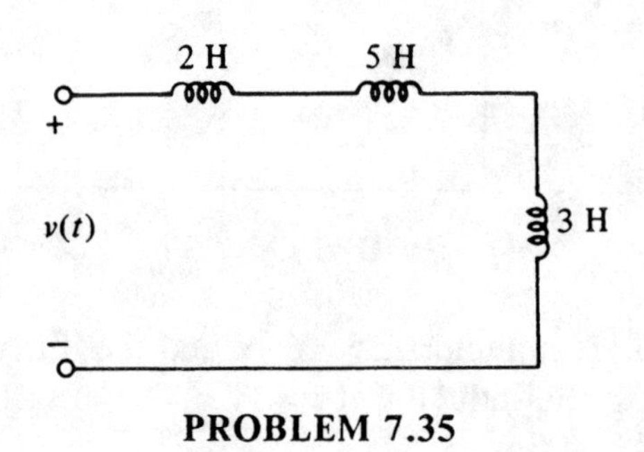

PROBLEM 7.35

7.36 Find the equivalent inductance.

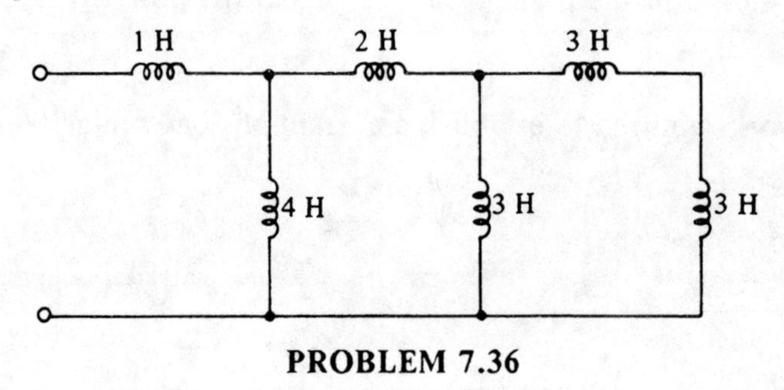

PROBLEM 7.36

7.37 Determine L_{eq}.

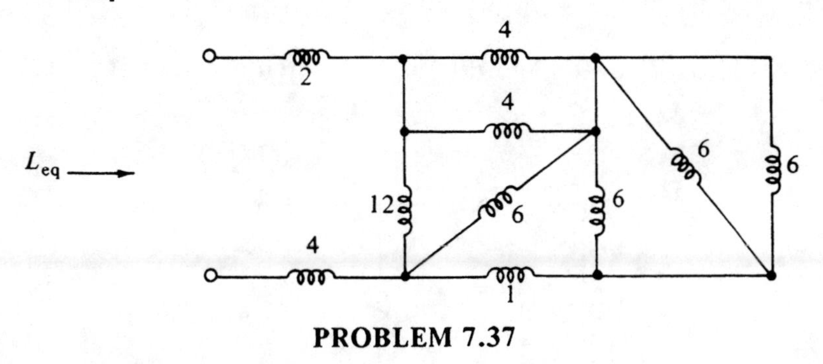

PROBLEM 7.37

7.8 Practical Capacitors and Inductors

7.38 Mylar capacitors have a resistance-capacitance product of $10^5 \Omega$. Find the equivalent parallel resistor in the following figure for the following capacitors: (a) 33pF, (b) 0.001μF, and (c) 2.7μF.

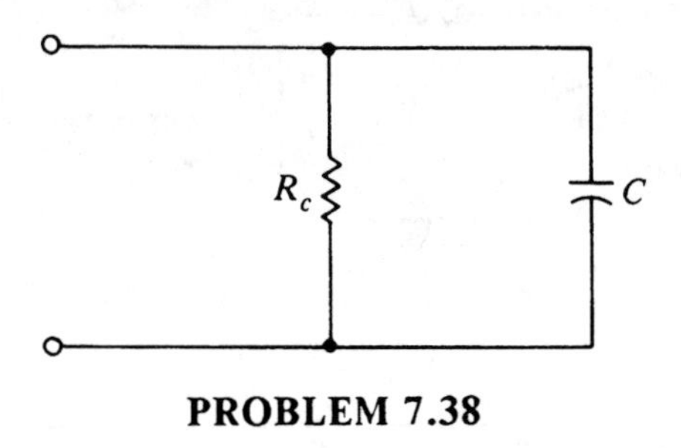

PROBLEM 7.38

7.9 Duality and Linearity

7.39 Determine a dual circuit for the network of Prob. 7.19.

7.40 Determine a dual circuit for the network of Prob. 7.37.

7.10 Singular Circuits

7.41 Find $v_1(0^+)$, $v_2(0^+)$, $i(0^+)$ and $i(0^-)$ if the switch is closed at $t = 0$, and $v_1(0^-) = 14$V and $v_2(0^-) = 6$V.

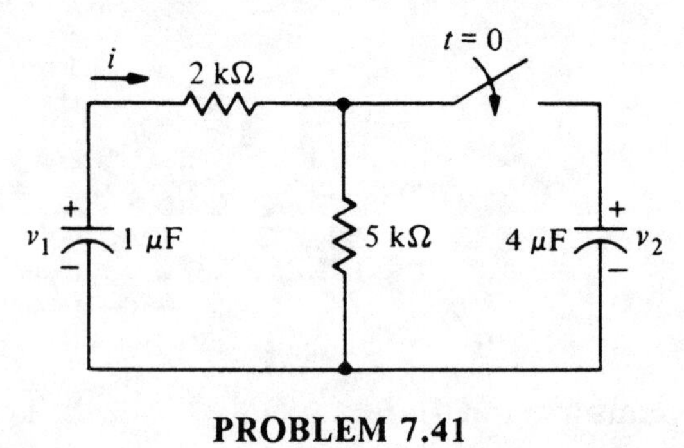

PROBLEM 7.41

7.42 If $v(0^-) = 6$V and the switch is opened, as indicated, at $t = 0$, find $i_1(0^-)$, $i_2(0^-)$, $i_1(0^+)$, and $i_2(0^+)$.

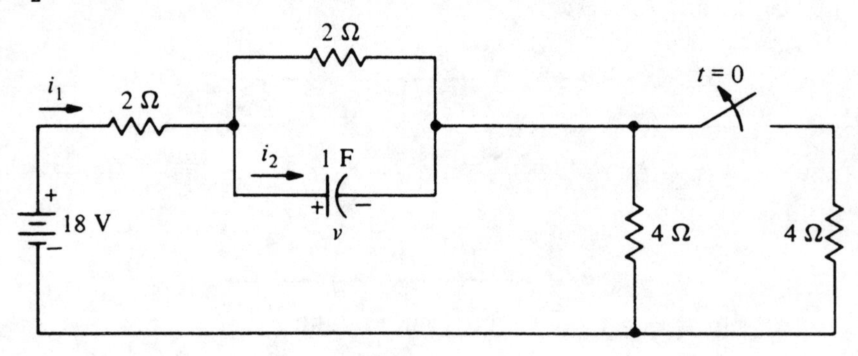

PROBLEM 7.42

7.43 If $i_1(0^-) = 1$A and the switch is opened at $t = 0$, find $i_1(0^+)$, $i_2(0^-)$, $i_2(0^+)$, $v(0^-)$, and $v(0^+)$.

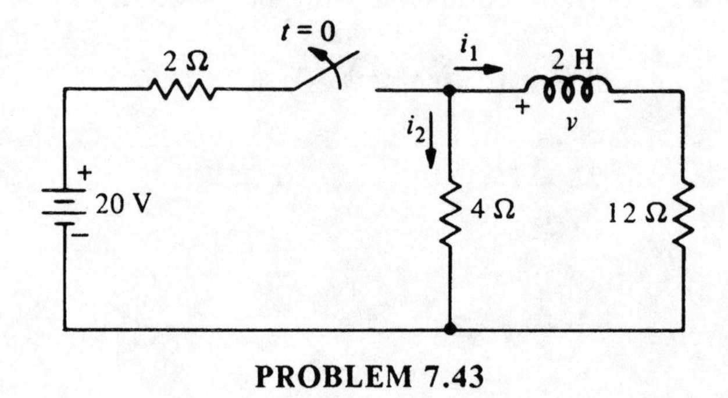

PROBLEM 7.43

7.44 If $v_1(0^-) = 9\text{V}$, $i_1(0^-) = 3\text{A}$, and the switch is opened at $t = 0$, find $v_1(0^+)$, $i_1(0^+)$, $i_2(0^-)$, $i_2(0^+)$, $v_2(0^-)$, and $v_2(0^+)$.

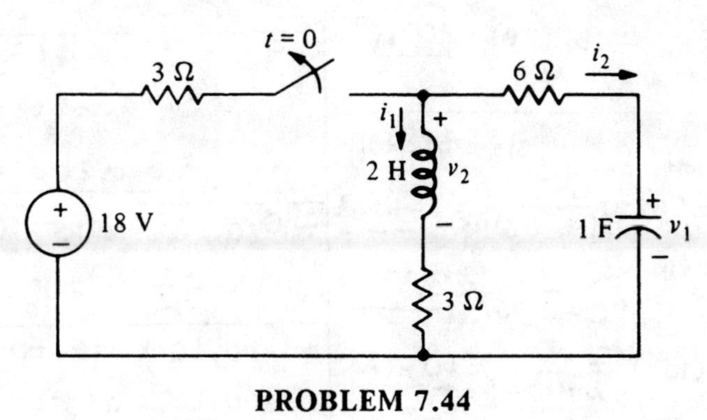

PROBLEM 7.44

7.1 $q = \int_0^{20} i\,dt + q(0) = \int_0^{20} 10^{-5}\,dt = 10^{-5}(20) = \underline{200\,\mu C}$

$v = q/C = \frac{200\times10^{-6}}{5\times10^{-6}} = \underline{40V}$.

7.2 $i = C\frac{dv}{dt}$; $\frac{dv}{dt}$ is the slope of graph.

$t=-7ms: i = (10^{-5})\frac{0-2}{[-6-(-8)]10^{-3}} = \underline{-10mA}$

$t=-3ms: i=(10^{-5})\frac{0+2}{[0+6]10^{-3}} = \underline{3.33mA}$

$t=1ms: i=(10^{-5})\frac{2-0}{[2-0]10^{-3}} = \underline{10\,mA}$

$t=3ms: i=(10^{-5})(0) = \underline{0mA}$

$t=7ms: i=(10^{-5})\frac{0-2}{[8-6]10^{-3}} = \underline{-10mA}$.

7.3 $v = \frac{1}{C}\int_{-\infty}^{t} i\,dt + v(t_0)$;

area under the graph is the integral and $v(t_0)=0$ at less than $-8ms$.

$v(-6ms) = \frac{1}{10^{-6}}\left[\frac{1}{2}(2ms)(-2mA)\right]+0 = \underline{-2V}$.

$v(0) = \frac{1}{10^{-6}}\left[\frac{1}{2}(6ms)(2mA)\right]-2 = \underline{-8V}$.

$v(2ms) = \frac{1}{10^{-6}}\left[\frac{1}{2}(2ms)(2mA)\right]-8 = \underline{-6V}$.

$v(8ms) = \frac{10^{-6}}{10^{-6}}\left[(2)(4)+\frac{1}{2}(2)(2)\right]-6 = \underline{4V}$.

7.4 $v = \frac{1}{C}\int_0^t i\,dt + v(0)$

$= \frac{1}{10^{-7}}\int_0^t [3\cos 2000t]\times10^{-3}\,dt - 15$

$= \underline{15\sin(2000t) - 15\,V}$

7.5 $q = Cv = (2\times10^{-6})(50) = \underline{100\,\mu C}$ on each plate (±)

$v = q/C = 100\times10^{-6}/10^{-6} = \underline{100V}$.

7.6 $v = q/C = 100\times10^{-6}/2\times10^{-6} = \underline{50V}$

$q = \int_{t_0}^{t_1} i\,dt = i(t_1-t_0)$

$\therefore (t_1-t_0)$ = time to deliver charge $= q/i = \frac{100\times10^{-6}}{50\times10^{-3}} = \underline{2ms}$

7.7 $i = C\frac{dv}{dt}$; $\frac{dv}{dt}$ is the slope of graph

$i(5ms) = (10^{-5})\left[\frac{-12-0}{(10-0)10^{-3}}\right] = \underline{-12mA}$.

$i(20ms) = (10^{-5})\left[\frac{0+12}{(40-10)10^{-3}}\right] = 4mA$

$v(20ms) = \frac{2}{3}(12) = 8V$; $p = vi = \underline{32mW}$

7.8 $q = it$ or $t = \frac{100\times10^{-6}}{25\times10^{-3}} = \underline{4ms}$

7.9 $v = Ri + \frac{1}{C}\int_0^t i\,dt + v(0)$, $v(0)=0$

$v = 2(4\cos 2t) + \frac{1}{0.5}\int_0^t 4\cos 2t\,dt$

$= \underline{8\cos 2t + 4\sin 2t}$ V

7.10 $q = Cv = (10^{-5})(20) = \underline{200\,\mu C}$

$W = \frac{1}{2}Cv^2 = \frac{10^{-5}}{2}(20)^2 = \underline{2mJ}$.

7.11 $v_c(0^-) = 1R_2 = 1(3) = 3V = v_c(0^+)$

(a) $q(0^-) = q(0^+) = Cv(0^-) = \frac{1}{3}(3) = \underline{1C}$

(b) $i_{R_1}(0^-) = \frac{v_1(0^-)}{R_1} = \frac{V - v_c(0^-)}{R_1} = \frac{9-3}{3} = \underline{2A}$

$i_{R_1}(0^+) = \underline{0}$ since the circuit is open

(c) $i_c(0^-) = i_{R_1}(0^-) - \frac{v_c(0^-)}{R_2} = 2 - \frac{3}{3} = \underline{1A}$

$i_c(0^+) = i_{R_1}(0^+) - \frac{v_c(0^+)}{R_2} = 0 - \frac{3}{3} = \underline{-1A}$

7.12 $v = \frac{1}{C}\int_0^{3\pi/4} i\,dt + v(0)$

$= \frac{1}{0.25}\int_0^{3\pi/4} 6\cos 2t\,dt + 4 = -12+4 = -8V$

$W = \frac{1}{2}Cv^2 = \frac{1}{2}(.25)(-8)^2 = \underline{8J}$

7.13 $v = \frac{1}{C}\int_0^1 i\,dt + v(0)$

$= \frac{1}{0.5}\int_0^1 6t\,dt + 2 = 6+2 = 8V$

$W = \frac{1}{2}Cv^2 = \frac{1}{2}(0.5)(8)^2 = \underline{16J}$

7.14 $W = \frac{1}{2}Cv^2 = \frac{1}{2}(10^{-7})(200)^2 = \underline{2mJ}$

7.15 $i = \frac{v}{R} + C\frac{dv}{dt} = \frac{4e^{-t}}{1} + \frac{1}{4}\frac{d}{dt}(4e^{-t})$

$= 3e^{-t}A$

$p = vi = (4e^{-t})(3e^{-t}) = \underline{12e^{-t}W}$.

7.16 $v_1(0^+) = v_1(0^-) = 18V$;

$v_2(0^+) = v_2(0^-) = 6V$,

$i_2(0^-) = \frac{v_2(0^-)}{4K\Omega} = \underline{1.5mA}$

$i_1(0^-) = \frac{v_1(0^-) - v_2(0^-)}{2K\Omega} = \underline{6mA}$

At $t=0^+$, the switch is open and $i_1(0^+) = i_2(0^+)$.

$\therefore i_1(0^+) = i_2(0^+) = \frac{v_1(0^+)}{(2+4)K\Omega} = \underline{3mA}$

7.17 $\frac{1}{C_s} = \frac{1}{1} + \frac{1}{1/3} + \frac{1}{1/5}$ or $C_s = \underline{.1F}$

$v(t_0) = 6+3-7 = \underline{2V}$

7.18 (circuit reduction: 12, 6, 3 series; 1, 4, 6 shunt ⇒ 12, 6; 1, 4, 2 ⇒ 12, 6; 1, 6 ⇒ 12; 1, 3 ⇒ 12; 4 ⇒) $3\,\mu F$

7.19 (circuit reduction ⇒ ... ⇒ 3, 12 ⇒) $15\,\mu F$

7.20 $v = L\frac{di}{dt}$; slope of graph equals $\frac{di}{dt}$

$v(-3ms) = (600\times10^{-3})\frac{(0+2)}{(0+6)} = \underline{0.2V}$

$v(1ms) = (600\times10^{-3})\left[\frac{2-0}{2-0}\right] = \underline{0.6V}$

$v(3ms) = (600\times10^{-3})(0) = \underline{0\,V}$

$v(7ms) = (600\times10^{-3})\left[\frac{0-2}{8-6}\right] = \underline{-0.6V}$

7.21 $i = \frac{1}{L}\int_{-\infty}^{t} v\,dt + i_o$,

$i(-6) = \frac{1}{10^{-3}}\left[-\frac{1}{2}(2mV)(2ms)\right] + 0 = \underline{-2mA}$

$i(0) = \frac{1}{10^{-3}}\left[-\frac{1}{2}(2mV)(6ms)\right] - 2mA = \underline{-8mA}$

$i(8) = \frac{1}{10^{-3}}\left[(2mV)(6ms)\right] - 8mA = \underline{4mA}$

7.22 $v = L\,di/dt$

$v = (0.1)\,d/dt(100\sin 100t)$

$= \underline{\cos 100t\ V}$.

7.23 $i = \frac{1}{L}\int_0^t v\,dt + i_o$

$= \frac{1}{0.1}\int_0^t 10e^{-100t}\,dt + 0$

$= \underline{1 - e^{-100t}\,A}$

7.24 $\lambda = Li$,

$\lambda(10ms) = (50\times10^{-3})(\frac{12}{2}mA) = \underline{300\mu Wb}$

$\lambda(40ms) = (0.05)[\frac{3}{4}(12mA)] = \underline{450\mu Wb}$

$\lambda(80ms) = (0.05)(\frac{12}{4}mA) = \underline{150\mu Wb}$

7.25 $v = L\,di/dt$, di/dt equals slope,

$v(10ms) = (10)\frac{12}{20} = \underline{6V}$

$v(40ms) = (10)\frac{(-12)}{(100-20)} = \underline{-1.5V}$

$v(80ms) = (10)(-\frac{12}{80}) = \underline{-1.5V}$

7.26 $i = \frac{1}{L}\int_0^t v\,dt + i_o$

$i(20ms) = \frac{1}{120}\left[\frac{1}{2}(20ms)12\right] + 2mA = \underline{3mA}$

$i(60ms) = \frac{1}{120}\left[\frac{1}{2}(12+\frac{12}{2})(40ms)\right] + 3mA$

$= \underline{6mA}$

$i(100ms) = \frac{1}{120}\left[\frac{1}{2}(12)(80ms)\right] + 3mA$

$= \underline{7mA}$

7.27 $v = L\,di/dt$

(a) $v = (0.1)\,d/dt(5) = \underline{0V}$.

(b) $v = (0.1)\,d/dt(100t) = \underline{10V}$.

(c) $v = (0.1)\,d/dt(10\sin 200t)$

$= \underline{200\cos 200t}\ V$

(d) $v = (0.1)\,d/dt(10e^{-50t}) = \underline{-50e^{-50t}}\,V$.

7.28 $i = \frac{1}{L}\int_0^t v\,dt + i_o$

$= \frac{1}{10^{-2}}\int_0^t 2\sin 1000t\,dt + 0$

$= \underline{200(1-\cos 1000t)}\,mA$

7.29 $\lambda = Li = (0.005)(0.2) = \underline{1\,mWb}$

$W = \frac{1}{2}Li^2 = \frac{1}{2}(0.005)(0.2)^2 = \underline{100\mu J}$

7.30 $i_L(0^-) = I - i_1(0^-) = 3 - 2 = \underline{1A}$

The inductor current is continuous so that $i_L(0^+) = i_L(0^-) = \underline{1A}$. $v_{R_1}(0^+) = 0$ = voltage across a short ciruit ∴ $i_1(0^+) = \frac{v_{R_1}}{R_1} = \underline{0}$.

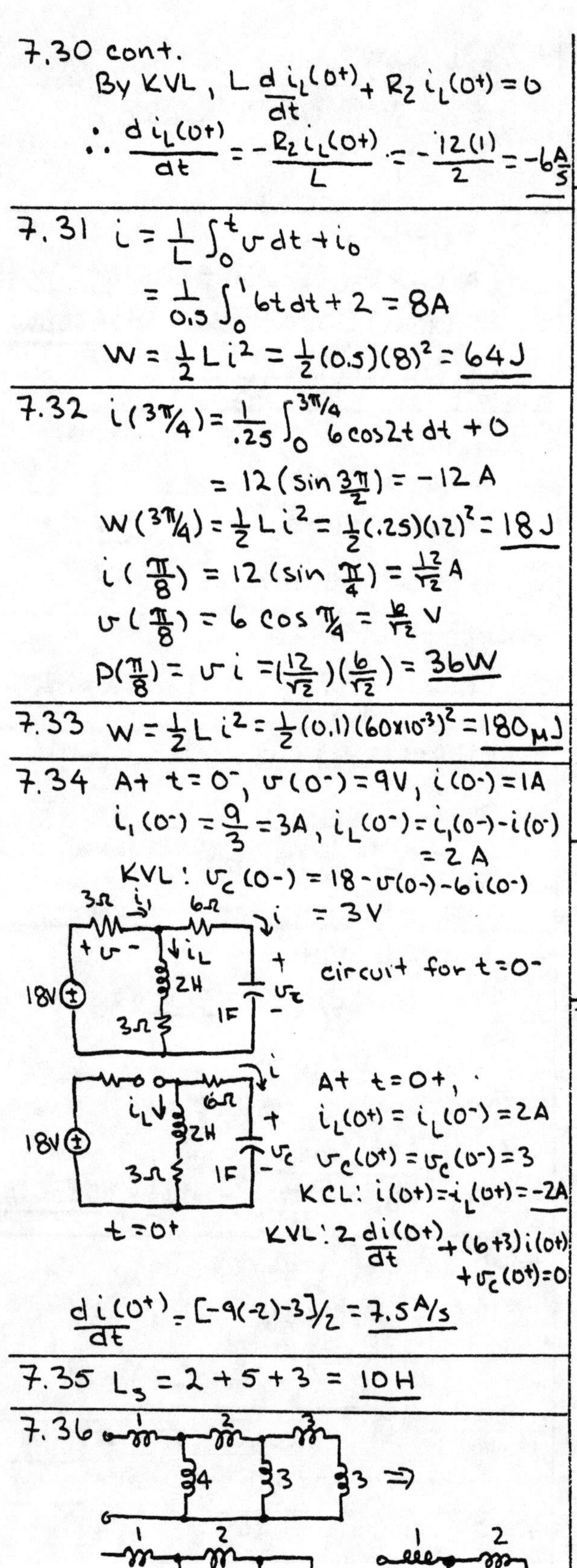

7.30 cont.

By KVL, $L\frac{di_L(0^+)}{dt} + R_2 i_L(0^+) = 0$

$\therefore \frac{di_L(0^+)}{dt} = -\frac{R_2 i_L(0^+)}{L} = -\frac{12(1)}{2} = -6\frac{A}{s}$

7.31 $i = \frac{1}{L}\int_0^t v\,dt + i_0$

$= \frac{1}{0.5}\int_0^1 6t\,dt + 2 = 8A$

$W = \frac{1}{2}Li^2 = \frac{1}{2}(0.5)(8)^2 = 64J$

7.32 $i(3\pi/4) = \frac{1}{.25}\int_0^{3\pi/4} 6\cos 2t\,dt + 0$

$= 12(\sin\frac{3\pi}{2}) = -12A$

$W(3\pi/4) = \frac{1}{2}Li^2 = \frac{1}{2}(.25)(12)^2 = 18J$

$i(\frac{\pi}{8}) = 12(\sin\frac{\pi}{4}) = \frac{12}{\sqrt{2}}A$

$v(\frac{\pi}{8}) = 6\cos\frac{\pi}{4} = \frac{6}{\sqrt{2}}V$

$p(\frac{\pi}{8}) = vi = (\frac{12}{\sqrt{2}})(\frac{6}{\sqrt{2}}) = 36W$

7.33 $W = \frac{1}{2}Li^2 = \frac{1}{2}(0.1)(60\times10^{-3})^2 = 180\mu J$

7.34 At $t = 0^-$, $v(0^-) = 9V$, $i(0^-) = 1A$

$i_1(0^-) = \frac{9}{3} = 3A$, $i_L(0^-) = i_1(0^-) - i(0^-) = 2A$

KVL: $v_c(0^-) = 18 - v(0^-) - 6i(0^-) = 3V$

circuit for $t = 0^-$

At $t = 0^+$,

$i_L(0^+) = i_L(0^-) = 2A$

$v_c(0^+) = v_c(0^-) = 3$

KCL: $i(0^+) = -i_L(0^+) = -2A$

KVL: $2\frac{di(0^+)}{dt} + (6+3)i(0^+) + v_c(0^+) = 0$

$t = 0^+$

$\frac{di(0^+)}{dt} = [-9(-2)-3]/2 = 7.5A/s$

7.35 $L_s = 2 + 5 + 3 = 10H$

7.36

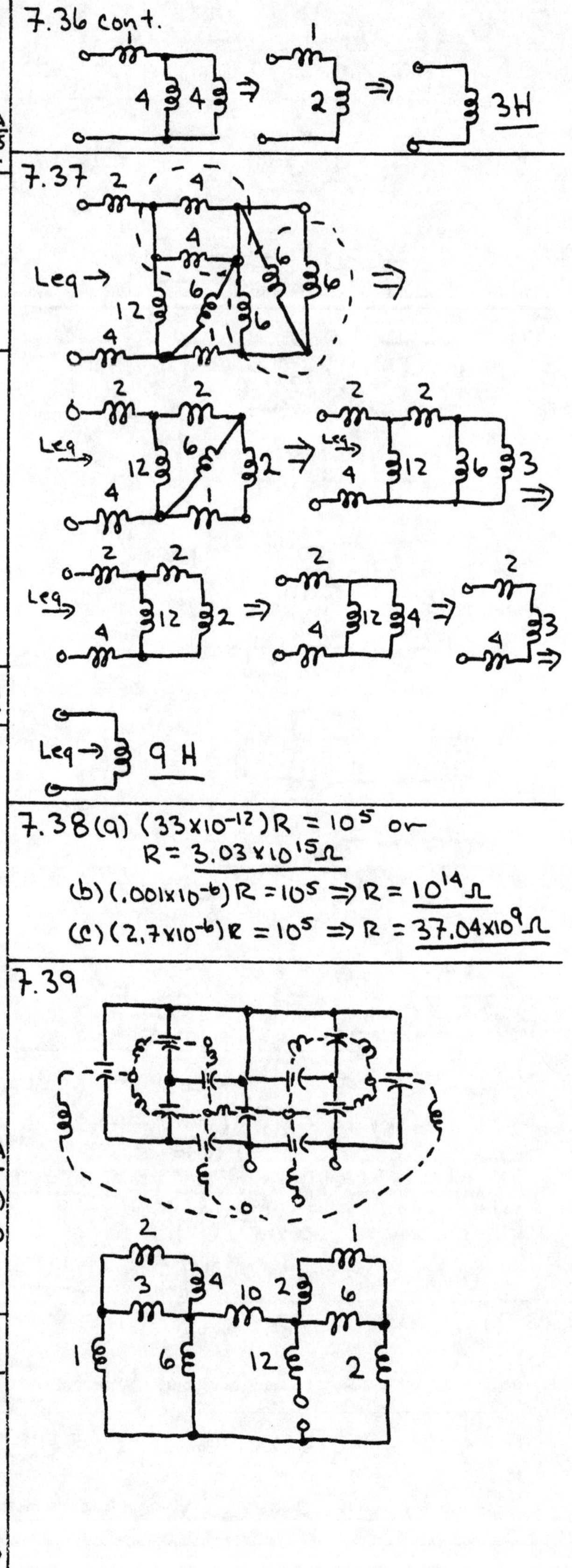

7.36 cont.

3H

7.37

$L_{eq} \rightarrow$ 9 H

7.38 (a) $(33\times10^{-12})R = 10^5$ or $R = 3.03\times10^{15}\Omega$

(b) $(.001\times10^{-6})R = 10^5 \Rightarrow R = 10^{14}\Omega$

(c) $(2.7\times10^{-6})R = 10^5 \Rightarrow R = 37.04\times10^9\Omega$

7.39

7.40

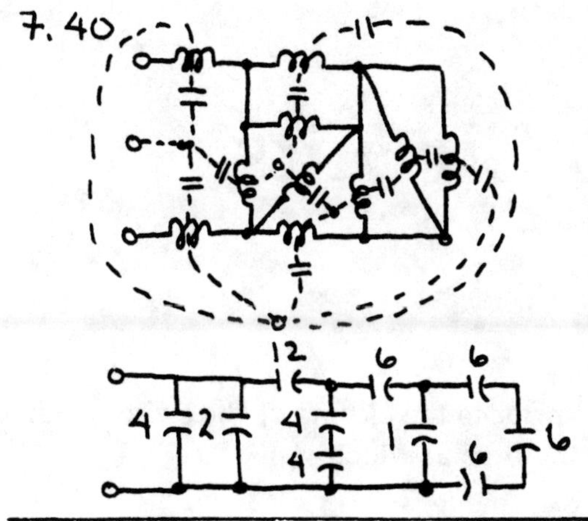

7.41 $v_1(0^+) = v_1(0^-) = \underline{14V}$

$v_2(0^+) = v_2(0^-) = \underline{6V}$

$i_1(0^-) = \dfrac{v_1(0^-)}{(2+5)k\Omega} = \underline{2mA}$.

$i_1(0^+) = \dfrac{v_1(0^+) - v_2(0^+)}{2\,k\Omega} = \dfrac{14-6}{2k\Omega} = \underline{4mA}$

7.42 At $t = 0^-$; KVL gives

$i_1(0^-)2 + v(0^-) + i_1(0^-)\left[\frac{4}{2}\right] = 18$ or

$4i_1(0^-) = 12$ or $i_1(0^-) = \underline{3A}$

KCL gives:

$i_2(0^-) = i_1(0^-) - \dfrac{v(0^-)}{2} = 3 - \dfrac{6}{2} = \underline{0A}$

$v(0^-) = v(0^+) = 6V$ the At $t = 0^+$

KVL gives

$i_1(0^+)2 + v(0^+) + i_1(0^+)4 = 18$ or

$6i_1(0^+) = 12$ or $i_1(0^+) = \underline{2A}$

KCL gives

$i_2(0^+) = 2 - \frac{6}{2} = \underline{-1A}$

7.43

$t = 0^-$

By KCL,

$i_1(0^-) + i_2(0^-) + \dfrac{v_1(0^-) - 20}{2} = 0$ or

$2i_2(0^-) + v_1(0^-) = 18$, since

$v_1(0^-) = i_2(0^-)4$ then

$6i_2(0^-) = 18$ or $i_2(0^-) = \underline{3A}$

$v(0^-) = v_1(0^-) - i_1(0^-)12 = 12 - 12 = \underline{0V}$

At $t = 0^+$; $i_1(0^+) = i_1(0^-) = \underline{1A}$,

$i_2(0^+) = -i_1(0^+) = \underline{-1A}$

7.43 cont.

$v(0^+) = i_2(0^+)(12+4) = \underline{-16V}$

7.44 At $t = 0^-$,

$v_1(0^-) = 9V$,

$i_1(0^-) = 3A$,

By KCL,

$t = 0^-$

$\dfrac{v(0^-) - 18}{3} + i_1(0^-) + \dfrac{v(0^-) - v_1(0^-)}{6} = 0$ or

$3v(0^-) = 27$ or $v(0^-) = 9V$

$i_2(0^-) = \dfrac{v(0^-) - v_1(0^-)}{6} = \dfrac{9-9}{6} = \underline{0A}$

$v_2(0^-) = v(0^-) - i_1(0^-)3 = 9 - 9 = \underline{0V}$

At $t = 0^+$,

$i_1(0^+) = i_1(0^-) = \underline{3A}$

$v_1(0^+) = v_1(0^-) = \underline{9V}$

$i_2(0^+) = -i_1(0^+) = \underline{-3A}$

$t = 0^+$

By KVL,

$V_2(0^+) = 6i_2(0^+) + v_1(0^+) + 3i_2(0^+)$

$= \underline{-18V}$.

Chapter 8

Simple RC and RL Circuits

8.1 Source-Free RC Circuit

8.1 The capacitor is charged to a voltage of 100V prior to the closing of the switch. For $t > 0$, find (a) $v(t)$, (b) $i(t)$, (c) $w_c(t)$, and (d) the time at which $v(t) = 50$V.

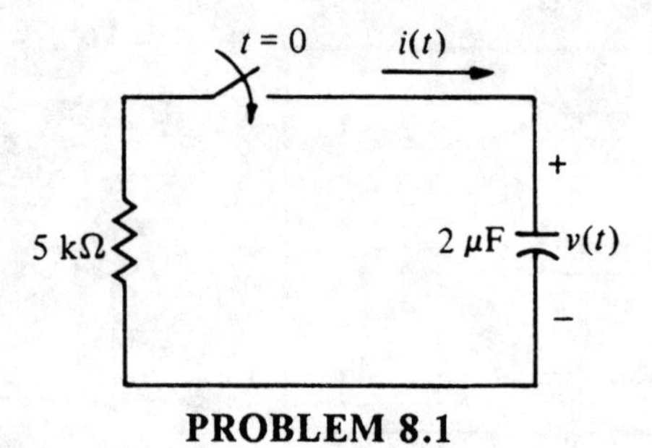

PROBLEM 8.1

8.2 The switch in the network opens at $t = 1$s, at which time $v = 10$V. For $t > 1$, find $v(t)$ and $w_c(t)$.

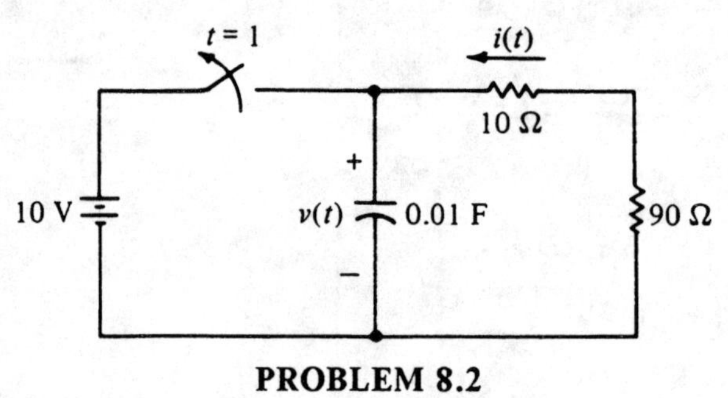

PROBLEM 8.2

8.3 Find v for $t > 0$ if $i(0) = 2A$.

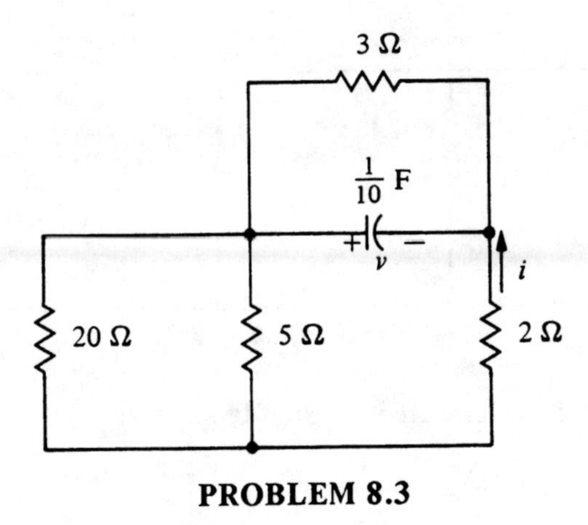

PROBLEM 8.3

8.4 Find v for $t > 0$ if the circuit is in steady state at $t = 0^-$.

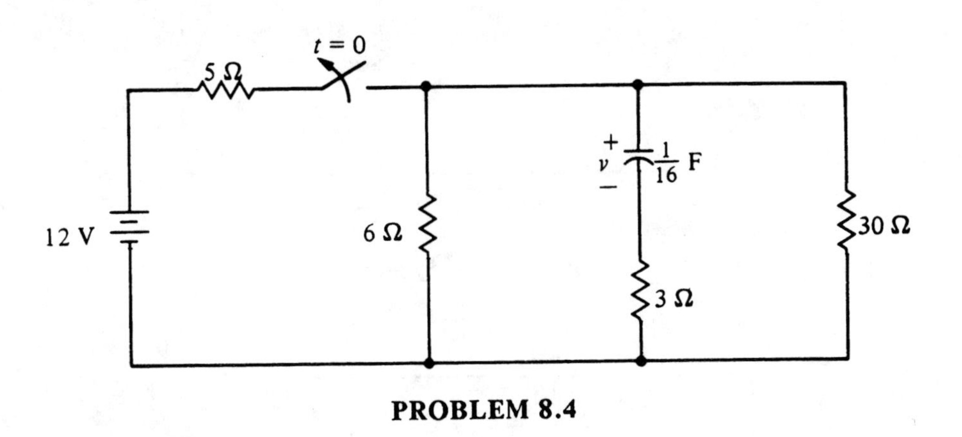

PROBLEM 8.4

8.2 Time Constants

8.5 In a series RC circuit, determine (a) for $R = 2k\Omega$ and $C = 5$mF, (b) C for $R = 10k\Omega$ and $\tau = 10$ms, and (c) R for $v(t)$ to half every 2ms on a 0.1μF capacitor.

8.6 A series RC circuit consists of a $2k\Omega$ and 0.1μF capacitor. It is desired to increase the current in the network by a factor of 3 without changing the capacitor voltage. Find the necessary values of R and C.

8.7 Find i for $t > 0$ if the circuit is in steady state at $t = 0^-$.

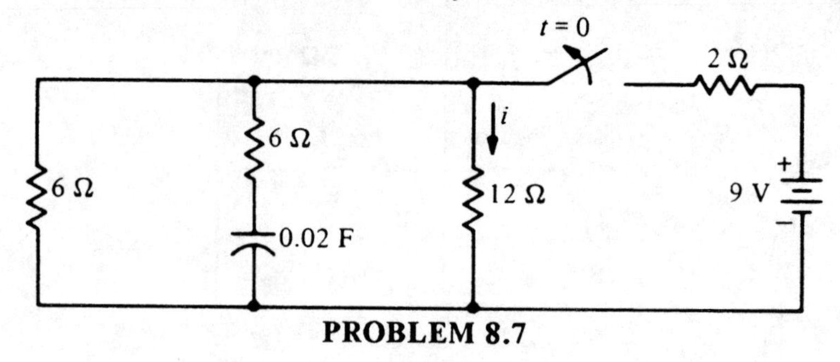

PROBLEM 8.7

8.8 Find v for $t > 0$ if the circuit is in steady state at $t = 0^-$.

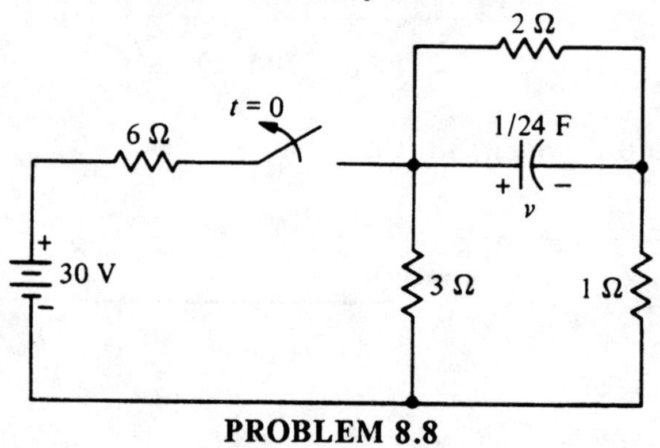

PROBLEM 8.8

8.9 Find v for $t > 0$ if the circuit is in steady state at $t = 0^-$.

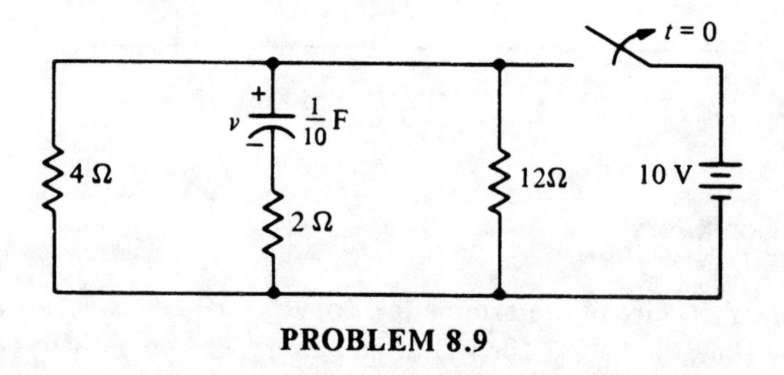

PROBLEM 8.9

8.10 Find i for $t > 0$ if the circuit is in steady state at $t = 0^-$.

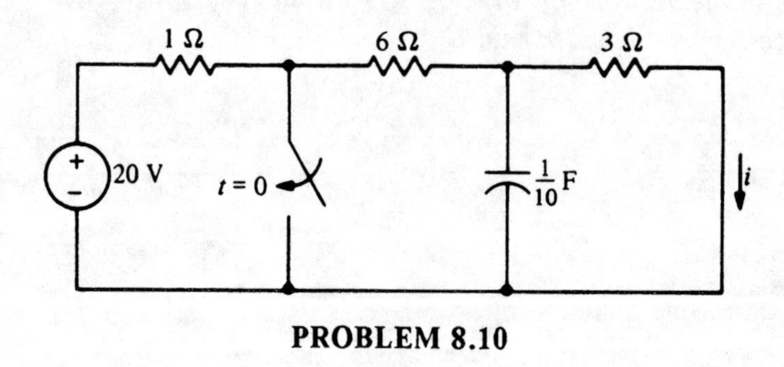

PROBLEM 8.10

8.11 Find v for $t > 0$ if $v(0) = 4\text{V}$.

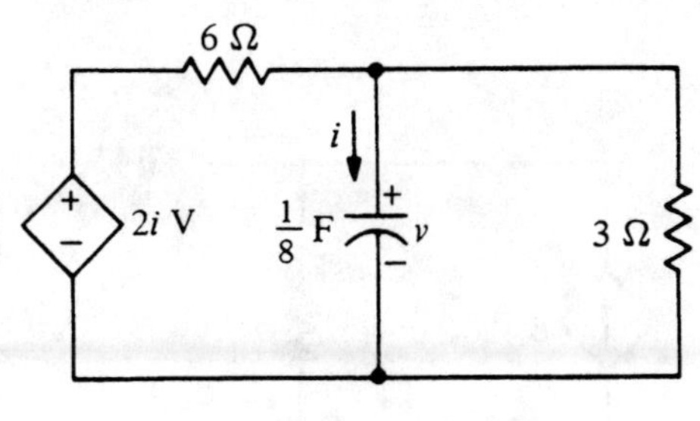

PROBLEM 8.11

8.3 Source-Free RL Circuit

8.12 In a series RL circuit, determine (a) τ for $R = 1k\Omega$ and $L = 15\text{mH}$, (b) L for $R = 10k\Omega$ and $\tau = 40\text{ms}$, and (c) R for the stored energy in a 5mH inductor to halve every 4ms.

8.13 A series RL circuit has a time constant of 5ms with an inductance of 4H. It is desired to halve the inductor voltage without changing the current response. Find the new values of inductance and resistance required.

8.14 The circuit shown is in a dc steady-state condition at $t = 0^-$. Find v for $t > 0$.

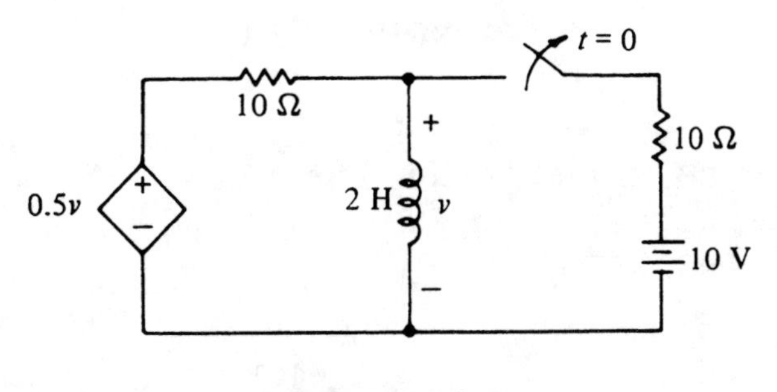

PROBLEM 8.14

8.15 Find v for $t > 0$ if the circuit is in steady state at $t = 0^-$.

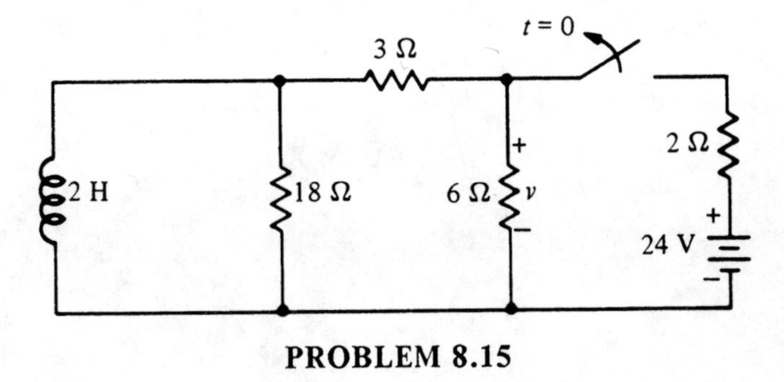

PROBLEM 8.15

8.16 Find i for $t > 0$ if $i(0) = 2\text{A}$.

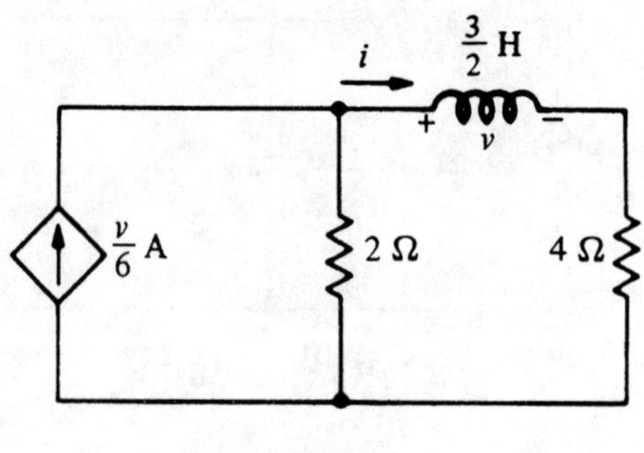

PROBLEM 8.16

8.17 Find i for $t > 0$ if the circuit is in steady state at $t = 0^-$.

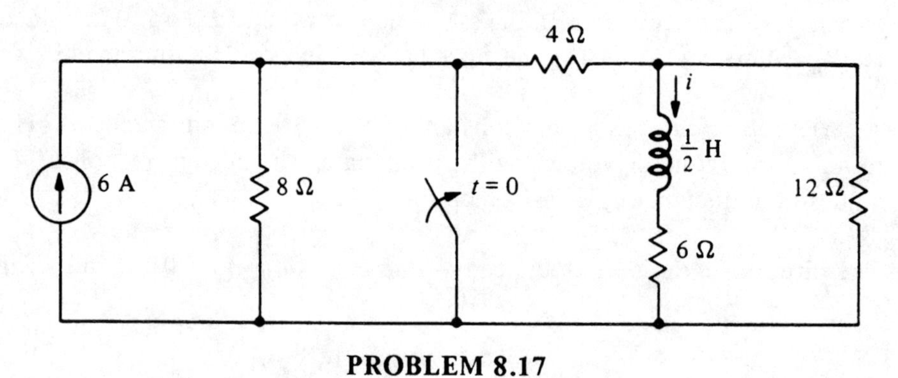

PROBLEM 8.17

8.4 Response to a Constant Forcing Function

8.18 Find v for $t > 0$ if $v(0^-) = 5\text{V}$.

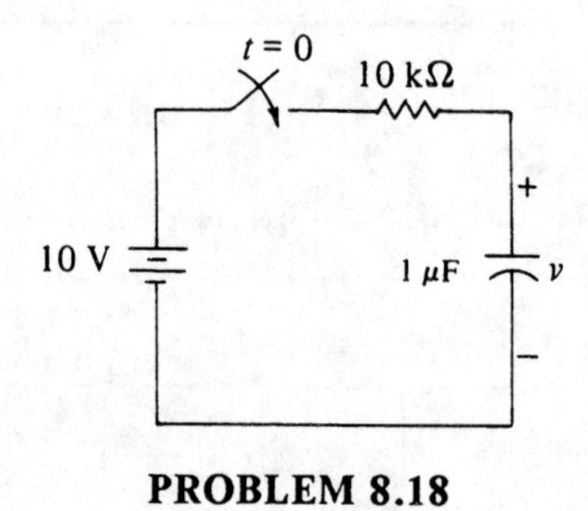

PROBLEM 8.18

8.19 The circuit is in steady state at $t = 0^-$. Find i for $t > 0$ if the switch is moved from position 1 to position 2 at $t = 0$.

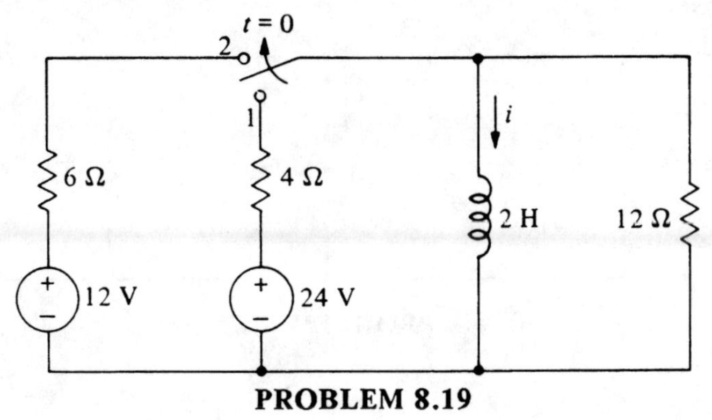

PROBLEM 8.19

8.20 Find the current in the inductor for $t > 0$ if the circuit is in steady state at $t = 0^-$.

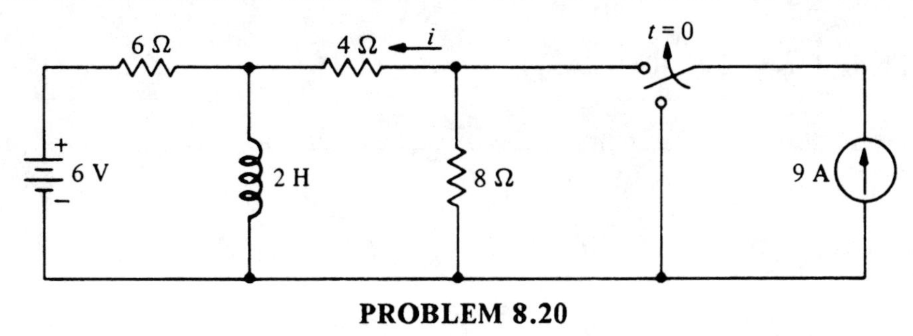

PROBLEM 8.20

8.21 Find i for $t > 0$ if the circuit is in steady state at $t = 0^-$.

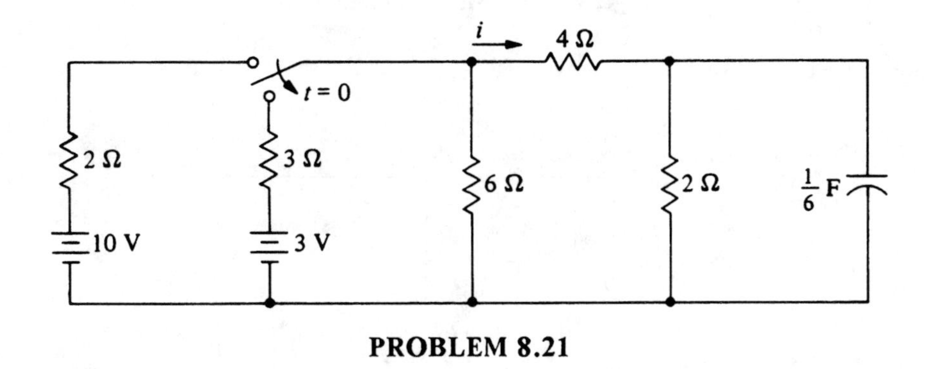

PROBLEM 8.21

8.22 Find i for $t > 0$ if the circuit is in steady state at $t = 0^-$.

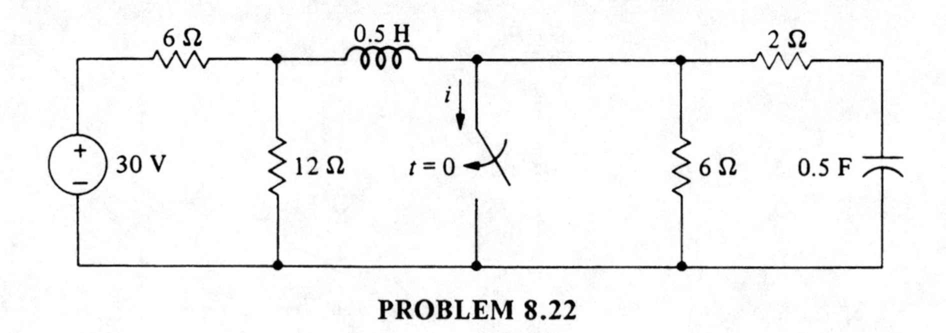

PROBLEM 8.22

8.5 The General Case

8.23 Find i for $t > 0$ if $i(0) = 2\text{A}$.

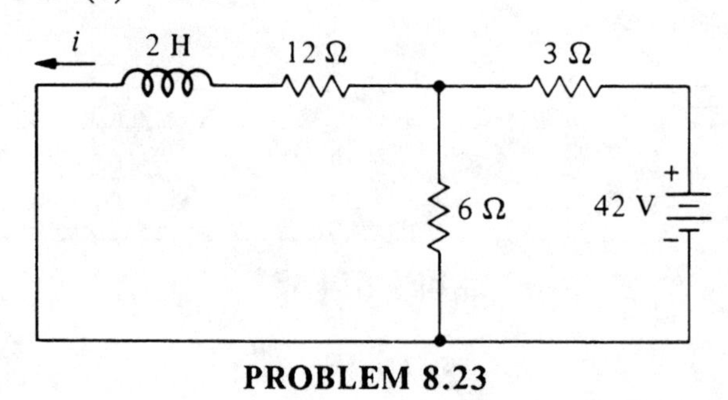

PROBLEM 8.23

8.24 Solve Prob. 8.23 if the 42V source is replaced by a source of $18e^{-4t}\text{V}$.

8.25 Find i for $t > 0$ if $v(0) = 12\text{V}$.

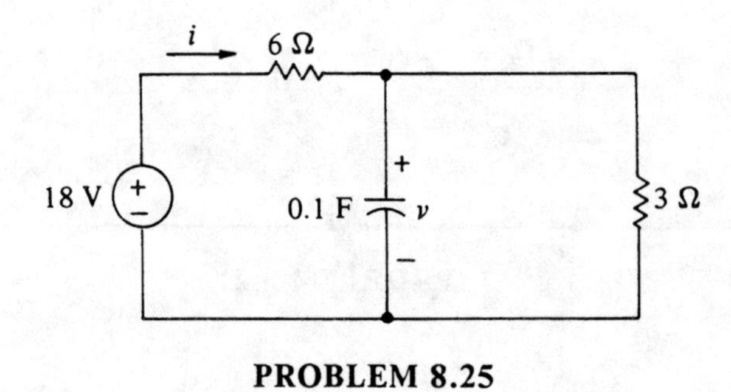

PROBLEM 8.25

8.26 Find v for $t > 0$ if $v(0) = 12\text{V}$.

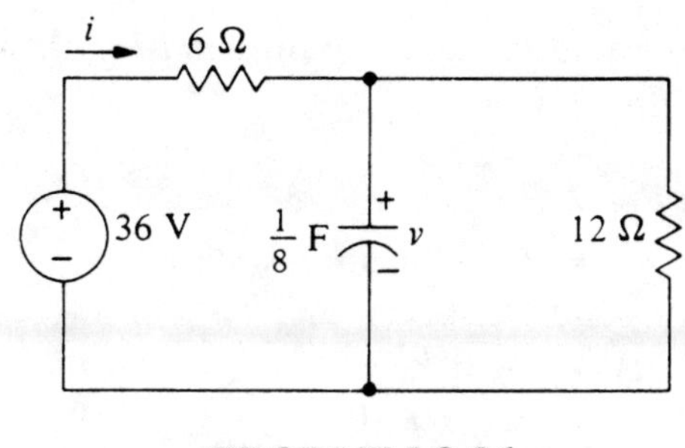

PROBLEM 8.26

8.27 Repeat Prob. 8.26 if the 36V source is replaced by a $18e^{-t}\text{V}$ source with the same polarity.

8.6 A Shortcut Procedure

8.28 Find v on the capacitor to $t > 0$ if the circuit is in steady state at $t = 0^-$.

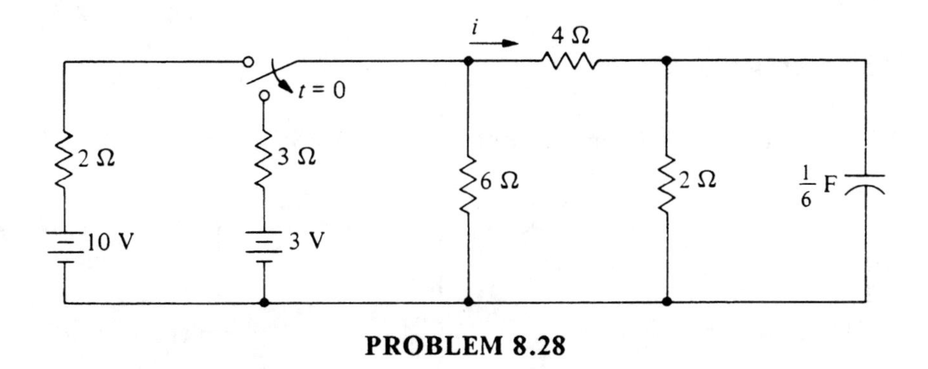

PROBLEM 8.28

8.29 Find v on the capacitor for $t > 0$ if the circuit is in steady state at $t = 0^-$.

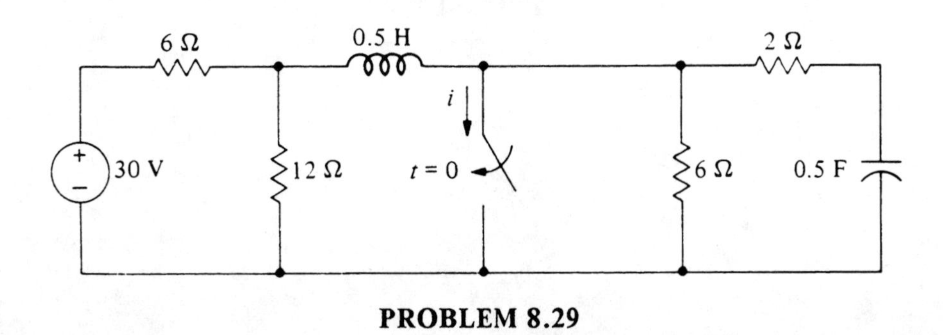

PROBLEM 8.29

8.7 The Unit Step Function

8.30 Using unit step functions, write an expression for the current $i(t)$ which satisfies

$$\text{(a) } i(t) = 0,\ t < 0$$
$$= 5\text{A},\ t > 0$$
$$\text{(b) } i(t) = -1\text{A},\ t < -20\text{ms}$$
$$= 2\text{A}, -20\text{ms} < t < 40\text{ms}$$
$$= 3\text{A},\ 40\text{ms} < t$$
$$\text{(c) } i(t) = 6\text{A},\ t < 10\text{s}$$
$$= -6\text{A},\ t > 10\text{s}$$

8.31 Sketch the voltage given by $3u(t - 3) + tu(t) - 3u(t + 3)$.

8.32 Express $v(t)$ in terms of unit step functions.

$$v(t) = 0\text{V},\ t < -10$$
$$= -10\text{V}, -10 < t < 0$$
$$= 20\text{V},\ 0 < t < 10$$
$$= 15\text{V},\ 10 < t$$

8.8 The Step Response

8.33 Find the response $v_1(t)$ to the voltage step $v(t) = 10u(t)$ V.

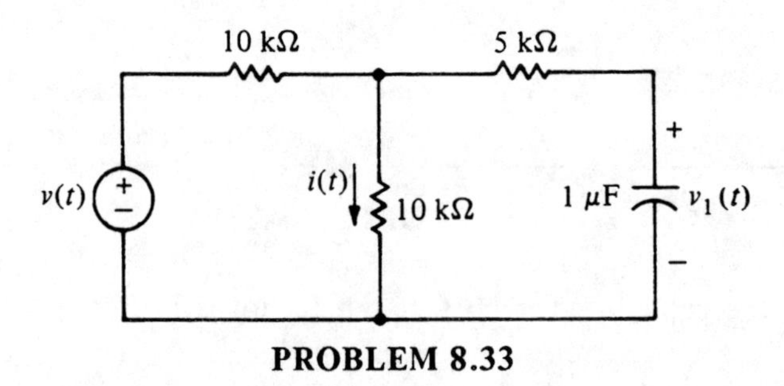

PROBLEM 8.33

8.34 Find v if $v_g = 3e^{-3(t+4)}u(t+4)$ V and there is no intital stored energy.

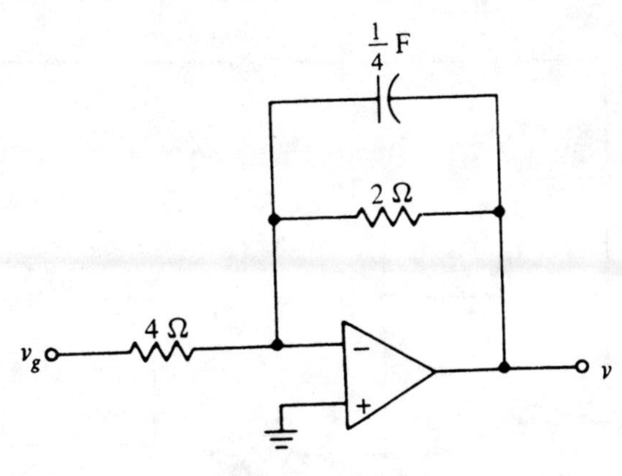

PROBLEM 8.34

8.35 Find the step response i. [Note that this means $v_g = u(t)$ V.]

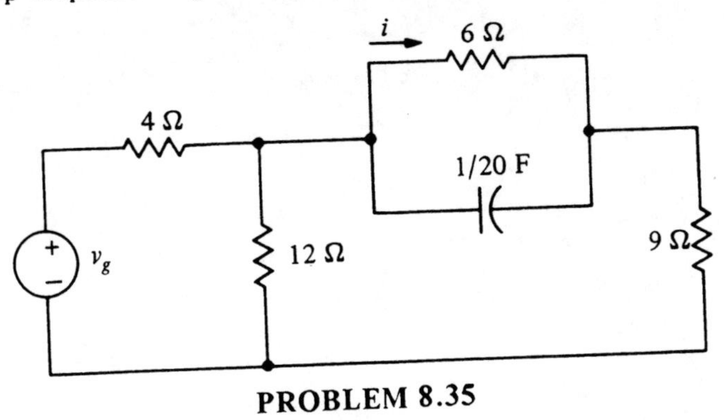

PROBLEM 8.35

8.9 Application of Superposition

8.36 Find v for $t > 0$.

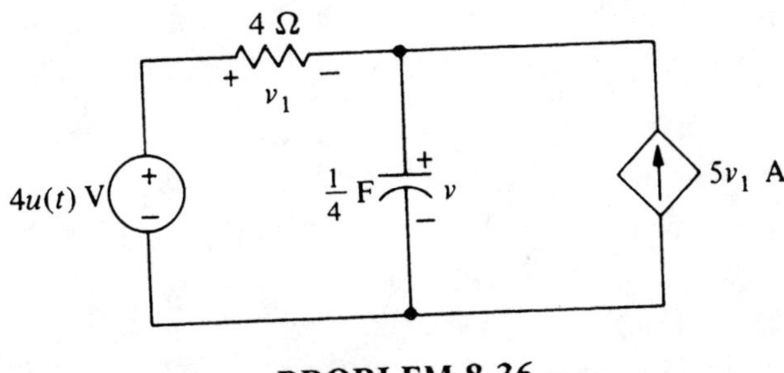

PROBLEM 8.36

8.37 Find v_1 for $t > 0$ if $v_1(0) = 0$ and $v_g = 5u(t)$ V.

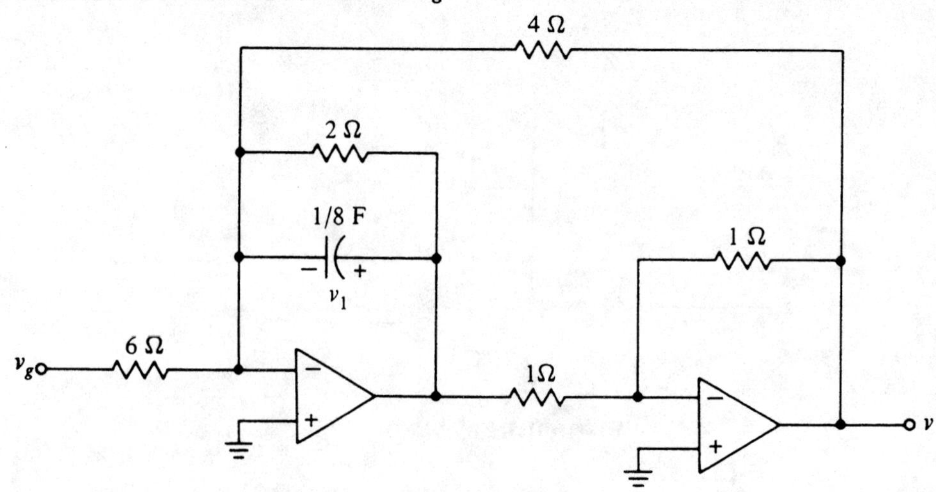

PROBLEM 8.37

8.1 (a) $v(t) = v(0)e^{-t/RC}$, where
$RC = (5\times10^3)(2\times10^{-6}) = 10^{-2}\ s$
$v(0) = 100V$, $v(t) = \underline{100\ e^{-100t}\ V}$

(b) $i(t) = -v(t)/5k\Omega = \underline{20e^{-100t}\ mA}$

(c) $w_C(t) = \frac{1}{2}Cv^2 = \frac{1}{2}(2\mu F)v^2(t)$
$= \underline{10\ e^{-200t}\ mJ}$

(d) $50 = 100\ e^{-100t}$, solve for t,
$t = \frac{\ln(50/100)}{-100} = \underline{6.93\ s}$

8.2 $v(t) = v(1)e^{-(t-1)/RC}$, where
$RC = (0.01)(10+90) = 1\ s$, $v(1) = 10V$,
$\underline{v(t) = 10e^{1-t}\ V}$
$w_C(t) = \frac{1}{2}Cv^2 = \frac{1}{2}(0.01)(10e^{1-t})^2$
$= \underline{0.5\ e^{2(1-t)}\ J}$

8.3 Replacing the 20Ω and 5Ω parallel combination by 4Ω, we see that
$v(0) = (4+2)\,i(0) = 12V$.
Req = resistance seen by capacitor
$= (4+2)(3)/_{4+2+3} = 2\Omega$.
$1/\tau = 1/R_{eq}C = \frac{1}{2(1/10)} = 5\ s^{-1}$
$\therefore v = \underline{12\ e^{-5t}\ V}$.

8.4 At $t=0^-$ the capacitor is open-circuited and it's voltage is the voltage across the 6Ω and 30Ω parallel combination. By voltage div.
$v(0^+) = v(0^-) = \frac{30(6)/30+6}{5+5}(12) = 6V$
For $t>0$ the resistance seen by the capacitor is
$R_{eq} = 3 + \frac{30(6)}{30+6} = 8\Omega$; $\frac{1}{\tau} = \frac{1}{R_{eq}C} = \frac{1}{8(\frac{1}{16})}$
$= 2s^{-1}$
$\therefore \underline{v = 6e^{-2t}\ V}$.

8.5 (a) $\tau = RC = (2\times10^3)(5\times10^{-3}) = \underline{10\ s}$
(b) $C = \tau/R = (10\times10^{-3})/(10\times10^3) = \underline{1\mu F}$
(C) $v(t_0) = V_0$; $v(t_0 + 2\times10^{-3}) =$
$= V_0e^{-2\times10^{-3}/\tau}$
$\therefore \tau = RC = \frac{-2\times10^{-3}}{\ln 0.5} \Rightarrow R = \frac{-2\times10^{-3}}{(0.1\times10^{-6})\ln 0.5}$
$= \underline{28.85\ K\Omega}$

8.6 Since $i(t) = \frac{V_0}{R}e^{-(t-t_0)/\tau}$, we must have R replaced by $R/3$. Therefore
$\tau = RC = (\frac{R}{3})(3C)$
$\therefore R_{new} = \frac{R}{3} = \underline{666.7\Omega}$, $C_{new} = 3C = \underline{0.3\mu F}$

8.7 At $t=0^-$, since $i_C(0^-) = 0$, the capacitor voltage is that across the parallel combination of 12Ω and 6Ω, equivalently 4Ω. By voltage division
$v_C(0^-) = \frac{4}{4+2}(9) = 6V$
The resistance seen by the capacitor is $R_{eq} = \frac{6(12)}{6+12} + 6 = 10\Omega$
$\tau = R_{eq}C = 10(0.02) = 0.2\ s$
$\therefore v_C = 6\ e^{-5t}\ V$; $i_C = C\frac{dv}{dt} = (0.02)(-30e^{-5t})$
$= -0.6e^{-5t}\ A$
By current division,
$i = -\frac{6}{6+12}i_C = \underline{0.2\ e^{-5t}\ A}$

8.8 At $t=0^-$, the capacitor is open-circuited and its voltage is the voltage across the 2Ω resistor.

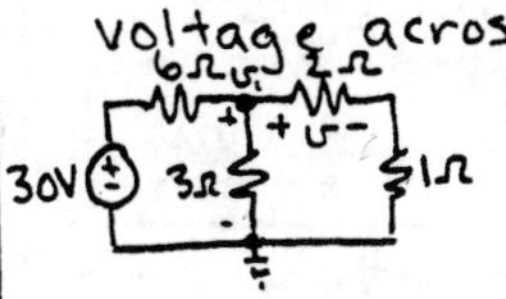

By $v_1 = \frac{\frac{(2+1)3}{3+3}}{6+3/2}(30) = 6V$

$v(0^-) = \frac{2}{1+2}(6) = 4V$; for $t>0$,
Req = resistance seen by capacitor
$= \frac{(3+1)2}{3+1+2} = 1.33\Omega$
$\frac{1}{\tau} = \frac{1}{R_{eq}C} = \frac{1}{(\frac{1}{24})(1.33)} = 18\ s^{-1}$
$\therefore \underline{v = 4\ e^{-18t}\ V}$

8.9 At $t=0^-$, since $i_C(0^-) = 0$, the capacitor voltage is that of 10V source. The resistance seen by the capacitor $R_{eq} = \frac{(4)(12)}{4+12} + 2 = 5\Omega$
$\frac{1}{\tau} = \frac{1}{R_{eq}C} = \frac{1}{(5)(1/10)} = 2\ s^{-1}$
$\therefore v = \underline{10\ e^{-2t}\ V}$

8.10 At $t=0^-$ the capacitor is open-circuited. By voltage division
$v_C(0^-) = \frac{3}{1+6+3}(20) = 6V$. For $t>0$
Req = resistance seen by capacitor
$= \frac{(6)(3)}{6+3} = 2\Omega$
$\frac{1}{\tau} = \frac{1}{R_{eq}C} = \frac{1}{2(\frac{1}{10})} = 5\ s^{-1}$
$\therefore v_C = 6e^{-5t}\ V$; $i = \frac{v_C}{3} = \underline{2\ e^{-5t}\ A}$

8.11 KCL gives $\frac{1}{8}\frac{dv}{dt} + \frac{v}{3} + \frac{v-2i}{6} = 0$;
$i = \frac{1}{8}\frac{dv}{dt}$ $\therefore$ $\frac{dv}{dt} + 6v = 0$ and
$\frac{dv}{v} = -6dt$, $\ln v = -6t + \ln K$
or $v = Ke^{-6t}$, $v(0) = 4 = K$,
$\therefore \underline{v = 4\ e^{-6t}\ V}$

8.12 (a) $\tau = L/R = \frac{15\times10^{-3}}{10^3} = \underline{15\,\mu s}$

(b) $L = R\tau = (10^4)(40\times10^{-3}) = \underline{400\,H}$

(c) $w(t_0+4\times10^{-3}) = \frac{1}{2}w(t_0)$

$= \frac{1}{2}\left[\frac{1}{2}LI_0^2e^{-2Rt_0/L}\right]$

$\therefore \frac{1}{2}LI_0^2e^{-2R(t_0+4\times10^{-3})/L} =$

$(\frac{1}{2})\left[\frac{1}{2}LI_0^2e^{-2Rt_0/L}\right]$

Then $e^{-2R(4\times10^{-3})/L} = \frac{1}{2}$ and

$-2R(4\times10^{-3})/5\times10^{-3} = \ln 0.5 = -\ln 2$

$R = \frac{5}{8}\ln 2 = \underline{0.433\,\Omega}$.

8.13 $R = L/\tau = 4/5\times10^{-3} = 800\Omega$

$v = RI_0e^{-(t-t_0)/\tau}$

If R is replaced by $\frac{R}{2} = \underline{400\Omega}$, the voltage is halved. Then

$\tau = \frac{L}{R} = \frac{L/2}{R/2} \Rightarrow L_{new} = \frac{L}{2} = \underline{2\,H}$.

8.14 i_L = inductor current downward.

$v(0^-) = 0$ (short circuit)

$i_L(0^-) = \frac{10V}{10\Omega} = 1A$, For $t>0$, KVL,

$v + 10i_L - \frac{v}{2} = 0$; $v = 2\frac{di_L}{dt}$. Then

$\frac{di_L}{dt} + 10i_L = 0 \therefore i_L(t) = i_L(0+)e^{-10t}$

$= e^{-10t}A$

Then $v = 2\frac{d}{dt}(e^{-10t}) = \underline{-20e^{-10t}V}$

8.15 i_L = inductor current downward.

$v_L(0^-) = 0$ (short circuit). For the circuit shown, voltage division gives,

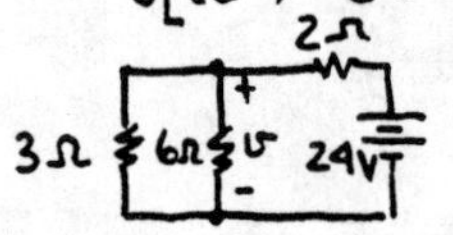

$t = 0^-$ $\quad v(0^-) = \frac{\frac{(3)(6)}{3+6}}{2+2}(24) = 12V$

the $i_L(0^-) = \frac{v(0^-)}{3} = 4A$, for $t>0$

$R_{eq} = \frac{18(3+6)}{18+3+6} = 6\Omega$ (seen by inductor)

$\frac{1}{\tau} = \frac{R_{eq}}{L} = \frac{6}{2} = 3s^{-1} \therefore i_L(0^+) = 4e^{-3t}A$

$v_L(0^+) = L\frac{d}{dt} = 2(4)(-3)e^{-3t} = -24e^{-3t}V$

By voltage division,

$v(0^+) = \frac{6}{6+3}v_L = \underline{-16e^{-3t}V}$

8.16 KVL around the right mesh gives

$2(i - \frac{v}{6}) + v + 4i = 0$; $v = L\frac{di}{dt}$ then

$\frac{di}{dt} + 6i = 0 \qquad \frac{di}{i} = -6dt$.

$\therefore i = ke^{-6t} = i(0)e^{-6t} = \underline{2e^{-6t}A}$

8.17 At $t=0^-$, the inductor is a short circuit. Thus for the circuit shown current division gives

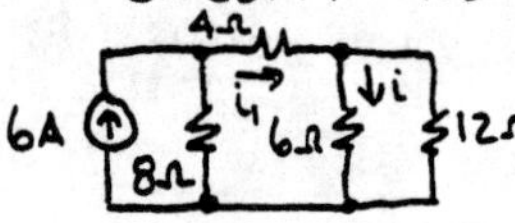

$t = 0^-$

$i_1(0^-) = \frac{8\ (6)}{4+8+\frac{6(12)}{6+12}} = 3A$;

$i(0^-) = \frac{12}{6+12}(i_1) = 2A$; For $t>0$

$R_{eq} = 6 + \frac{12(4)}{4+12} = 9\Omega$ (seen by inductor)

$\frac{1}{\tau} = \frac{R_{eq}}{L} = \frac{9}{1/2} = 18s^{-1}$

$\therefore \underline{i = 2e^{-18t}A}$

8.18 capacitor current downward $= i_c$

$i_c = 10^{-6}\frac{dv}{dt}$. By KVL

$10k\Omega(i_c) + v = 10$ or $\frac{dv}{dt} + 100v = 10^3$

separating the variables

$\frac{dv}{v-10} = -100dt$ or $\ln(v-10) = -100t + \ln K$

$v - 10 = Ke^{-100t}$, $v(0^+) = v(0^-) = 5V$

then $5 - 10 = K = -5$

$\therefore v = \underline{10 - 5e^{-100t}V}$

8.19 At $t=0^-$, the inductor is a short circuit. thus $i(0^-) = 24/4 = 6A$.

For $t>0$, KCL gives for v_L

$\frac{v_L - 12}{6} + i + \frac{v_L}{12} = 0$ or $3v_L + 12i = 24$

setting $v_L = 2\frac{di}{dt}$ gives $\frac{di}{dt} + 2i = 4$.

separation of variables yeilds

$\frac{di}{i-2} + 2dt = 0$ and $\ln(i-2) = -2t + \ln K$

$\therefore i - 2 = Ke^{-2t}$, since $i(0^-) = i(0^+)$

$i(0^+) - 2 = K \Rightarrow K = 6 - 2 = 4$.

then $i = \underline{4e^{-2t} + 2A}$ for $t>0$

8.20 At $t=0^-$, the inductor voltage is zero, thus $i_L(0^-) = \frac{6V}{6\Omega} = 1A$, down.

For $t>0$, for the circuit shown

$R_{eq} = \frac{6(4+8)}{6+4+8} = 4\Omega$ (seen by inductor)

$\frac{1}{\tau} = \frac{R_{eq}}{L} = \frac{4}{2} = 2s^{-1}$;

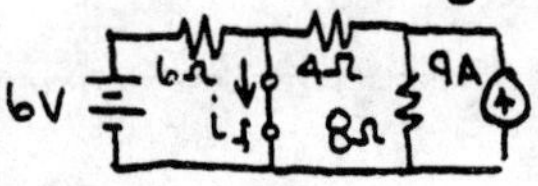

$i_f = \frac{6}{6} + \frac{8}{4+8}(9) = 7A$, since there are essentially two circuits. Therefore

$i_L = i_n + i_f = Ke^{-R_{eq}t/L} + 7$

$i_L(0^+) = i_L(0^-) = 1 = K + 7 \Rightarrow K = -6$

$\therefore i_L = \underline{7 - 6e^{-2t}}\,A$.

8.21 At $t=0^-$, the capacitor is open-circuited. By voltage division $v_c(0-) = 2/(2+4)\,(v_{6\Omega})$, where $v_{6\Omega}$ is the voltage across the 6Ω resistor. Since $v_{6\Omega} = 3v_c(0^-)$, KVL at the top of 6Ω resistor yeilds.

$$3v_c(0^-)\left(\frac{1}{6}+\frac{1}{6}+\frac{1}{2}\right) = \frac{10}{2} \Rightarrow v_c(0-) = 2V$$

Now $i = i_n + i_f$, where i_f is the dc steady-state value and i_n is an exponential with $\tau = R_{eq}C$ with the source dead, the resistance seen by the capacitor is

$$R_{eq} = \frac{2[4+(6)(3)/(6+3)]}{2+4+2} = \frac{3}{2}\Omega$$

$$\tau = \left(\frac{3}{2}\right)\left(\frac{1}{6}\right) = \frac{1}{4}\ s^{-1}$$

with the circuit in dc steady-state, the capacitor is open, and KCL

$$v_{6\Omega}\left(\frac{1}{6}+\frac{1}{6}+\frac{1}{3}\right) = \frac{3}{3} \Rightarrow v_{6\Omega} = \frac{3}{2}$$

$$\therefore\ i_f = \frac{3/2}{(4+2)} = \frac{1}{4}A$$

$$i = \frac{1}{4} + Ke^{-4t}$$

To determine K we need $i(0^+)$:

$$v_{6\Omega}(0^+) = v_c(0^+) + 4i(0^+) = 4i(0^+) + 2$$

KCL at top of 6Ω resistor yeilds

$$\frac{v_{6\Omega}(0^+)-3}{3} + \frac{v_{6\Omega}(0^+)}{6} + i(0^+) = 0 \text{ or}$$

$$\frac{4i(0^+)+2-3}{3} + \frac{4i(0^+)+2}{6} + i(0^+) = 0$$

$$\therefore\ i(0^+) = 0 = \frac{1}{4} + k \Rightarrow k = -1/4$$

$$i = \frac{1}{4}(1-e^{-4t})A$$

8.22

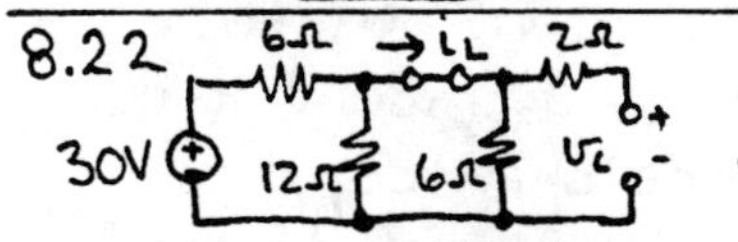

At $t=0^-$, by voltage division

$$v_c(0^-) = \frac{6(12)/6+12}{6+4}(30) = 12V, \text{ then}$$

$$i_L(0^-) = \frac{v_c(0^-)}{6} = 2A.$$ At $t=0^+$, the closing of the switch seperates the circuit into two independent circuits due to the short-circuit.

For $t>0$, $i = i_L - i_c$ as shown above In the capacitor circuit.

8.22 cont.

$$-i_c = -0.5\frac{dv_c}{dt} = -0.5\,v_c(0^+)\frac{d}{dt}\left(e^{-\frac{t}{RC}}\right)$$

$$= (0.5)(12)\left(\frac{1}{2(0.5)}\right)e^{-t} = 6e^{-t}A$$

In the inductor circuit $i_L = i_{Ln} + i_{Lf}$ where $i_{Lf} = 30/6 = 5A$ (inductor is a short-circuit). with the source dead the inductor sees the resistance $R_{eq} = \frac{6(12)}{6+12} = 4\Omega$. Then $\frac{R_{eq}}{L} = \frac{4}{1/2} = 8$ and $i_L = 5 + ke^{-8t}$;

$$i_L(0^+) = i_L(0^-) = 2 = 5 + K \Rightarrow K = -3$$

and $i_L = 5 - 3e^{-8t}A$.

$$\therefore\ i = 5 - 3e^{-8t} + 6e^{-t}A \quad (i_L - i_c = i)$$

8.23 For counter clockwise mesh currents i and i_1, KVL gives

$$2\frac{di}{dt} + 12i + 6(i - i_1) = 0;$$

$6(i-i_1) + 3i_1 = 42$; Eliminating i_1 gives $\frac{di}{dt} + 7i = 14$. Since $P = 7$ and $Q = 14$, the solution becomes

$$i = Ae^{-7t} + 2,\ i(0) = 2 = A + 2 \Rightarrow A = 0$$

$\therefore\ i = 2A.\ t > 0$

8.24 Replacing 42V in Prob. 8.23 by $18e^{-4t}$ gives $\frac{di}{dt} + 7i = 6e^{-4t}$

$P = 7$, $Q = 6e^{-4t}$, then

$$i = Ae^{-7t} + e^{-7t}\int 6e^{-4t}(e^{7t})dt$$

$$= Ae^{-7t} + e^{-7t}\left(\frac{6}{3}\right)e^{3t} = Ae^{-7t} + 2e^{-4t}$$

$i(0) = 2 = A + 2 \Rightarrow A = 0$,

$\therefore\ i = 2e^{-4t}A$.

8.25 KCL gives

$$\frac{v-18}{6} + 0.1\frac{dv}{dt} + \frac{v}{3} = 0 \text{ or}$$

$\frac{dv}{dt} + 5v = 30$. Since $P = 5$, $Q = 30$

$$v = Ae^{-5t} + 6,$$

$v(0) = 12 = A + 6 \Rightarrow A = 6$ then

$$v = 6e^{-5t} + 6;\ i = \frac{18-v}{6}$$

$\therefore\ i = 2 - e^{-5t}$ A

8.26 KCL gives

$$\frac{v-36}{6} + \frac{1}{8}\frac{dv}{dt} + \frac{v}{12} = 0 \text{ or}$$

$$\frac{dv}{dt} + 2v = 48$$

8.26 cont.

since $P = 2$, $Q = 48$; then

$v = Ae^{-2t} + 24$, therefore

$v(0) = 12 = A + 24 \Rightarrow A = -12$

$\therefore v = 24 - 12e^{-2t}$ V

8.27 By KCL

$$\frac{v - 18e^{-t}}{6} + \frac{1}{8}\frac{dv}{dt} + \frac{v}{12} = 0 \text{ or}$$

$$\frac{dv}{dt} + 2v = 24e^{-t}, \text{ therefore}$$

$$ve^{2t} = A + \int 24e^{-t}(e^{2t})dt$$

$$= A + 24e^{t}$$

$$v = Ae^{-2t} + 24e^{-t}$$

$$v(0) = 12 = A + 24 \Rightarrow A = -12$$

$$\therefore v = 24e^{-t} - 12e^{-2t} \text{ V}$$

8.28 at $t = 0^-$ KCL yields

$$v_1\left(\frac{1}{2} + \frac{1}{6} + \frac{1}{4}\right) - v_c\left(\frac{1}{4}\right) = \frac{10}{2} \text{ or}$$

$$11v_1 - 3v_c = 60$$

$$v_c\left(\frac{1}{4} + \frac{1}{2}\right) - v_1\left(\frac{1}{4}\right) = 0 \text{ or } -v_1 + 3v_c = 0$$

Multiply second equation by 11 and Adding gives $30v_c = 60$ or

$v_c(0^-) = 2\text{V} = v_c(0^+)$

At $t > 0$ And the capeitor open circuited, KCL yields

$$v_1\left(\frac{1}{3} + \frac{1}{6} + \frac{1}{4}\right) + v_{cf}\left(\frac{1}{4}\right) = \frac{3}{3} \text{ or } 9v_1 - 3v_{cf} = 12$$

$-v_1 + 3v_{cf} = 0$, Adding 9 times the second equation $24v_{cf} = 12$ or $v_{cf} = \frac{1}{2}$;

When the voltage source is shorted

$$R_{eq} = \frac{\left[\frac{3(6)}{3+6} + 4\right]2}{2 + 4 + 2} = \frac{3}{2}\,\Omega \text{ (seen by the capacitor)}$$

$$\frac{1}{\tau} = \frac{1}{R_{eq}C} = \frac{1}{(\frac{3}{2})(\frac{1}{6})} = 4\text{ s}^{-1}$$

$$\therefore v_{cn} = Ae^{-4t}, \quad v_c = v_{cf} + v_{cn},$$

$$v_c = Ae^{-4t} + \frac{1}{2}$$

$$v_c(0) = 2 = A + \frac{1}{2} \Rightarrow A = \frac{3}{2}$$

$$v_c = \frac{3}{2}e^{-4t} + \frac{1}{2} \text{ V}$$

8.29 At $t = 0^-$, By voltage div.

$$v_c(0^-) = \frac{\frac{(6)(12)}{6+12}}{6 + 4}(30) = 12\text{V}$$

At $t > 0$, $v_c(0^+) = v_c(0^-)$,

$$\frac{1}{\tau} = \frac{1}{R_{eq}C} = \frac{1}{2(\frac{1}{2})} = 1\text{ s}^{-1}$$

since $v_{cf} = 0$ then

$v_c = Ae^{-t} \Rightarrow v_c(0^+) = 12 \Rightarrow A = 12\text{V}$

$v_c = 12e^{-t}$ V

8.30 (a) $i(t) = 5u(t)$ A.

(b) $i(t) = -u(-20\text{ms} - t) + 3u(t + 20\text{ms}) + u(t - 40\text{ms})$ A.

(c) $i(t) = 6u(10 - t) - 12u(t - 10)$ μA.

8.31

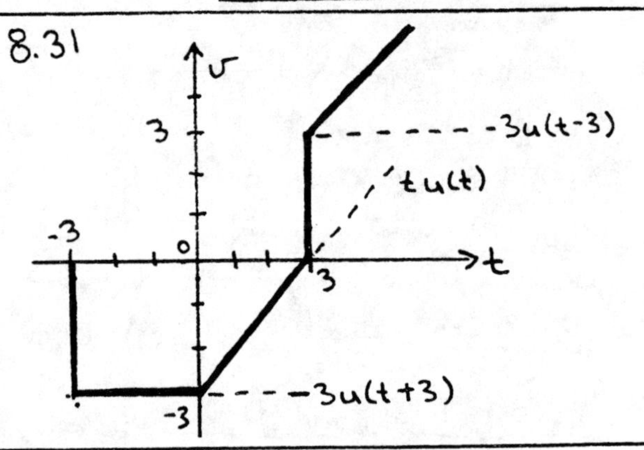

8.32

$$v(t) = -10u(t + 10) + 30u(t) - 5u(t - 10) \text{ V}$$

8.33 For $t < 0$, $v_1(t) = 0$. For $t > 0$, $v_1 = v_n + v_f$. with the voltage source shorted, the capacitor sees the resistance

$R_{eq} = 5 + \frac{10}{2} = 10\text{k}\Omega$

$\therefore v_n = Ae^{-t/R_{eq}C} = Ae^{-100t}$

v_f = dc steady state $= \frac{10}{2} = 5\text{V}$

Since $v_1(0^-) = v_1(0^+) = 0$

$v_1 = Ae^{-100t} + 5$

$v_1(0) = A + 5 = 0 \Rightarrow A = -5$

$v_1(t) = (5 - 5e^{-100t})u(t)$ V

8.34 KCL gives at the inverting op amp terminals,

$$\frac{1}{4}\frac{dv}{dt} + \frac{v}{2} + \frac{v_g}{4} = 0 \text{ or}$$

$$\frac{dv}{dt} + 2v = -3e^{-3(t+4)}u(t+4)$$

For $t > -4$,

$$ve^{2t} = A + \int(-3e^{-3(t+4)})(e^{2t})dt$$

$$= A + 3e^{-(t+12)}$$

$$v = Ae^{-2t} + 3e^{-3(t+4)}$$

$$v(-4) = Ae^{8} + 3 = 0 \Rightarrow A = -3e^{-8}$$

$$\therefore v = \left[3e^{-3(t+4)} - 3e^{-2(t+4)}\right]u(t+4) \text{ V.}$$

8.35 For $t < 0$, $v_g = 0$ and therefore the capacitor voltage is zero. Hence $i(0+) = 0$. With the source dead the resistance seen by the capacitor is

$$R_{eq} = \frac{6\left[9 + \frac{4(12)}{4+12}\right]}{6 + 9 + 3} = 4\Omega$$

$$R_{eq}C = 4\left(\frac{1}{20}\right) = \frac{1}{5}\text{ s}$$

Now $i = i_n + i_f = A_1e^{-5t} + i_f$, where i_f is the dc steady-state current. With the capacitor open circuited the voltage across the $12\,\Omega$ resistor is $(6+9)i_f = 15i_f$, By voltage div.

$$15i_f = \frac{15(12)/(15+12)}{4 + \frac{(15)(12)}{15+12}}(1) \Rightarrow i_f = \frac{1}{24}\text{ A}$$

$$\therefore i = A_1e^{-5t} + 1/24, \; t > 0$$

$$i(0^+) = 0 = A_1 + 1/24 \Rightarrow A_1 = -1/24$$

$$\therefore i = \frac{1}{24}(1 - e^{-5t})u(t)\text{ A}$$

8.36 For $t < 0$, the voltage soure is zero and therefore the capacitor voltage is also zero. Since there is only one independent source. KCL gives for $t > 0$,

$$\frac{1}{4}\frac{dv}{dt} + \frac{v-4}{4} - 5v_1 = 0,$$

$v_1 = 4 - v$, therefore

$\frac{dv}{dt} + 21v = 84$, therefore

8.36 cont.

$$v = Ae^{-21t} + 4$$

$$v(0) = 0 = A + 4 \Rightarrow A = -4$$

$$\therefore v = 4(1 - e^{-21t})u(t)\text{ V}$$

8.37 For $t > 0$ KCL at the inverting op amp input terminals yields

$$\frac{5}{6} + \frac{1}{8}\frac{dv_1}{dt} + \frac{v_1}{2} + \frac{v}{4} = 0, \quad v_1 + v = 0$$

Note that v_1 is the output voltage of the first op amp. Eliminating v results in $\frac{dv_1}{dt} + 2v_1 = -\frac{20}{3}$

$$\therefore v = v_n + v_f = Ae^{-2t} - \frac{10}{3}$$

$$v_1(0^+) = v_1(0^-) = 0 \Rightarrow A = \frac{10}{3}$$

$$\therefore v = \frac{10}{3}(e^{-2t} - 1)\text{ V} \quad t > 0$$

Chapter 9

Second-Order Circuits

9.1 Circuits with Two Storage Elements

9.1 Find the equation satisfied by the mesh current i_1.

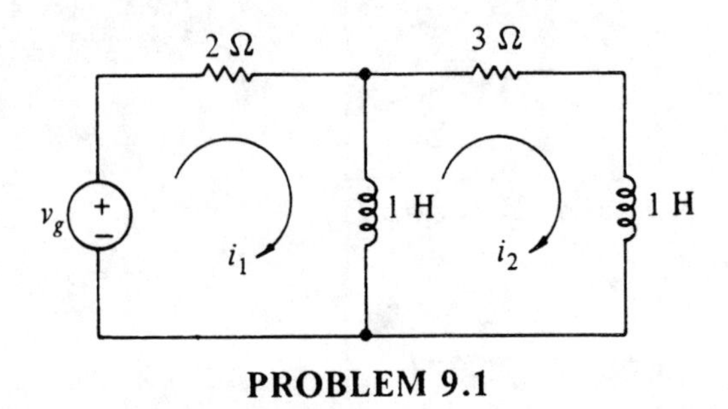

PROBLEM 9.1

9.2 Find i for $t > 0$ if $i(0) = 4$A and $v(0) = 6$V.

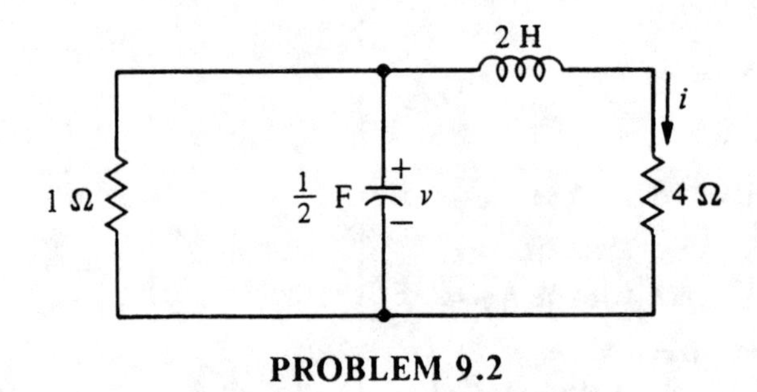

PROBLEM 9.2

9.3 Find i for $t > 0$ if the circuit is in steady state at $t = 0^-$.

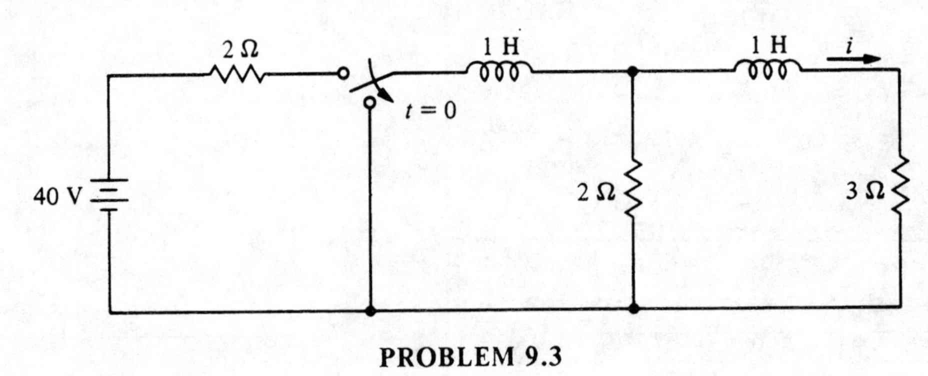

PROBLEM 9.3

9.4 Show that $x_1 = A_1e^{-t}$ and $x_2 = A_2e^{-2t}$ are solutions of $\frac{d^2x}{dt^2} + 3\frac{dx}{dt} + 2x = 0$ regardless of the values of the constants, A_1 and A_2.

9.5 Show that $x = x_1 + x_2 = A_1 e^{-t} + A_2 e^{-2t}$ is a solution of the differential equation $\frac{d^2x}{dt^2} + 3\frac{dx}{dt} + 2x = 0$.

9.6 Show that if the right member of the differential equation of Ex. 9.7 is changed from 0 to 6, then $x = A_1 e^{-t} + A_2 e^{-2t} + 3$ is a solution.

9.3 The Natural Response

9.7 Given $\frac{d^2x}{dt^2} + 5\frac{dx}{dt} + 4x = 0$ find the characteristic equation and the natural frequencies.

9.8 Given $\frac{d^2x}{dt^2} + 6\frac{dx}{dt} + 9x = 0$ find the characteristic equation and the natural frequencies.

9.4 Types of Natural Frequencies

9.9 Find the natural frequencies of a circuit described by $\frac{d^2x}{dt^2} + a_1\frac{dx}{dt} + a_0 x = 0$ if (a) $a_1 = 6$, $a_0 = 8$; (b) $a_1 = 4$, $a_0 = 5$; and (c) $a_1 = 2$, $a_0 = 1$.

9.10 Find x if $\frac{d^2x}{dt^2} + 9x = 0$.

9.5 The Forced Response

9.11 Find the forced response if $\frac{d^2x}{dt^2} + 4\frac{dx}{dt} + 3x = f(t)$ is given by (a) $6t^2$, (b) $4e^t$.

9.12 If $x(0) = 8$ and $dx(0)/dt = -1$, find the complete solution in Prob. 9.11.

9.6 Excitation at a Natural Frequency

9.13 Find the forced response if $\frac{d^2x}{dt^2} + 4\frac{dx}{dt} + 3x = f(t)$ is given by $2e^{-3t} - e^{-t}$.

9.14 Find the complete response if $\frac{d^2x}{dt^2} + 9x = \sin(3t)$ and $x(0) = dx(0)/dt = 0$.

9.7 The Complete Response

9.15 Find i_1 in Prob. 9.1 for $t > 0$, if $v_g = 12\text{V}$, $i_1(0) = -1\text{A}$, and $i_2(0) = 2\text{A}$.

9.16 Find i for $t > 0$, if $v(0) = 0\text{V}$, $i(0) = 0\text{A}$, $L = 2\text{H}$ and $R = 2\Omega$.

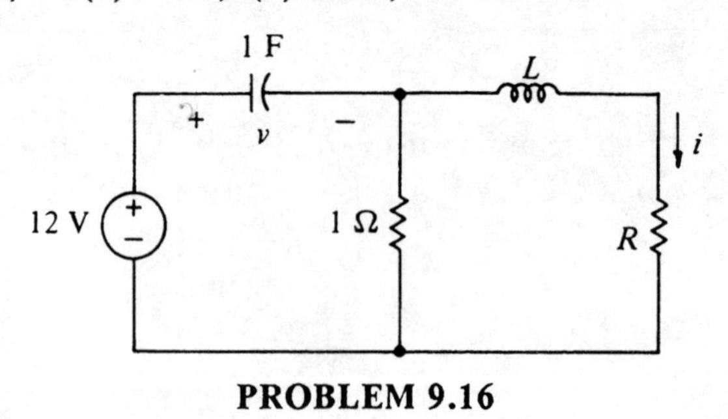

PROBLEM 9.16

9.17 Find i in Prob. 9.16 for $L = 2\text{H}$ and $R = 0\Omega$.

9.18 Find i for $t > 0$ if the circuit is in steady state at $t = 0^-$.

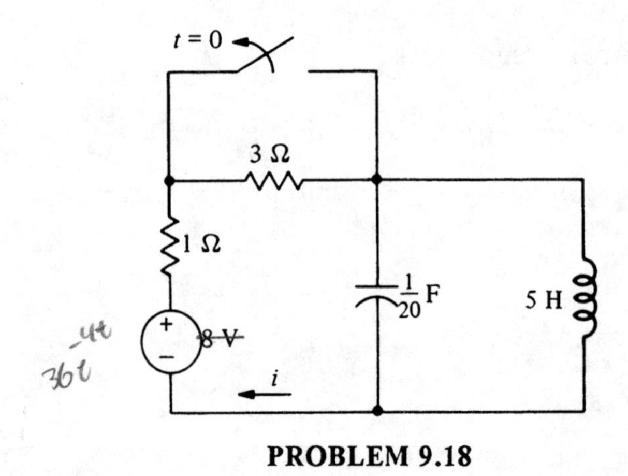

PROBLEM 9.18

9.19 Solve Prob. 9.18 if the 8V source is replaced by a source of $36e^{-4t}\text{V}$ with the same polarity, and there is no initial stored energy.

9.20 Find v for $t > 0$ if the circuit is in steady state at $t = 0^-$.

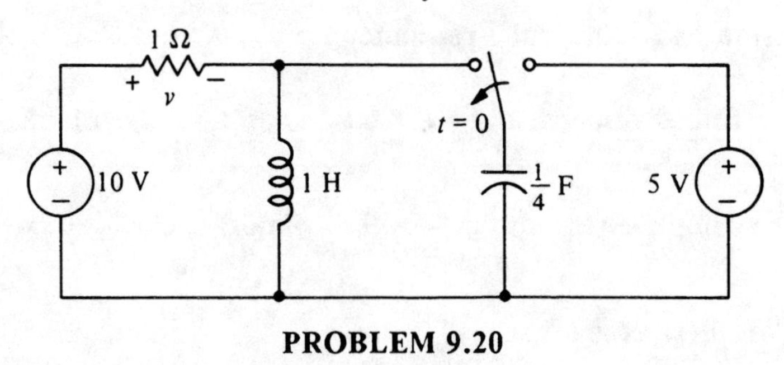

PROBLEM 9.20

9.21 Find v for $t > 0$ if $i_1(0) = -1\text{A}$ and $i_2(0) = 0$.

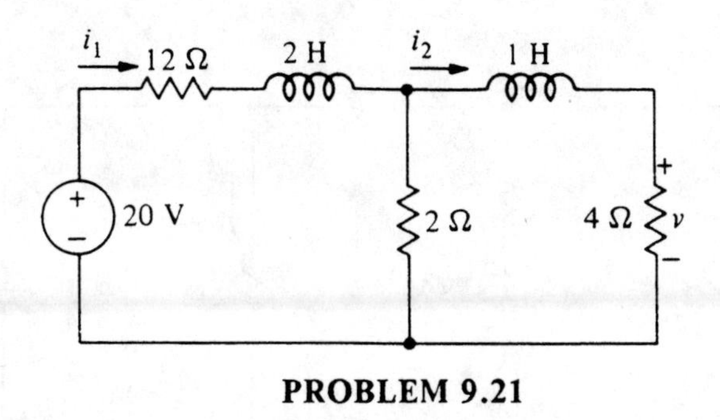

PROBLEM 9.21

9.22 Find v for $t > 0$ if $v(0) = 4\text{V}$ and $i(0) = 3\text{A}$.

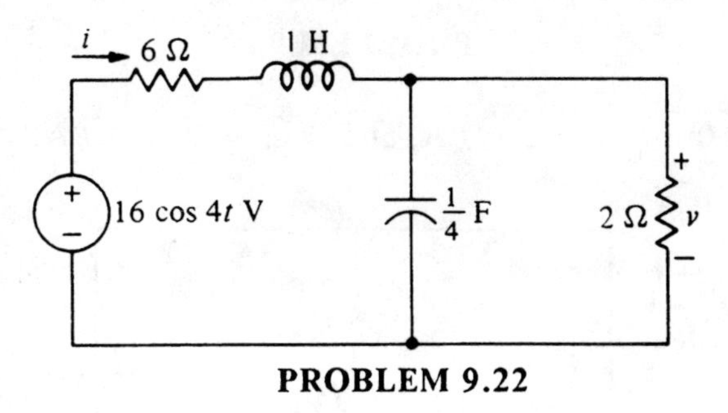

PROBLEM 9.22

9.23 Find v for $t > 0$ if there is no initial stored energy.

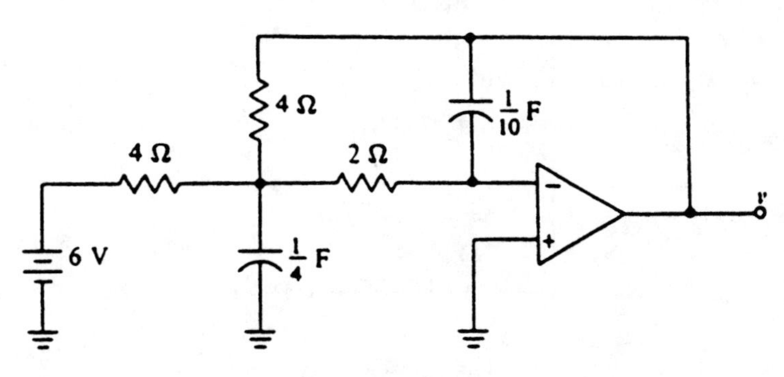

PROBLEM 9.23

9.24 Find i for $t > 0$ if the circuit is in steady state at $t = 0^-$.

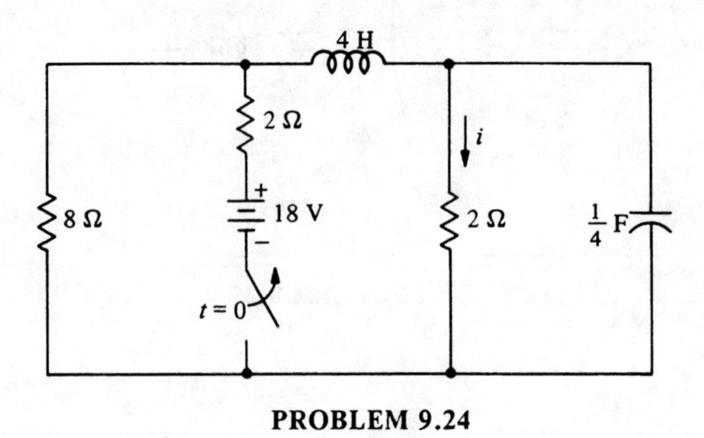

PROBLEM 9.24

9.25 Find v for $t > 0$ if (a)$i_g = 2u(t)$ A, and (b) $i_g = 2e^{-t}u(t)$ A.

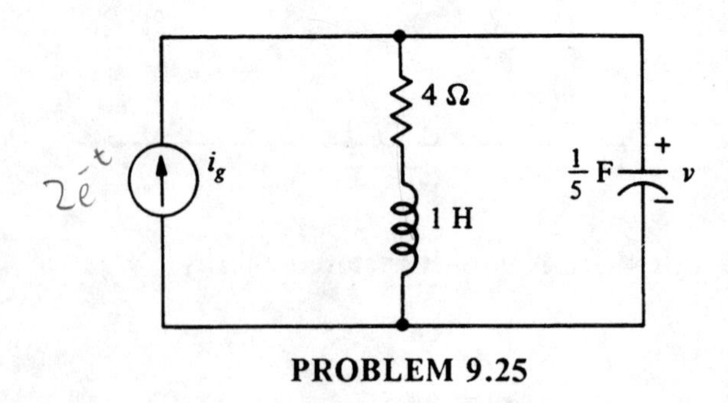

PROBLEM 9.25

9.26 Find v, $t > 0$, if $v(0) = 4$V and $i(0) = 2$A.

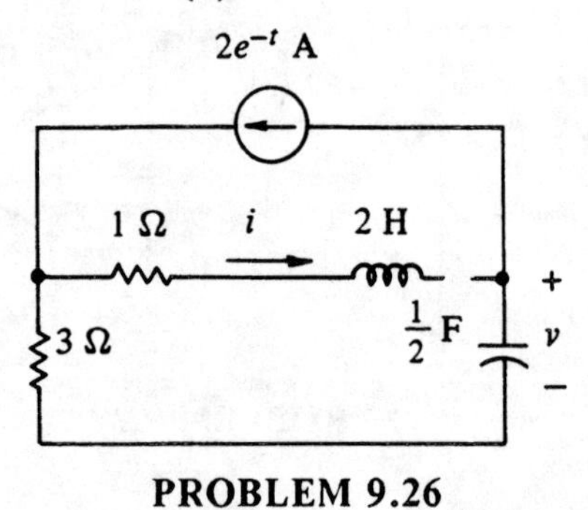

PROBLEM 9.26

9.8 The Parallel RLC Circuit

9.27 In a source-free parallel RLC circuit, $R = 4k\Omega$ and $C = 0.1\mu F$. Find L so that the circuit is critically damped.

9.28 In a source-free parallel RLC circuit, $R = 4k\Omega$ and $C = 0.1\mu F$. Find L so that the circuit is underdamped with $\omega_d = 250$.

9.29 In a source-free parallel *RLC* circuit, $R = 4k\Omega$ and $C = 0.1\mu F$. Find L so that the circuit is overdamped with $s_{1,2} = 2000, 500$.

9.9 The Series RLC Circuit

9.30 Let $R = 2\Omega$, $L = 4H$, $v_g = 0$, $v(0) = 4V$, and $i(0) = 0$. Find i for $t > 0$ if C is $\frac{1}{8}F$.

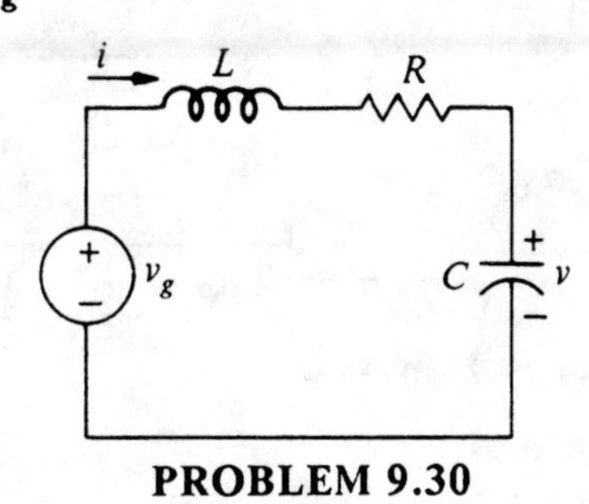

PROBLEM 9.30

9.31 Find i for $t > 0$ if $L = \frac{8}{3}H$.

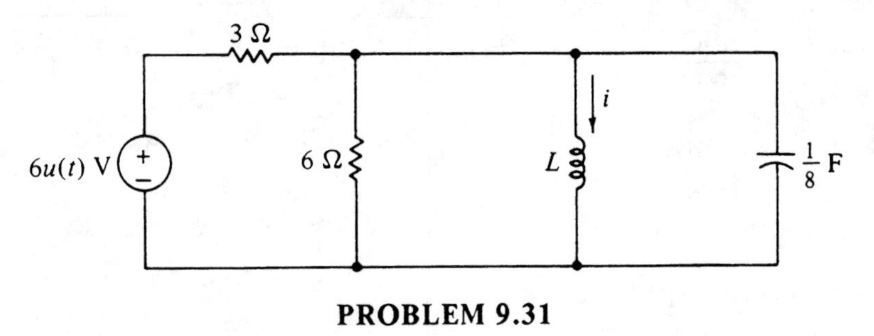

PROBLEM 9.31

9.32 Find i in Prob. 9.31 for $t > 0$ if $L = 2H$.

9.10 Alternative Methods for Obtaining the Describing Equations

9.33 Solve Prob. 9.23 using the first method of this section applied to the node equations.

9.34 Find the describing equation for Prob. 9.16.

9.1 KVL yields

(1) $\frac{di_1}{dt} + 2i_1 - \frac{di_2}{dt} = v_g$; (2) $-\frac{di_1}{dt} + 2\frac{di_2}{dt} + 3i_2 = 0$

Adding 2 times the first equation

(3) $\frac{di_1}{dt} + 4i_1 + 3i_2 = 2v_g$, or

$\frac{d^2 i_1}{dt} + 4\frac{di_1}{dt} + 3\frac{di_2}{dt} = 2\frac{dv_g}{dt}$

Substituting for $\frac{di_2}{dt}$ from (1) yields

(4) $\frac{d^2 i_1}{dt} + 7\frac{di_1}{dt} + 6i_1 = 2\frac{dv_g}{dt} + 3v_g$

9.2 Let i and i_1 be the clockwise mesh currents. KVL for left mesh:

$i_1 + 2\int_0^t (i_1 - i)dt + 6 = 0$ or

$\frac{di_1}{dt} + 2i_1 - 2i = 0$, KVL for outer loop:

$2\frac{di_1}{dt} + 4i + i_1 = 0 \Rightarrow i_1 = -2\frac{di}{dt} - 4i$

Eliminating i_1: $\frac{d^2 i}{dt} + 4\frac{di}{dt} + 5i = 0$

$s^2 + 4s + 5 = 0 \Rightarrow s_{1,2} = -2 \pm j1$

$i = e^{-2t}(A_1 \cos t + A_2 \sin t)$; $i(0) = 4 = A_1$

KVL for right mesh at $t = 0$;

$v(0) = 6 = 2\frac{di(0)}{dt} + 4i(0)$

$\therefore \frac{di(0)}{dt} = -5 = A_2 - 2A_1 \Rightarrow A_2 = 3$

$i = e^{-2t}(4\cos t + 3\sin t)$ A

9.3 Let i_1 = current in left 1-H inductor to right. At $t = 0^-$ the inductors are short circuits.

$\therefore i_1(0^-) = \frac{40}{2 + \frac{2(3)}{5}} = \frac{25}{2}$ A;

$i(0^-) = \frac{2}{2+3} i_1(0^-) = 5$ A

For $t > 0$ the mesh equations are

(1) $\frac{di_1}{dt} + 2(i_1 - i) = 0$; (2) $\frac{di}{dt} + 3i + 2(i - i_1) = 0$

From (2) $i_1 = \frac{1}{2}\frac{di}{dt} + \frac{5}{2}i$. Substitution into (1): $\frac{d^2 i}{dt^2} + 7\frac{di}{dt} + 6i = 0 \Rightarrow$

$s^2 + 7s + 6 = 0 \Rightarrow s_{1,2} = -1, -6$

$\therefore i = A_1 e^{-t} + A_2 e^{-6t}$

$i(0^+) = A_1 + A_2$; From (2) $\frac{di(0^+)}{dt} =$

$2i_1(0^+) - 5i(0^+) = 0$

$\frac{di(0^+)}{dt} = -A_1 - 6A_2 = 0$; $\therefore A_1 = 6, A_2 = -1$

$\therefore i = 6e^{-t} - e^{-6t}$ A

9.4 $x_1 = A_1 e^{-t}$, $x_2 = A_2 e^{-2t}$

$\frac{d^2 x_1}{dt^2} + 3\frac{dx_1}{dt} + 2x_1 = A_1(-1)^2 e^{-t} + A_1 3(-1)e^{-t} + 2A_1 e^{-t}$
$= 0$

$\frac{d^2 x_2}{dt^2} + 3\frac{dx_2}{dt} + 2x_2 = A_2(-2)^2 e^{-2t} + 3(-2)A_2 e^{-2t} + 2A_2 e^{-2t}$
$= 0$

9.5 $\frac{d^2(x_1 + x_2)}{dt^2} + 3\frac{d(x_1 + x_2)}{dt} + 2(x_1 + x_2) =$
$A_1(-1)^2 e^{-t} + A_2(-2)^2 e^{-2t} + 3(A_1(-1)e^{-t} + A_2(-2)e^{-2t}) +$
$2(A_1 e^{-t} + A_2 e^{-2t}) = 0$

9.6 $\frac{d^2 x}{dt^2} + 3\frac{dx}{dt} + 2x = (A_1(-1)^2 e^{-t} + A_2(-2)^2 e^{-2t})$
$+ 3(A_1(-1)e^{-t} + A_2(-2)e^{-2t})$
$+ 2(A_1 e^{-t} + A_2 e^{-2t} + 3) = 6$

9.7 $s^2 + 5s + 4 = 0$ or $(s+1)(s+4) = 0$

$s_{1,2} = -1, -4$

9.8 $s^2 + 6s + 9 = 0$ or $(s+3)(s+3) = 0$

$s_{1,2} = -2, -2$

9.9 $s^2 + a_1 s + a_0 = 0$

(a) $s^2 + 6s + 8 = 0$ or $(s+2)(s+4) = 0$

$s_{1,2} = -2, -4$

(b) $s^2 + 4s + 5 = 0$ or $(s+2)^2 + 1 = 0$

$s_{1,2} = -2 \pm j$

(c) $s^2 + 2s + 1 = 0$ or $(s+1)^2 = 0$

$s_{1,2} = -1, -1$

9.10 characteristic equation:

$s^2 + 9 = 0$; $s_{1,2} = \pm j3$

$\therefore x = A_1 \cos 3t + A_2 \sin 3t$

9.11 (a) $x_f = At^2 + Bt + C$:

$\frac{dx_f}{dt} = 2At + B$; $\frac{d^2 x_f}{dt^2} = 2A$;

$\frac{d^2 x_f}{dt^2} + 4\frac{dx}{dt} + 3x = 6t^2$

$(2A) + 4(2At + B) + 3(At^2 + Bt + C) =$

$3At^2 + (8A + 3B)t + (2A + 4B + 3C) = 6t^2$

$3A = 6 \Rightarrow A = 2$; $8A + 3B = 0 \Rightarrow B = -\frac{16}{3}$

$2A + 4B + 3C = 0 \Rightarrow C = \frac{52}{9}$

$\therefore x_f = 2t^2 - \frac{16}{3}t + \frac{52}{9}$

(b) $x_f = Ae^t$: $\frac{dx_f}{dt} = \frac{d^2 x_f}{dt^2} = Ae^t$;

$Ae^t + 4(Ae^t) + 3(Ae^t) = 8Ae^t = 4e^t$

$8A = 4$; $A = \frac{1}{2}$ $\therefore x_f = \frac{1}{2}e^t$

9.12 The characteristic equation is $s^2+4s+3=0$; $s_{1,2}=-1,-3$ and $x_n = A_1e^{-t}+A_2e^{-3t}$

(a) $x = x_n + x_f$; using x_f of Prob. 9.11

$= A_1e^{-t}+A_2e^{-3t}+2t^2-\frac{16}{3}t+52/9$

$x(0)=8=A_1+A_2+52/9$ $\Big\}$ $A_1+A_2=20/9$

$\frac{dx(0)}{dt}=-1=-A_1-3A_2-\frac{16}{3}$ $\Big\}$ $-A_1-3A_2=13/3$

Adding, $-2A_2=\frac{59}{9} \Rightarrow A_2=\frac{-59}{18}$, $A_1=\frac{11}{2}$.

$\therefore x = \frac{11}{2}e^{-t}-\frac{59}{18}e^{-3t}+2t^2-\frac{16}{3}t+\frac{52}{9}$.

(b) $x = A_1e^{-t}+A_2e^{-3t}+\frac{1}{2}e^{t}$

$x(0)=A_1+A_2+\frac{1}{2}=8$ $\Big\}$ $A_1+A_2=\frac{15}{2}$

$\frac{dx(0)}{dt}=-A_1-3A_2+\frac{1}{2}=-1$ $\Big\}$ $-A_1-3A_2=-\frac{3}{2}$

Adding, $-2A_2=6 \Rightarrow A_2=-3$, $A_1=\frac{21}{2}$

$\therefore x = 2\frac{1}{2}e^{-t}-3e^{-3t}+\frac{1}{2}e^{t}$

9.13 Since $s_{1,2}=-1,-3$ and $f(t)=2e^{-3t}-e^{-t}$, try

$x_f = Ate^{-3t}+Bte^{-t}$

$\frac{dx_f}{dt}=-3Ate^{-3t}+Ae^{-3t}-Bte^{-t}+Be^{-t}$

$\frac{d^2x_f}{dt^2}=9Ate^{-3t}-6Ae^{-3t}+Bte^{-t}-2Be^{-t}$

$\therefore (9Ate^{-3t}-6Ae^{-3t}+Bte^{-t}-2Be^{-t})+$
$4(-3Ate^{-3t}+Ae^{-3t}-Bte^{-t}+Be^{-t})+$
$3(Ate^{-3t}+Bte^{-t})=$
$-2Ae^{-3t}+2Be^{-t}=2e^{-3t}-e^{-t}$

Equating coefficients

e^{-3t}: $-2A=2 \Rightarrow A=-1$

e^{-t}: $2B=-1 \Rightarrow B=-\frac{1}{2}$

$\therefore x_f = -t(e^{-3t}+\frac{1}{2}e^{-t})$

9.14 Since $s_{1,2}=\pm j3$ and $f(t)=\sin 3t$ try $x_f = t(A\cos 3t + B\sin 3t)$

$\frac{dx_f}{dt}=t(-3A\sin 3t+3B\cos 3t)+A\cos 3t + B\sin 3t$,

$\frac{d^2x_f}{dt^2}=t(-9A\cos 3t-9B\sin 3t)-6A\sin 3t + 6B\cos 3t$,

$\frac{d^2x_f}{dt^2}+9x_f=-6A\sin 3t+6B\cos 3t=\sin 3t$

Equating coefficients

$A=-1/6$, $B=0$; $x_f=-\frac{1}{6}t\cos 3t$

9.14 cont. $x = x_n + x_f$

$x = A_1\cos 3t + A_2\sin 3t - \frac{1}{6}t\cos 3t$

$x(0)=0=A_1$

$\frac{dx}{dt}=-3A_1\sin 3t+3A_2\cos 3t-\frac{1}{6}\cos 3t - \frac{t}{2}\sin 3t$

$\frac{dx(0)}{dt}=3A_2-\frac{1}{6}=0 \Rightarrow A_2=\frac{1}{18}$

$\therefore x = \frac{1}{18}\sin 3t - \frac{1}{6}t\cos 3t$.

9.15 From Prob. 9.1 equation (3)

$\frac{di_1}{dt}=2v_g-3i_2-4i_1$, then

$\frac{di_1(0)}{dt}=2(12)-3(2)-4(-1)=22$

using equation (4) of Prob. 9.1 to find i_{1f}; $\frac{d^2i_{1f}}{dt^2}+7\frac{di_{1f}}{dt}+6i_{1f}=3v_g+2\frac{dv_g}{dt}$

since $\frac{di_{1f}}{dt}=\frac{d^2i_{1f}}{dt^2}=\frac{dv_g}{dt}=0$,

$6i_{1f}=3v_g \Rightarrow i_{1f}=6A$.

The characteristic equation of (4) is $s^2+7s+6=0$; $s_{1,2}=-1,-6$

$i_{1n}=A_1e^{-t}+Ae^{-6t}$, the

$i_1=i_{1n}+i_{1f}=A_1e^{-t}+A_2e^{-6t}+6$

$i_1(0)=A_1+A_2+6=-1$ or $A_1+A_2=-7$

$\frac{di_1(0)}{dt}=-A_1-6A_2=22$; Adding,

$-5A_2=15 \Rightarrow A_2=-3$, $A_1=-4$

$\therefore i_1 = -4e^{-t}-3e^{-6t}+6$ A.

9.16 Since the capacitor is an open circuit in dc steady-state, $i_f=0$, and $i(0^+)=0$. Let i_1 and i be the clockwise mesh currents. KVL gives

$\int_0^t i_1 dt + i_1 - i = 12$ or $\frac{di}{dt}=\frac{di_1}{dt}+i_1$

$L\frac{di}{dt}+Ri+i-i_1=0 \Rightarrow i_1 = L\frac{di}{dt}+(R+1)i$

$\therefore L\frac{di}{dt}+(R+1)i+\frac{d}{dt}\left[L\frac{di}{dt}+(R+1)i\right]-\frac{di}{dt}=0$

$L\frac{d^2i}{dt^2}+(L+R)\frac{di}{dt}+(R+1)i=0$

KVL around the outside loop at $t=0^+$ gives $-12+v(0^+)+L\frac{di(0^+)}{dt}+Ri(0^+)=0$

$\therefore \frac{di(0^+)}{dt}=\frac{12}{L}$

9.16 cont. for $L=2H$, $R=2\Omega$

The characteristic equation is

$s^2+2s+3/2=0$; $s_{1,2}=-1\pm j\,1/\sqrt{2}$

$i=e^{-t}(A_1\cos t/\sqrt{2}+A_2\sin t/\sqrt{2})$

$i(0)=A_1=0$, then setting $A_1=0$

$\frac{di(0)}{dt}=\frac{A_2}{\sqrt{2}}=\frac{12}{2}\Rightarrow A_2=6\sqrt{2}$

$\therefore i=e^{-t}(6\sqrt{2})\sin t/\sqrt{2}$ A.

9.17 From Prob. 9.16 the characteristic equation is

$s^2+s+1/2=0$; $s_{1,2}=-1/2\pm j1/2$

$i=e^{-t/2}(A_1\cos t/2+A_2\sin t/2)$

$i(0)=A_1=0$, then setting $A_1=0$

$\frac{di(0)}{dt}=\frac{A_2}{2}=\frac{12}{2}\Rightarrow A_2=12$

$\therefore i=12e^{-t/2}\sin t/2$ A

9.18 At $t=0^-$, the capacitor is an open circuit; the 3Ω resistor and the inductor are shorted.

$\therefore i_L(0^-)=\frac{8}{1}=8A$, $v_c(0^-)=0$

For $t>0$, $v_{cf}=0$, KCL gives

(1) $\frac{v_c-8}{4}+\frac{1}{20}\frac{dv_c}{dt}+\frac{1}{5}\int_0^t v_c\,dt+i_L(0^-)=0$

or $\frac{d^2v_c}{dt^2}+5\frac{dv_c}{dt}+4=0$

From equation (1) $\frac{dv_c(0)}{dt}=-120$

the characteristic equation is

$s^2+5s+4=0$; $s_{1,2}=-1,-4$

$v_c=A_1e^{-t}+A_2e^{-4t}$

$v_c(0)=A_1+A_2=0$

$\frac{dv_c(0)}{dt}=-A_1-4A_2=-120$, Adding,

$-3A_2=-120\Rightarrow A_2=40$, $A_1=-40$

$\therefore v_c=40(e^{-4t}-e^{-t})$ V

$i=\frac{8-v_c}{4}=2+10(e^{-t}-e^{-4t})$ A

9.19 At $t>0$, KCL gives

(1) $\frac{v_c-36e^{-4t}}{4}+\frac{1}{20}\frac{dv_c}{dt}+\frac{1}{5}\int_0^t v_c\,dt=0$

, $i_L(0^+)=0$, then

(2) $\frac{d^2v_c}{dt^2}+5\frac{dv_c}{dt}+4v_c=-720e^{-4t}$

9.19 cont. Try $v_{cf}=Ate^{-4t}$

$\frac{dv_{cf}}{dt}=-4Ate^{-4t}+Ae^{-4t}$

$\frac{d^2v_{cf}}{dt^2}=16Ate^{-4t}-8Ae^{-4t}$

substituting into (2) gives

$(16Ate^{-4t}-8Ae^{-4t})+5(-4Ate^{-4t}+Ae^{-4t})$

$+4(Ate^{-4t})=-3Ae^{-4t}=-720e^{-4t}$

therefore $A=\frac{720}{3}=240$ and

$v_{cf}=240te^{-4t}$ and

$v_c=A_1e^{-t}+A_2e^{-4t}+240te^{-4t}$

$v_c(0)=0$; $\frac{dv_c(0)}{dt}=-5v_c(0)+180=180$

$v_c(0)=A_1+A_2=0$

$\frac{dv_c(0)}{dt}=-A_1-4A_2+240=180$ or $-A_1-4A_2=-60$

Adding, $-3A_2=-60\Rightarrow A_2=20$, $A_1=-20$

$v_c=20e^{-4t}-20e^{-t}+240te^{-4t}$

$i=\frac{36e^{-4t}-v_c}{4}=5e^{-t}+4e^{-4t}-60te^{-4t}$ A

9.20 At $t=0^-$, $i_L(0^-)=\frac{10}{1}=10A$, $v_c(0^-)=5V$; KCL gives

$\frac{v_c-10}{1}+\frac{1}{4}\frac{dv_c}{dt}+\int_0^t v_c\,dt+i_L(0)=0$ or

$\frac{d^2v_c}{dt^2}+4\frac{dv_c}{dt}+4=0$; $s_{1,2}=-2,-2$

$v_c=(A_1+A_2t)e^{-2t}$;

$\frac{dv_c(0)}{dt}=-4v_c(0)=-20=-2A_1+A_2$

$v_c(0)=A_1=5$; $A_2=-10$

$\therefore v_c=(5-10t)e^{-2t}$

$v=10-v_c=10-(5-10t)e^{-2t}$ V

9.21 for $t>0$, KVL gives

$12i_1+2\frac{di_1}{dt}+2(i_1-i_2)=20$

$\frac{di_2}{dt}+4i_2+2(i_2-i_1)=0$ or

$i_1=\frac{1}{2}\frac{di_2}{dt}+3i_2$, Eliminate i_1 from first equation and

$\frac{d^2i_2}{dt^2}+13\frac{di_2}{dt}+40i_2=20$,

$\frac{di_2(0)}{dt}=-6i_2(0)+2i_1(0)=-2$, $i_2(0)=0$

9.21 cont. Try $i_{2f} = A$ then

$40 i_{2f} = 20 \Rightarrow i_{2f} = 1/2$

The characteristic equation is

$s^2 + 13s + 40 = 0$; $s_{1,2} = -5, -8$

$i_2 = A_1 e^{-5t} + A_2 e^{-8t} + 1/2$

$i_2(0) = A_1 + A_2 + 1/2 = 0$

$\frac{di_2(0)}{dt} = -5A_1 - 8A_2 = -2$, Adding times the first gives $3A_1 = -6 \Rightarrow A_1 = -2$, $A_2 = 3/2$ $\therefore i_2 = -2e^{-5t} + \frac{3}{2}e^{-8t} + 1/2$,

$v = 4i = 6e^{-8t} - 8e^{-5t} + 2$ V.

9.22 KVL around the left mesh yields

$6i + \frac{di}{dt} + v = 16\cos 4t$

KCL at the top node of capacitor

$i = \frac{1}{4}\frac{dv}{dt} + \frac{v}{2}$

Eliminating i results in

$\frac{d^2v}{dt^2} + 8\frac{dv}{dt} + 16v = 64\cos 4t$

$s^2 + 8s + 16 = 0 \Rightarrow v_n = (A_1 + A_2 t)e^{-4t}$

$v_f = A\cos 4t + B\sin 4t$ results in

$32(-A\sin 4t + B\cos 4t) = 64\cos 4t$

$\therefore A = 0, B = 2$

$v = (A_1 + A_2 t)e^{-4t} + 2\sin 4t$

$v(0^+) = 4 = A_1$

$\frac{1}{4}\frac{dv(0^+)}{dt} = \frac{-v(0^+)}{2} + i(0^+) = 3 - \frac{4}{2} = 1$

$\frac{dv(0)}{dt} = 4 = -4A_1 + A_2 + 8 \Rightarrow A_2 = 12$

$v = (4 + 12t)e^{-4t} + 2\sin 4t$ V.

9.23 Let v_1 be the nodal voltage at the top of 1/4 capacitor, then KCL gives

$v_1\left(\frac{1}{4} + \frac{1}{4} + \frac{1}{2}\right) + \frac{1}{4}\frac{dv_1}{dt} - \frac{v}{4} - \frac{6}{4} = 0$ or

$4v_1 + \frac{dv_1}{dt} - v = 6$

node analysis at inverting opamp terminal gives.

$\frac{v_1}{2} + \frac{1}{10}\frac{dv}{dt} = 0 \Rightarrow v_1 = -\frac{1}{5}\frac{dv}{dt}$

substituting for v_1 in first equation

$\frac{d^2v}{dt^2} + 4\frac{dv}{dt} + 5v = -30$

since $\frac{d^2v_f}{dt^2} = \frac{dv_f}{dt} = 0$, $5v_f = -30$;

9.23 cont. $v_f = -6$ V

the characteristic equation is

$s^2 + 4s + 5 = 0$; $s_1 = -2 \pm j$

$v = e^{-2t}(A_1\cos t + A_2\sin t) - 6$

$\frac{dv(0)}{dt} = -5v_1(0) = 0$

$v(0) = 0 = A_1 - 6 \Rightarrow A_1 = 6$

$\frac{dv(0)}{dt} = -2A_1 + A_2 = 0 \Rightarrow A_2 = 12$

$\therefore v = e^{-2t}(6\cos t + 12\sin t) - 6$ V

9.24 At $t = 0^-$, the capacitor is open-circuited and the inductor is a short circuit. By current division and Ohm's law

$i(0^-) = \frac{18}{2 + \frac{2(8)}{10}} \cdot \frac{8}{2+8} = 4$A; $v_c(0^-) = 4(2) = 8$V

Let i_L be the inductor current to the right and i_c be the capacitor current downward. For $t > 0$,

$v_c = 2i$, $i_c = \frac{1}{4}\frac{dv_c}{dt} = \frac{1}{2}\frac{di}{dt}$ and

$i_L = i + i_c = i + \frac{1}{2}\frac{di}{dt}$. KVL around the left mesh yields

$8i_L + 4\frac{di_L}{dt} + 2i = 0$ or

$8\left(i + \frac{1}{2}\frac{di}{dt}\right) + 4\frac{d}{dt}\left(i + \frac{1}{2}\frac{di}{dt}\right) + 2i = 0$ or

$\frac{d^2i}{dt^2} + 4\frac{di}{dt} + 5i = 0 \Rightarrow s^2 + 4s + 5 = 0 \Rightarrow$

$s_{1,2} = -2 \pm j$, $i = e^{-2t}(A_1\cos t + A_2\sin t)$

$i(0^+) = \frac{1}{2}v_c(0^+) = \frac{1}{2}v_c(0^-) = \frac{8}{2} = 4 = A_1$

$\frac{di(0^+)}{dt} = 2[i_L(0^+) - i(0^+)] = 2i(0^-) - v_c(0^+)$

$= 2i(0^-) - v_c(0^+) = 2(4) - 8 = 0$

$= A_2 - 2A_1 \Rightarrow A_2 = 8$

$\therefore i = e^{-2t}(4\cos t + 8\sin t)$ A

9.25 v_L = inductor voltage, positive top.

KCL gives (1) $\frac{1}{5}\frac{dv}{dt} + \frac{v - v_L}{4} = i_g$

(2) $-\frac{v - v_L}{4} + \int_0^t v_L\,dt = 0$

Add (1) and (2), differentiate and substitute for v_L from (1):

$\frac{d^2v}{dt^2} + 4\frac{dv}{dt} + 5v = 5\frac{di_g}{dt} + 20i_g$

At $t = 0^-$, $v(0^-) = i_L(0^-) = 0$.

Therefore $v(0^+) = 0$ and $v_L(0^+) = 0$ by KVL. From (1) $\frac{dv(0^+)}{dt} = 5i_g(0^+)$

9.25 cont

(a) $i_g = 2u(t)$ A. For $t>0$, $\frac{di_g}{dt}=0$, $i_g = 2$ A:

$\frac{d^2v}{dt^2}+4\frac{dv}{dt}+5v=40 \Rightarrow s_{1,2}=-2\pm j$

$v = e^{-2t}(A_1\cos t + A_2 \sin t)+8$

$v(0^+)=0=A_1+8$; $\frac{dv(0^+)}{dt}=-2A_1+A_2=10$

$A_1=-8$; $A_2=-6 \Rightarrow$

$v = 8 - e^{-2t}(8\cos t + 6\sin t)$ V

(b) $i_g = 2e^{-t}u(t)$: For $t>0$; $i_g=2e^{-t}$ and $\frac{di_g}{dt}=-2e^{-t}$. Therefore

$\frac{d^2v}{dt^2}+4\frac{dv}{dt}+5v=30e^{-t}$

Try $v_f = Ae^{-t}$. Then

$A(1-4+5)e^{-t}=30e^{-t}\Rightarrow A=15$.

$v=e^{-2t}(A_1\cos t+A_2\sin t)+15e^{-t}$

$v(0)=0=A_1+15 \Rightarrow A_1=-15$.

$\frac{dv(0)}{dt}=5(2)=-2A_1+A_2-15\Rightarrow A_2=-5$

$\therefore v = 15e^{-t}-e^{-2t}(15\cos t+5\sin t)$ V $(t>0)$

9.26 KVL gives: $i - i_g = \frac{1}{2}\frac{dv}{dt}$

$i + 2\frac{di}{dt}+v-3(i_g-i)=0$;

$4(i_g+\frac{1}{2}\frac{dv}{dt})+2\frac{d}{dt}(i_g+\frac{1}{2}\frac{dv}{dt})+v=3i_g$

or $\frac{d^2v}{dt^2}+2\frac{dv}{dt}+v=-i_g-2\frac{di_g}{dt}=2e^{-t}$

$s^2+2s+1=0\Rightarrow s_{1,2}=-1,-1$; try

$v_f=At^2e^{-t}$ $\therefore 2Ae^{-t}=2e^{-t}\Rightarrow A=1$

$v=(A_1+A_2t)e^{-t}+t^2e^{-t}$

$=(A_1+A_2t+t^2)e^{-t}$

$v(0)=4=A_1$: $\frac{dv(0)}{dt}=2[i(0)-i_g(0)]$

$=2(2-2)=0$

$\frac{dv(0)}{dt}=0=-A_1+A_2\Rightarrow A_2=4$

$v=(4+4t+t^2)e^{-t}$ V

9.27 $L=4R^2C=4(4000)^2(10^{-7})=6.4$ H

9.28 $\sqrt{\frac{1}{LC}-(\frac{1}{2RC})^2}=\omega_d$

$L=\frac{1}{(10^{-7})[(661)^2+[\frac{1}{(2)(4\times10^3)(10^{-7})}]^2]}=5$ H

9.29 $s_{1,2}=-\frac{1}{2RC}\pm\sqrt{(\frac{1}{2RC})^2-\frac{1}{LC}}$

$2000=-1250+\sqrt{(1250)^2-\frac{10^7}{L}}\Rightarrow L=10$ H

9.30 The differential equation for the circuit is

$L\frac{di}{dt}+Ri+\frac{1}{C}\int_0^t i\,dt+V_0=v_g$

For $t=0^+$, $L\frac{di(0^+)}{dt}+Ri(0^+)+V_0=v_g(0^+)=0$

$\frac{di(0^+)}{dt}=-\frac{2}{4}(0)-\frac{4}{4}=-1$ A/s

$s_{1,2}=-\frac{R}{2L}\pm\sqrt{(\frac{R}{2L})^2-\frac{1}{LC}}=-\frac{2}{2(4)}\pm\sqrt{(\frac{1}{4})^2-\frac{8}{4}}$

$=-\frac{1}{4}\pm j1.392$

$\therefore i=e^{-t/4}(A_1\cos 1.39t+A_2\sin 1.39t)$

$i(0^+)=0=A_1$; $\frac{di(0^+)}{dt}=A_2 1.39=-1\Rightarrow$

$A_2=.7184$; $i=.718e^{-t/4}\sin 1.39t$ A

9.31

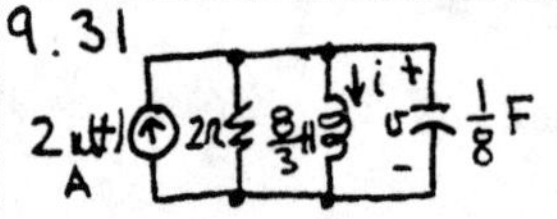

using the norton circuit of the first 3 elements, we obtain the parallel RLC circuit. The differential equation is

$\frac{d^2i}{dt}+\frac{1}{CR}di+\frac{1}{LC}i=\frac{i_g}{LC}$

For $t=0^+$; $i(0^+)=i(0^-)=0$,

$\frac{di(0^+)}{dt}=\frac{v(0^+)}{L}=\frac{v(0^-)}{L}=0$,

$s_{1,2}=-\frac{1}{2RC}\pm\sqrt{(\frac{1}{2RC})^2-\frac{1}{LC}}$, $\frac{1}{2RC}=2$

$=-2\pm\sqrt{4-3}=-1,-3$

$i_f=i_g=2$ A

$i=A_1e^{-t}+A_2e^{-3t}+2$

$i(0^+)=0=A_1+A_2+2$

$\frac{di(0^+)}{dt}=0=-A_1-3A_2$, Adding,

$-2A_2=-2\Rightarrow A_2=1$, $A_1=-3$

$\therefore i=e^{-3t}-3e^{-t}+2$ A $(t>0)$

9.32 Using the differential equation of Prob. 9.31

$s_{1,2}=-2\pm\sqrt{4-\frac{1}{2(\frac{1}{8})}}=-2$

$i_f=i_g=2$ A;

$i=(A_1+A_2t)e^{-2t}+2$

$i(0^+)=0=A_1+2\Rightarrow A_1=-2$

$\frac{di(0^+)}{dt}=0=-2A_1+A_2\Rightarrow A_2=-4$

$\therefore i=2-(2+4t)e^{-2t}$ A $(t>0)$

9.33 Node equations are

$$\frac{v_1-6}{4}+\frac{1}{4}\frac{dv_1}{dt}+\frac{v_1}{2}+\frac{v_1-v}{4}=0 \text{ or}$$

$$\frac{dv_1}{dt}+4v_1-v=6$$

$$\frac{v_1}{2}+\frac{1}{10}\frac{dv}{dt}=0 \text{ or } 5v_1+\frac{dv}{dt}=0$$

$$\left.\begin{aligned}(D+4)v_1-v&=6\\ 5v_1+Dv&=0\end{aligned}\right\}$$

$$\begin{vmatrix}D+4 & -1\\ 5 & D\end{vmatrix}v=\begin{vmatrix}D+4 & 6\\ 5 & 0\end{vmatrix}$$

$$\therefore [(D+4)D+5]v=-30 \text{ or}$$

$$\frac{d^2v}{dt}+4\frac{dv}{dt}+5v=-30;\quad s_{1,2}=-2\pm j$$

$$v=e^{-2t}(A_1\cos t+A_2\sin t)-6$$

$$v(0)=0=A_1-6 \Rightarrow A_1=6$$

$$\frac{dv(0)}{dt}=0=-2A_1+A_2 \Rightarrow A_2=12$$

$$\therefore v=\underline{e^{-2t}(6\cos t+12\sin t)-6\text{ V}}$$

9.34 The mesh equations with i_1 and i being the mesh currents going clockwise are.

$$\frac{1}{C}\int_0^t i_1\,dt+i_1(0)+(i_1-i)1=12 \text{ or}$$

$$\left.\begin{aligned}i_1+\frac{di_1}{dt}-\frac{di}{dt}&=0\\ L\frac{di}{dt}+iR+i-i_1&=0\end{aligned}\right\}\quad \text{with } L=2\text{H},\ R=2\,\Omega$$

$$\left.\begin{aligned}(D+1)i_1-Di&=0\\ -i_1+(2D+3&=0)\end{aligned}\right\}$$

$$\begin{vmatrix}(D+1) & -D\\ -1 & (2D+3)\end{vmatrix}i=\begin{vmatrix}D+1 & 0\\ -1 & 0\end{vmatrix}=0$$

$$[(D+1)(2D+3)+D]\,i=0$$

$$(2D^2+6D+3)\,i=0$$

$$\therefore \underline{2\frac{d^2i}{dt^2}+6\frac{di}{dt}+3i=0}$$

Chapter 10

Sinusoidal Excitation and Phasors

10.1 Properties of Sinusoids

10.1 Find the period of the following sinusoids: (a) $6\sin(5t + 17°)$, (b) $3\sin(4t + \frac{\pi}{4}) + 2\cos(4t + \pi)$, (c) $5\sin(6\pi t)$.

10.2 Find the amplitude and phase of the sinusoids in Prob. 10.1.

10.3 Find the frequency of the following sinusoids: (a) $2\cos(8t + \pi)$, (b) $3\sin(6{,}280t)$.

10.4 Given $v_1 = 5\sin(4t)$ V, $v_2 = 20[\cos(4t) + \sqrt{3}\sin(4t)]$ V determines if v_1 leads or lags v_2 and by what amount.

10.5 Find i_4, using only the properties of sinusoids, if $i_1 = 2\cos(3t)$ A, $i_2 = 25\cos(3t + 36.9°)$ A, and $i_3 = 13\sin(3t + 157.4°)$ A.

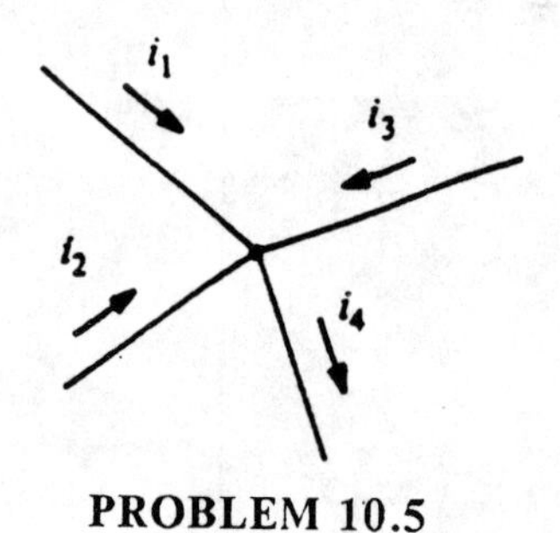

PROBLEM 10.5

10.2 An RL Circuit Example

10.6 If the source is $10\cos(3000t)$ mA and the output is $v = 8\cos(3000t - 36.9°)$ V, find R and C.

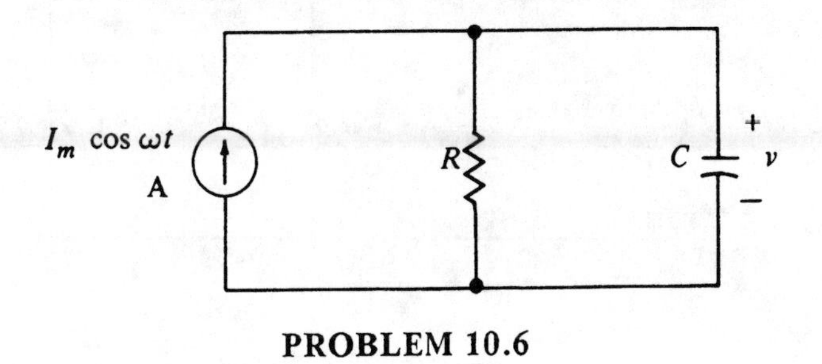

PROBLEM 10.6

10.3 An Alternative Method Using Complex Numbers

10.7 A complex voltage input $10e^{j(4t+20°)}$ V produces a current output $2e^{j(4t-25°)}$ A. Find the output current if the input voltage is (a) $45e^{(j4t+50°)}$ V, (b) $15\sin(4t + 60°)$ V, (c) $25\cos(4t)$ V.

10.4 Complex Excitations

10.8 (a) From the time-domain equations find the forced response v if $i_g = 10e^{j2t}$A. (b) Using the result in (a), find the forced response v if $i_g = 10\cos(2t)$ A.

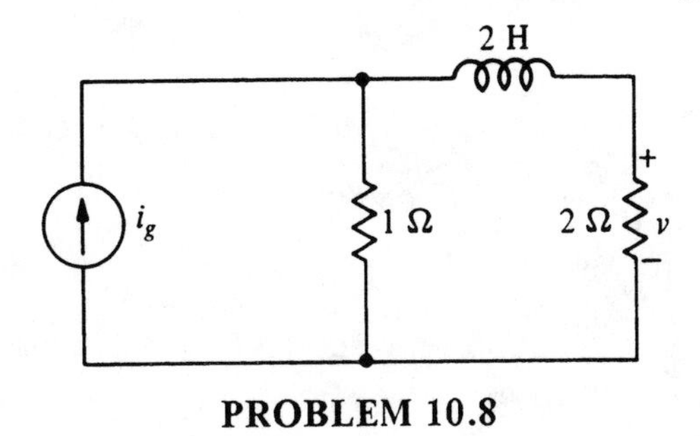

PROBLEM 10.8

10.9 Find the forced response v in Prob. 10.8 if $i_g = 10\sin(2t)$ A.

10.10 Find v_1 from the differential equation and use the result to find the forced response v to an input voltage of $34\cos(4t)$ V.

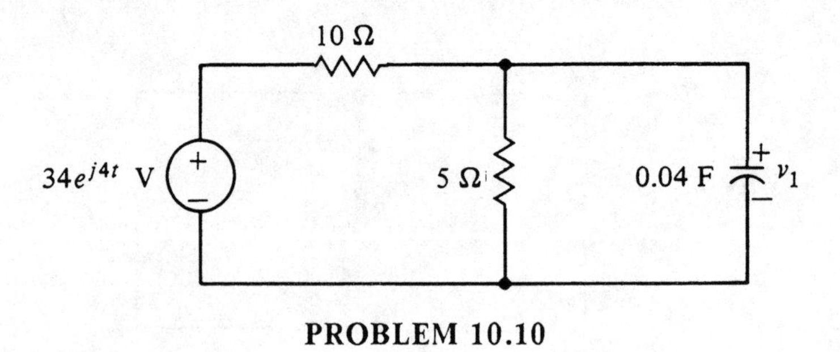

PROBLEM 10.10

10.11 Find the response v_1 to the source $2e^{j8t}$A and use the result to find the response v to (a) $2\cos(8t)$ A, and (b) $2\sin(8t)$ A.

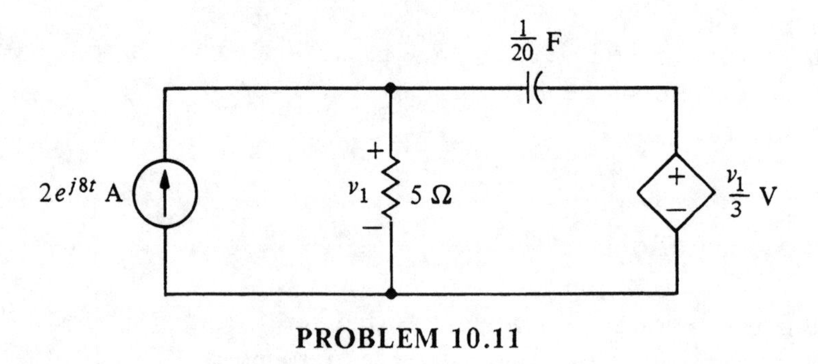

PROBLEM 10.11

10.5 Phasors

10.12 Find the phasor representation of (a) $5\cos(4t + 30°)$, (b) $-3\sin(3t) + 2\cos(3t)$, (c) $4\sin(10t - 20°)$.

10.13 Find the time-domain function represented by the phasors (a) $15\angle -45°$, (b) $5 + j12$, (c) $5 - j12$, $\omega = 5$ for all cases.

10.14 If $\omega = 5$ rad/s, find the time-domain functions represented by the phasors (a) $-6 - j6$, (b) $-5 + j12$, (c) $4 - j3$, (d) -10.

10.6 Voltage-Current Relationships for Phasors

10.15 Using phasors, find the ac steady-state current if $v = 12\cos(1000t + 30°)$ V, $R = 4k\Omega$.

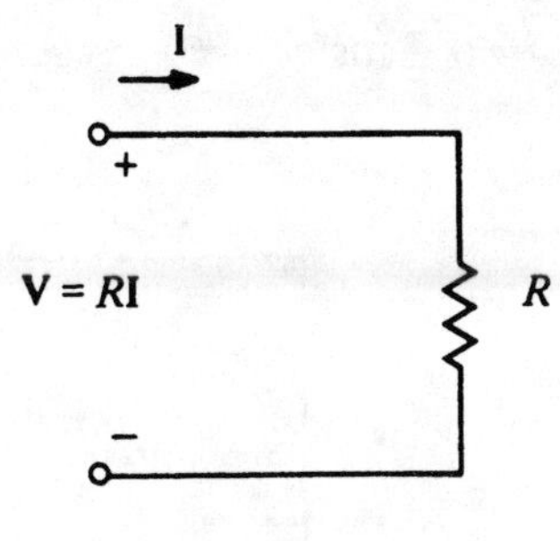

PROBLEM 10.15

10.16 Using phasors, find the ac steady-state current if $v = 10\sin(377t + 15°)$ V, L = 0.5H.

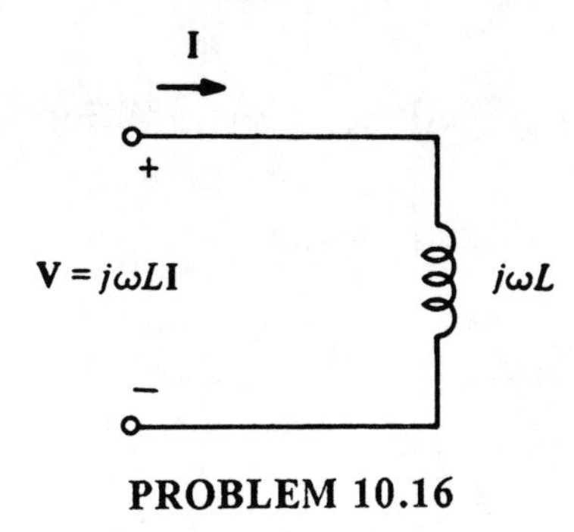

PROBLEM 10.16

10.17 Using phasors, find the ac steady-state current if $v = 12\sin(377t + 15°)$ V, $C = 10\mu$F.

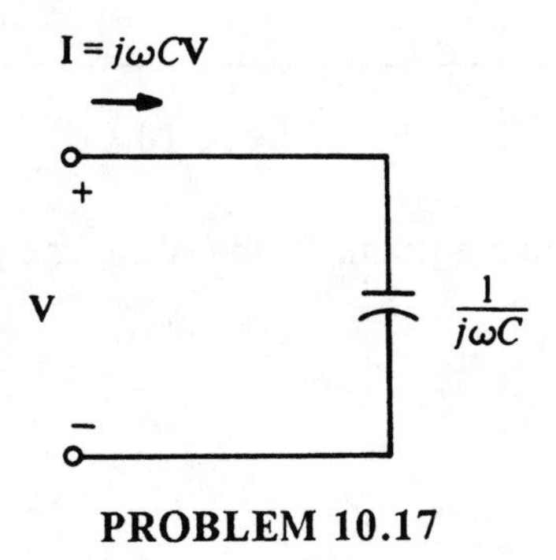

PROBLEM 10.17

10.7 Impedance and Admittance

10.18 Find the conductance and susceptance if **Z** is (a) $12 + j5$, (b) $3 - j3$, (c) $5\angle 30°$.

10.19 Find the resistance and reactance if **Y** is (a) $12 + j5$, (b) $3 - j3$, (c) $5\angle 30°$.

10.20 Find the impedance of the circuit shown if the time-domain functions represented by the phasors **V** and **I** are

(a) $v = -15\cos(2t) + 8\sin(2t)$ V, $i = 1.7\cos(2t + 40°)$ A.

(b) $v = \text{Re}[je^{j2t}]$ V, $i = \text{Re}[(1 + j)e^{j(2t+30°)}]$ mA.

(c) $v = aV_m\cos(\omega t + \theta)$ V, $i = V_m\cos(\omega t + \theta - \alpha)$ A.

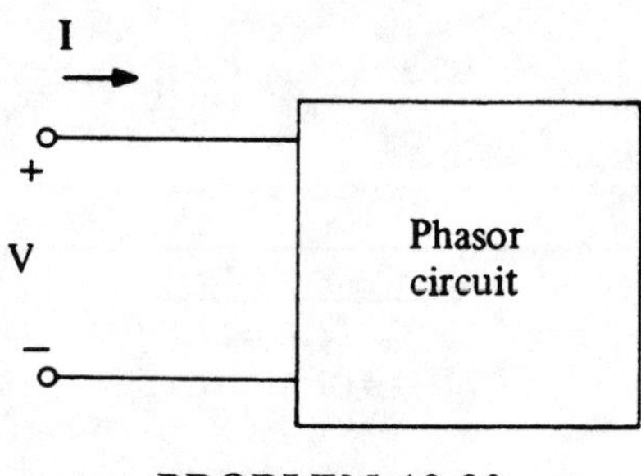

PROBLEM 10.20

10.8 Kirchoff's Laws and Impedance Combinations

10.21 Find the steady-state voltage v using phasors.

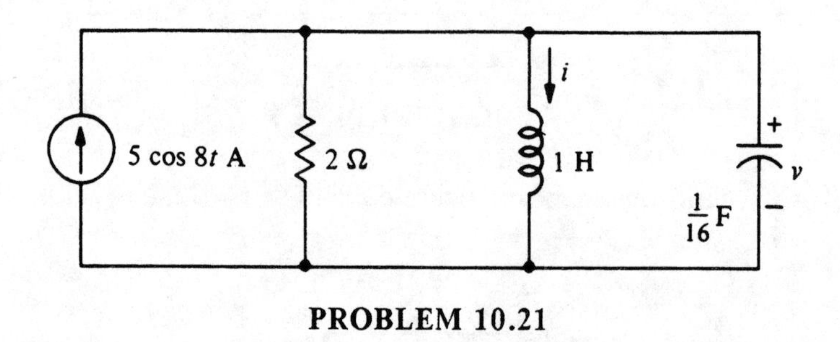

PROBLEM 10.21

10.22 Find the steady-state current i in Prob. 10.21 using phasors and current division.

10.23 Solve Prob. 10.5 using phasors.

10.9 Phasor Circuits

10.24 Find the steady-state voltage v using the phasor circuit.

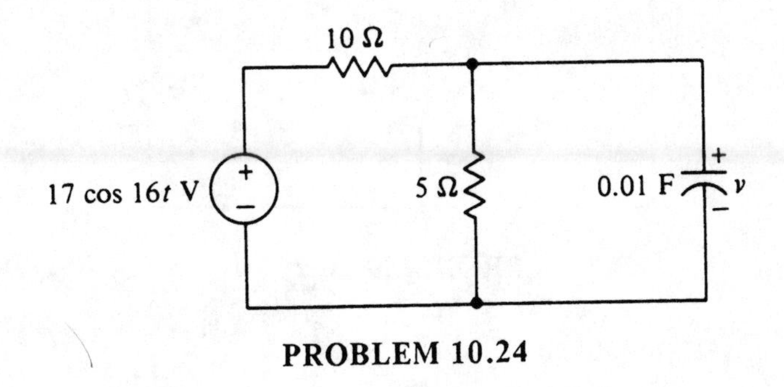

PROBLEM 10.24

10.25 Find the steady-state voltage v in Prob. 10.8 using the phasor circuit.

10.26 Find the steady-state current i.

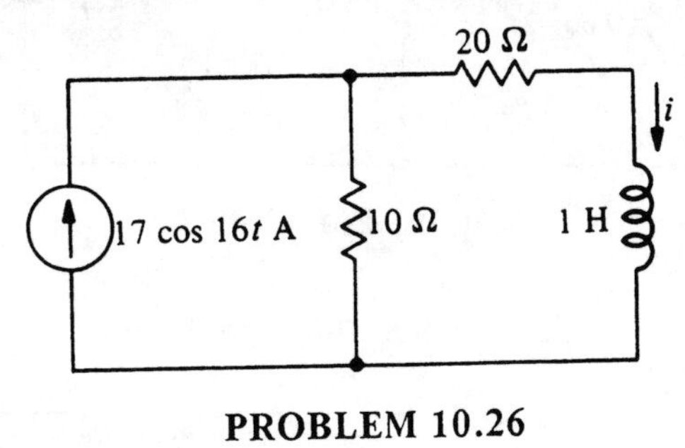

PROBLEM 10.26

10.27 Find the steady-state current i if $L = 1\text{H}$.

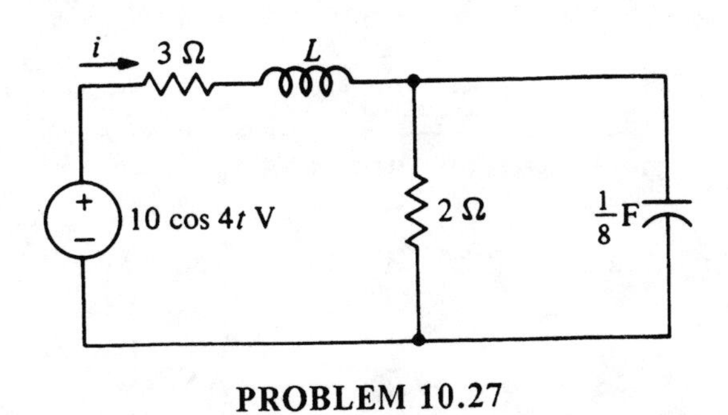

PROBLEM 10.27

10.28 Determine L in Prob. 10.27 so that the impedance seen by the source is real, and for this case find the power delivered by the source at $t = \pi/4$ s.

10.29 Find the forced response i using phasors and current division.

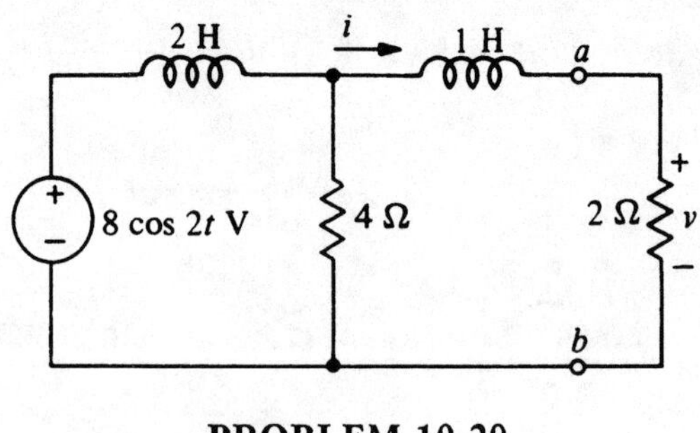

PROBLEM 10.29

10.30 Find the steady-state current i using phasors.

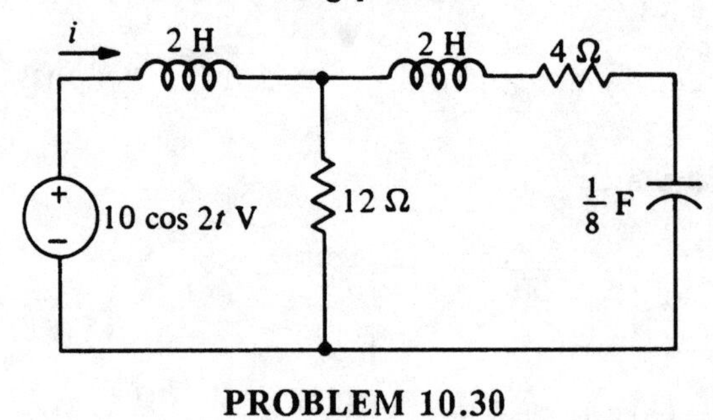

PROBLEM 10.30

10.31 Find the steady-state voltage v using phasors and voltage division.

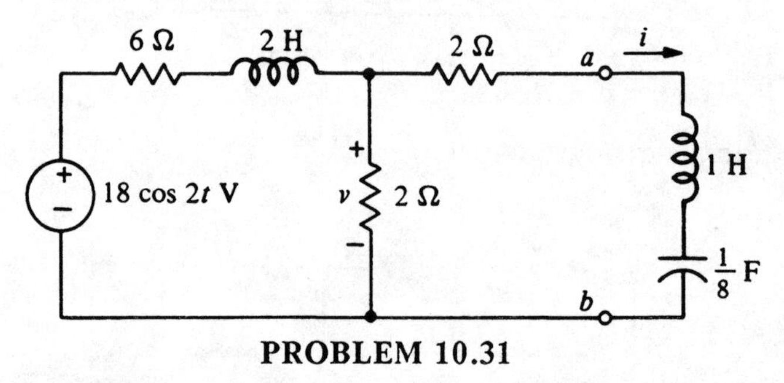

PROBLEM 10.31

10.32 Find the steady-state current of i if (a) $\omega = 4$ rad/s and (b) $\omega = 2$ rad/s.

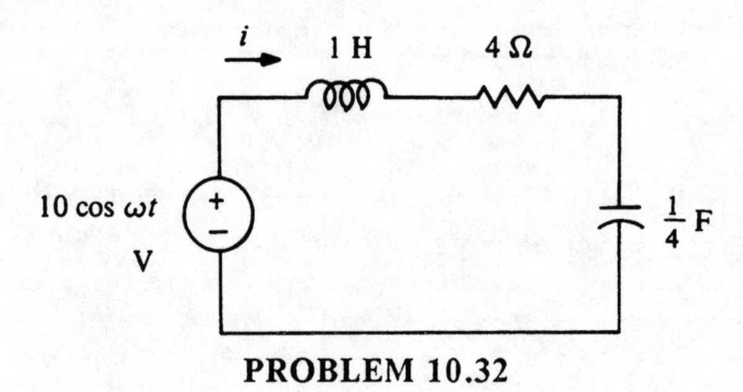

PROBLEM 10.32

10.33 Find the steady-state value of v_x.

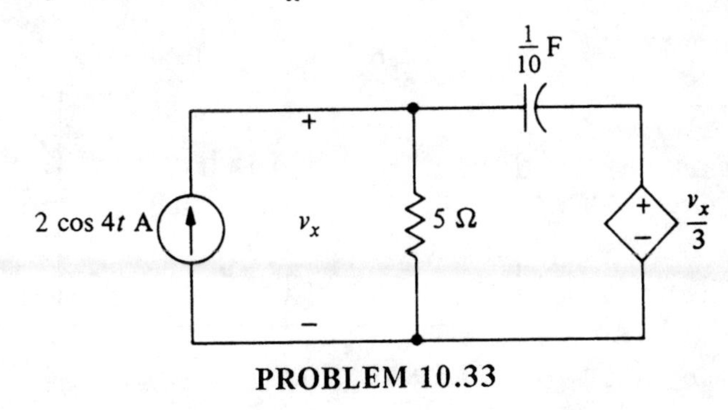

PROBLEM 10.33

10.34 Find the steady-state current i.

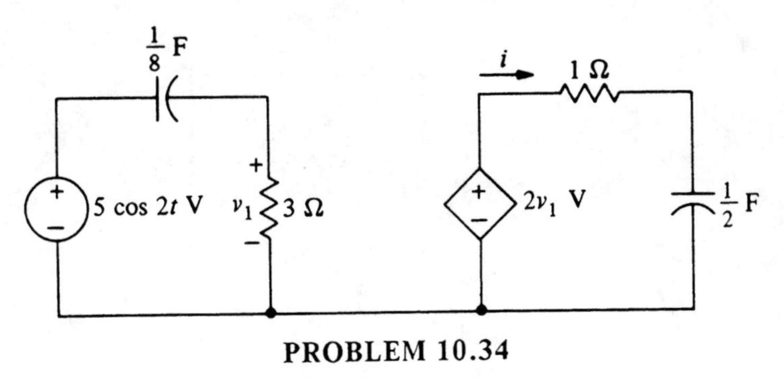

PROBLEM 10.34

10.35 Find the steady-state value of v.

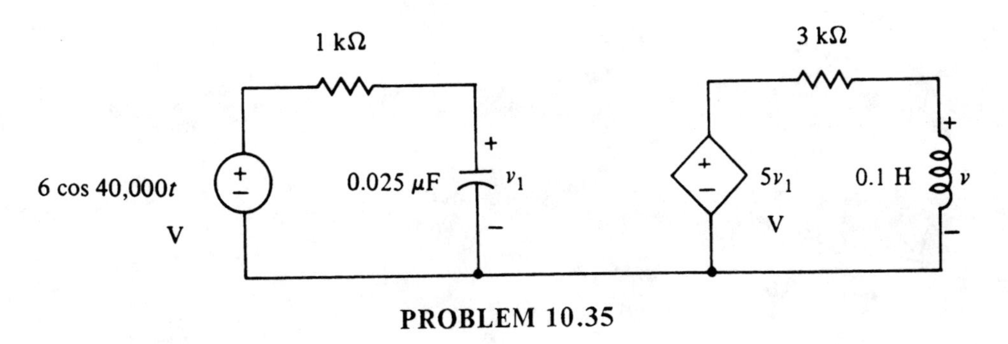

PROBLEM 10.35

10.36 Find the steady-state response i.

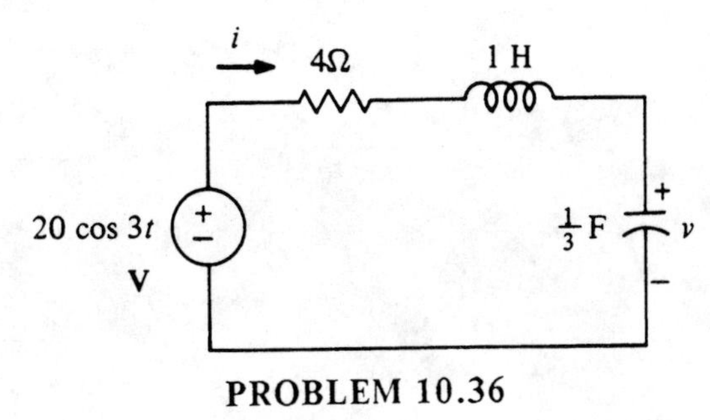

PROBLEM 10.36

10.37 Find the steady-state values of i and v.

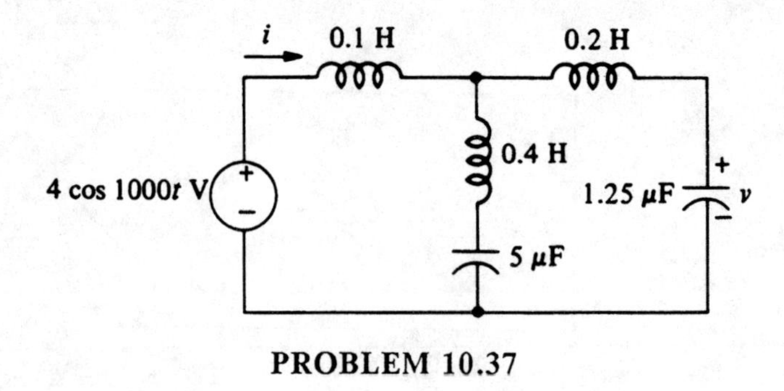

PROBLEM 10.37

10.38 Find the steady-state voltage v if $v_g = 10\cos(10{,}000t)$ V.

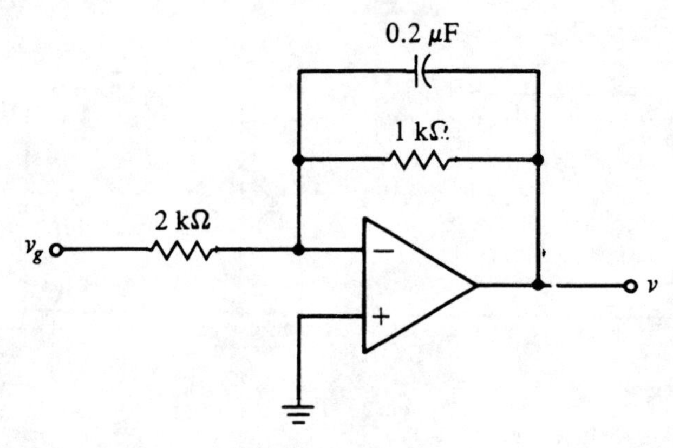

PROBLEM 10.38

10.1(a) $\omega = 5$ rad/s, $T = 2\pi/\omega = \underline{2\pi/5 \text{ s}}$

(b) $\omega = 4$ rad/s, $T = 2\pi/\omega = \underline{\pi/2 \text{ s}}$

(c) $\omega = 6$ rad/s, $T = 2\pi/\omega = \underline{\pi/3 \text{ s}}$

10.2(a) Amplitude $= \underline{6}$, $\phi = \underline{17°}$.

(b) $3\sin(4t + \frac{\pi}{4}) + 2\cos(4t + \pi)$

$= 3[\sin 4t \cos \pi/4 + \cos 4t \sin \pi/4]$

$+ 2[\cos 4t \cos \pi - \sin 4t \sin \pi]$

$= \sqrt{(\frac{3}{\sqrt{2}} - 2)^2 + \frac{3^2}{2}} \cos[4t - \tan^{-1}(\frac{3}{3 - 2\sqrt{2}})]$

$= 2.125 \cos(4t - 86.73°)$

Amplitude $= \underline{2.13}$, $\phi = \underline{86.7°}$.

(c) Amplitude $= \underline{5}$, $\phi = \underline{0°}$.

10.3(a) $f = \omega/2\pi = 8/2\pi = \underline{4/\pi \text{ Hz}}$.

(b) $f = \omega/2\pi = 6{,}280/2\pi = \underline{1 \text{ KHz}}$.

10.4 $v_1 = 5 \sin 4t = 5\cos(4t - 90°)$

$v_2 = 20\sqrt{1+3} \cos(4t - \tan^{-1}\frac{\sqrt{3}}{1})$

$= 40\cos(4t - 60°)$

$\therefore v_1$ lags v_2 by $90° - 60° = \underline{30°}$.

10.5 By KCL, $i_4 = i_1 + i_2 + i_3$

$i_4 = 25[\cos 3t \cos 36.9° - \sin 3t \sin 36.9°]$

$+ 13[\sin 3t \cos 157.4° + \cos 3t \sin 157.4°]$

$+ 2\cos 3t$

$= 27\cos 3t - 27 \sin 3t$

$= \underline{27\sqrt{2}\cos(3t + 45) \text{ A}}$

10.6 $C\frac{dv}{dt} + \frac{v}{R} = I_m \cos \omega t$

$\frac{dv}{dt} = 8(-3000 \sin(3000t - 36.9°))$

$-24{,}000C \sin(\omega t - 36.9°) + \frac{8}{R}\cos(\omega t - 36.9°)$

$= \sqrt{\frac{64}{R^2} + C^2 576 \times 10^6} \cos[\omega t - 36.9 - \tan^{-1}(-RC \cdot 3 \times 10^3)]$

$\therefore I_m = 10 \times 10^{-3} = \frac{\sqrt{64 + R^2C^2 \cdot 576 \times 10^6}}{R}$ or

$R = \frac{\sqrt{64 + (RC)^2 \times 576 \times 10^6}}{10 \times 10^{-3}}$

$\phi = 0 = 36.9° + \tan^{-1}[(-RC)(3 \times 10^3)]$ or

$RC = \frac{\tan(-36.9°)}{-3 \times 10^3} = 250 \times 10^{-6}$

$\therefore R = \underline{1 \text{ K}\Omega}$, $C = \frac{250 \times 10^{-6}}{R} = \underline{250 \text{ nF}}$

10.7 The output amplitude is 1/5 of the input amplitude. The phase is 45° less, therefore

(a) $i = 45/5 \, e^{j(4t + 50° - 45°)} = \underline{9e^{j(4t + 5°)} \text{ A}}$

(b) $i = 15/5 \sin(4t + 60° - 45°) = \underline{3\sin(4t + 15°) \text{ A}}$

(c) $i = 25/5 \cos(4t - 45°) = \underline{5\cos(4t - 45°) \text{ A}}$

10.8(a) $i = \frac{v}{2}$ = inductor current,

$L\frac{di}{dt} + v + (i - i_g) = 0$ or

$\frac{dv}{dt} + \frac{3}{2}v = i_g = 10e^{j2t}$

$v_f = Ae^{j2t}$, $\frac{dv_f}{dt} = jA2e^{j2t}$

$\frac{dv_f}{dt} + \frac{3}{2}v_f = jA2e^{j2t} + \frac{3}{2}Ae^{j2t} = 10e^{j2t}$

$A(\frac{3}{2} + j2) = 10 \Rightarrow A = \frac{20}{3 + j4} = 4e^{-j53.1°}$

$\therefore v_f = \underline{4e^{j(2t - 53.1°)} \text{ V}}$

(b) $v = \text{Re}\, v_f = \text{Re}[4e^{j(2t - 53.1°)}]$

$= \underline{4\cos(2t - 53.1°) \text{ V}}$

10.9 $v = \text{Im}\, v_f = \text{Im}(4e^{j(2t - 53.1°)})$

$= \underline{4\sin(2t - 53.1°) \text{ V}}$

10.10 KCL: $0.04\frac{dv_1}{dt} + \frac{v_1}{5} + \frac{v_1}{10} = \frac{34}{10}e^{j4t}$ or

$2\frac{dv_1}{dt} + 15v_1 = 34(5)e^{j4t}$, try $v_1 = Ae^{j4t}$

$Ae^{j4t}(j8 + 15) = 5(34)e^{j4t}$

$A = \frac{5(34)}{15 + j8} = 10e^{-j\theta}$; $\theta = \tan^{-1}\frac{8}{15} = 28.1°$

Then $v_1 = 10e^{j(4t - \theta)}$ and

$v = \text{Re}\, v_1 = 10\cos(4t - \theta) \text{ V}$

$= \underline{10\cos(4t - 28.1°) \text{ V}}$

10.11 KCL: $\frac{v_1}{5} + \frac{1}{20}\frac{d}{dt}(v_1 - \frac{v_1}{3}) = 2e^{j8t}$ or

$\frac{dv_1}{dt} + 6v_1 = 60e^{j8t}$. Try $v_1 = Ae^{j8t}$

then $Ae^{j8t}(j8 + 6) = 60e^{j8t}$

$A = \frac{60}{6 + j8} = 6e^{-j53.1°}$

$\therefore v_1 = \underline{6e^{j(8t - 53.1°)} \text{ V}}$

(a) $v = \text{Re}\, v_1 = \underline{6\cos(8t - 53.1°) \text{ V}}$

(b) $v = \text{Im}\, v_1 = \underline{6\sin(8t - 53.1°) \text{ V}}$

10.12 (a) amplitude = 5, $\phi = 30°$

phasor = $\underline{5\angle 30°}$

(b) $-3\sin 3t + 2\cos 3t = \sqrt{13}\cos(3t + 56.3°)$

phasor = $\underline{\sqrt{13}\angle 56.3°}$

(c) $4\sin(10t - 20°) = 4\cos(10t - 20° - 90° + 180°)$

$= 4\cos(10t + 70°)$

phasor = $\underline{4\angle 70°}$

10.13 (a) amplitude = 15, $\phi = -45°$

$\underline{v = 15\cos(5t - 45°)}$

(b) $5 + j12 = 13\angle 67.38°$

$\underline{v = 13\cos(5t + 67.38°)}$

(c) $5 - j12 = 13\angle -67.38°$

$\underline{v = 13\cos(5t - 67.38°)}$

10.14 (a) $-6 - j6 = 6\sqrt{2}\angle -135°$

$\underline{v = 6\sqrt{2}\cos(5t - 135°)}$

(b) $-5 + j12 = 13\angle 112.62°$

$\underline{v = 13\cos(5t + 112.62°)}$

(c) $4 - j3 = 5\angle -36.87°$

$\underline{v = 5\cos(5t - 36.87°)}$

(d) $-10 = 10\angle 180°$

$\underline{v = 10\cos(5t + 180°)}$

$\underline{= -10\cos 5t}$

10.15 $\underline{V} = 12\angle 30°$, $R = 4\,k\Omega$

$\underline{I} = \frac{\underline{V}}{R} = \frac{12\angle 30°}{4} = 3\angle 30°\,mA$

$\underline{i = 3\cos(1000t + 30°)\,mA}$

10.16 $\underline{V} = 10\angle 15°$ (sine conversion)

$\underline{I} = \frac{\underline{V}}{j\omega L} = \frac{10\angle 15°}{j(377)(.5)} = 53\angle -75°$

$\underline{i = 53\sin(377t - 75°)\,A}$

10.17 $\underline{V} = 12\angle 15°$ (sine conversion)

$\underline{I} = j\omega C\underline{V} = j(377)(10^{-5})(12\angle 15°)$

$= 45.24\angle 105°\,mA$

$i = 45.24\sin(377t + 105°)\,mA$

10.18 (a) $\underline{Y} = G + jB = \frac{1}{12 + j5} = \underline{0.071 - j0.030}$

(b) $\underline{Y} = G + jB = \frac{1}{3 - j3} = \underline{0.167 + j0.167}$

(c) $\underline{Y} = G + jB = \frac{1}{5\angle 30°} = \underline{0.173 - j0.100}$

10.19 (a) $\underline{Z} = R + jX = \frac{1}{12 + j5} = \underline{0.071 - j0.030}$

(b) $\underline{Z} = R + jX = \frac{1}{3 - j3} = \underline{0.167 + j0.167}$

(c) $Z = R + jX = \frac{1}{5\angle 30°} = \underline{0.173 - j0.100}$

10.20 (a) $\underline{Z} = \underline{V}/\underline{I} = \frac{-15 - j8}{1.7\angle 40°} = \frac{17\angle 151.9°}{1.7\angle 40°}$

$= \underline{10\angle 111.9\,\Omega}$

(b) $v = Re[j(\cos 2t + j\sin 2t] = -\sin 2t$

$= -\cos(2t - 90°) = \cos(2t + 90°)\,V$

$i = Re\{(1 + j1)[\cos(2t + 30°) + j\sin(2t + 30°)]\}$

$= \cos(2t + 30°) - \sin(2t + 30°)$

$= \sqrt{2}\cos(2t + 75°)\,mA$

$\underline{Z} = \frac{\underline{V}}{\underline{I}} = \frac{1\angle 90°}{\sqrt{2}\angle 75°} = \underline{\frac{1}{\sqrt{2}}\angle 15°\,k\Omega}$

(c) $\underline{Z} = \frac{aV_m\angle\theta}{V_m\angle\theta - \alpha} = \underline{a\angle\alpha\,\Omega}$

10.21 By KCL at top Node

$\underline{V}\left(\frac{1}{2} + \frac{1}{j\omega L} + j\omega C\right) = 5\angle 0°\,A$; $\omega = 8$ rad/s

$\underline{V} = \frac{5\angle 0°}{\frac{1}{2} + j\frac{3}{8}} = 8\angle -36.87°$

$\therefore v = \underline{8\cos(8t - 36.87°)\,V}$

10.22 the impedance of the $2\,\Omega$ resistor and capacitor are

$Z_{RC} = \frac{(2)\,1/j\omega C}{2 + 1/j\omega C} = \frac{2}{1 + j\omega 2/16} = 1 - j$

By current division

$\underline{I} = \frac{(1 - j)(5)}{1 - j + j8} = 1\angle -126.9°$

$\underline{i = \cos(8t - 126.9°)\,A}$

10.23 $\underline{I}_4 = \underline{I}_1 + \underline{I}_2 + \underline{I}_3$

$= 2 + 25\angle 36.9° + 13\angle$

$= 2 + 20 + j15 + 5 + j12$

$= 27 + j27 = 27\sqrt{2}\angle 45°$

$\therefore i = \underline{27\sqrt{2}\cos(3t + 45°)\,A}$

10.24 the impedance of the $5\,\Omega$ resistor and capacitor is

$Z_{RC} = \frac{(5)\,1/j16(0.01)}{5 + 1/j16(0.01)} = 3.9\angle -38.66°$.

By voltage division:

$\overline{V} = \frac{Z_{RC}}{Z_{RC} + 10}(17) = 5\angle -28.07°$

10.24 cont.

$\therefore v = \underline{5\cos(16t - 28.07^\circ)\text{ V}}$

10.25 Let $\underline{I}$ = phasor current of inductor then by current division

$$\underline{I} = \frac{\underline{I}g}{1+2+j\omega L} = \frac{10}{3+j(2)(2)} = 2\angle{-53.13^\circ}\text{A}$$

$$\underline{V} = \underline{I}R = (2)(2\angle{-53.13^\circ}) = 4\angle{-53.13}\text{ V}$$

$v = \underline{4\cos(2t - 53.13^\circ)\text{ V}}$

10.26 By current division

$$\underline{I} = \frac{10\ (17)}{10+20+j(16)(1)} = 5\angle{-28.07^\circ}$$

$i = \underline{5\cos(16t - 28.07^\circ)\text{ A}}$

10.27 the impedance seen by the source is:

$$\underline{Z} = 3 + j(4)(1) + \frac{2}{1 + j(4)(2)(1/8)} = 4 + j3 = 5\angle{36.9^\circ}$$

$$\underline{I} = \frac{\underline{V}}{\underline{Z}} = \frac{10}{5\angle{36.9^\circ}} = 2\angle{-36.9^\circ}\text{ A}$$

$i = \underline{2\cos(4t - 36.9^\circ)\text{ A}}$

10.28

$$Z = 3 + j4L + 1 - j = 4 + j(4L-1)$$

$\therefore 4L - 1 = 0 \Rightarrow L = \underline{1/4\text{ H}}$

$$\underline{I} = \frac{10}{4} = 2.5\angle 0\text{ A}$$

$\therefore i = 2.5\cos 4t\text{ A},\ v = 10\cos 4t$

$$p = vi = 2.5\cos[4(\tfrac{\pi}{4})] \times 10\cos[4(\pi/4)]$$

$= \underline{25\text{ W}}$ at $t = \pi/4$ s

10.29 the impedance seen by the source is.

$$\underline{Z} = j(2)(2) + \frac{4(2+j2)}{4+2+j2} = 1.6 + j4.8 = 5.06\angle{71.6^\circ}$$

Let $\underline{I}_1$ be the source current

$$\underline{I}_1 = \frac{8}{5.06\angle{71.6^\circ}} = 1.581\angle{71.57^\circ}$$

By current division

$$\underline{I} = \frac{4\ \underline{I}_1}{4+2+j2} = \frac{2-j6}{6+j2} = 1\angle{-90^\circ}$$

$i = \cos(2t - 90^\circ) = \underline{\sin 2t\text{ A}}$

10.30 the impedance seen by the source is

$$\underline{Z} = j(2)(2) + \frac{12[4 + j4 + \frac{1}{j2(1/8)}]}{12 + 4 + j4 - j4} = 3 + j4$$

$$\underline{I} = \underline{V}/\underline{Z} = 10/3+j4 = 2\angle{-53.13^\circ}$$

$i = \underline{2\cos(2t - 53.13^\circ)\text{ A}}$

10.31 By voltage division

$$\underline{V} = \frac{\frac{(2)(2+j2-j4)}{2+2-j2}}{6 + j4 + 1.2 - j0.4}(18) = 2\sqrt{2}\angle{-45^\circ}$$

$v = \underline{2\sqrt{2}\cos(2t - 45^\circ)\text{ V}}$

10.32

(a) $\underline{I} = \dfrac{\underline{V}}{\underline{Z}} = \dfrac{10}{4 + j\omega - j4/\omega} = \dfrac{10}{4+j3} = 2\angle{-36.9^\circ}$

$i = \underline{2\cos(4t - 36.9^\circ)\text{ A}}$

(b) $\underline{I} = \dfrac{10}{4 + j2 - j2} = 2.5\angle 0^\circ$

$i = \underline{2.5\cos(2t)\text{ A}}$

10.33 By KCL

$$\underline{V}_x\left(\tfrac{1}{5} + j\tfrac{4}{10}\right) - \tfrac{\underline{V}_x}{3}\left(j\tfrac{4}{10}\right) = 2 \text{ or}$$

$$\underline{V}_x = \frac{60}{6+j8} = 6\angle{-53.13^\circ}$$

$v_x = \underline{6\cos(4t - 53.13^\circ)\text{ V}}$

10.34 By voltage division

$$\underline{V}_1 = \frac{5(3)}{3-j4} = 3\angle{53.13^\circ}$$

then,

$$\underline{I} = \frac{2\underline{V}_1}{1 + \frac{1}{j\omega C_2}} = \frac{6\angle{53.13^\circ}}{1 - j(2)(\frac{1}{2})} = 3\sqrt{2}\angle{98.13^\circ}$$

$i = \underline{3\sqrt{2}\cos(2t + 98.13^\circ)\text{ A}}$

10.35 By voltage division

$$\underline{V}_1 = \frac{6}{1 + j(40{,}000)(10^3)(0.025\times 10^{-6})} = 3\sqrt{2}\angle{-45^\circ}\text{ , and,}$$

$$\underline{V} = \frac{j(40{,}000)(0.1)\,5\underline{V}_1}{3\times 10^3 + j4000} = 12\sqrt{2}\angle{-8.13^\circ}$$

$v = \underline{12\sqrt{2}\cos(40{,}000t - 8.13^\circ)\text{ V}}$

10.36

$$\underline{I} = \frac{20}{4 + j(3)(1) - j\frac{1}{(3)(1/3)}} = \frac{20}{4+j2} = 4.47\angle{-26.57^\circ}$$

$i = \underline{4.47\cos(3t - 26.57^\circ)\text{ A}}$

10.37 The impedance seen by the source is

$$\underline{Z} = j100 + \frac{(j400-j200)(j200-j800)}{j400-j200+j200-j800}$$

$$= j400\ \Omega$$

$$\underline{I} = \frac{4\angle 0^\circ}{\underline{Z}} = \frac{4}{j400}\ A = 10\angle -90^\circ \text{ mA}$$

$$i = 10\cos(1000t - 90^\circ) = \underline{10 \sin 1000t} \text{ mA}$$

The current through the 0.2H inductor is

$$\underline{I}_1 = \frac{j400-j200}{j400-j200+j200-j800}(10\angle -90^\circ)$$

$$= j5 \text{ mA}$$

$$\underline{V} = -j800\underline{I}_1 = 4\angle 0^\circ;$$

$$v = \underline{4\cos 1000t}\ V$$

10.38 KCL at the inverting input terminal of the op amp yields

$$(j2+1)\underline{V} + \frac{\underline{V}_g}{2} = 0$$

$$\underline{V} = \frac{1}{2(1+j2)}\underline{V}_g = \frac{5}{\sqrt{5}}\angle -63.4^\circ + 180^\circ$$

$$= \sqrt{5}\angle 116.6^\circ$$

$$v = \underline{\sqrt{5}\cos(10{,}000t + 116.6^\circ)}\,V$$

Chapter 11

AC Steady-State Analysis

11.1 Nodal Analysis

11.1 Find the forced response v using nodal analysis.

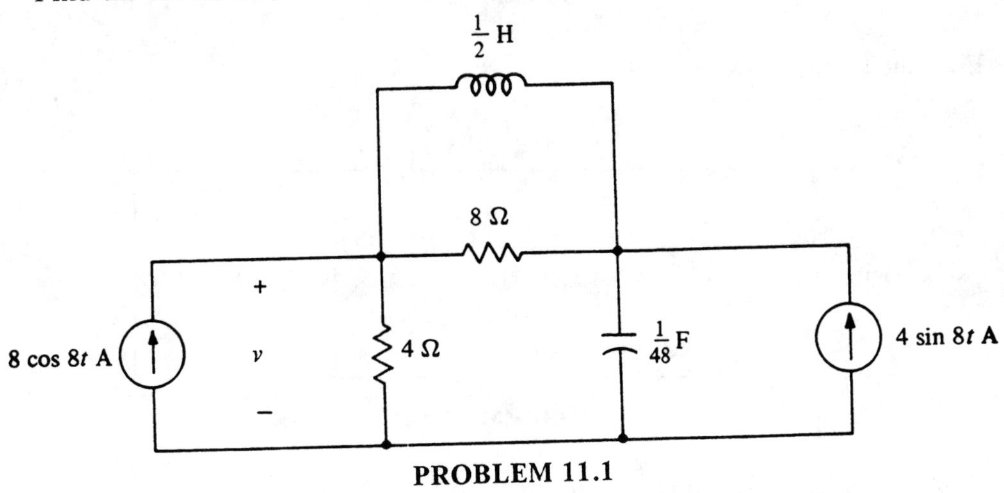

PROBLEM 11.1

11.2 Solve Prob. 10.31 using nodal analysis.

11.3 Find the steady-state voltage v using nodal analysis.

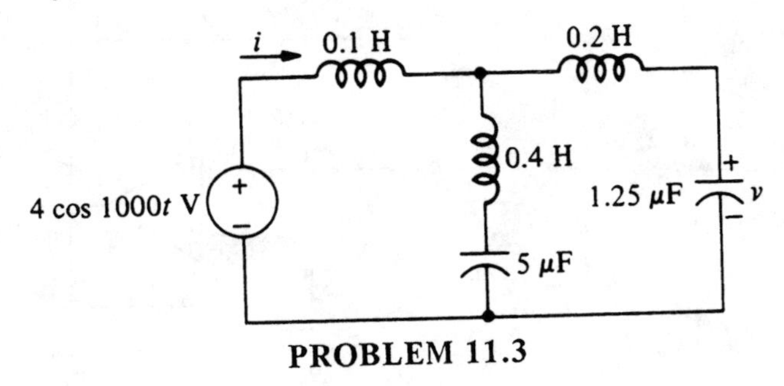

PROBLEM 11.3

11.4 Find the steady-state current i using nodal analysis.

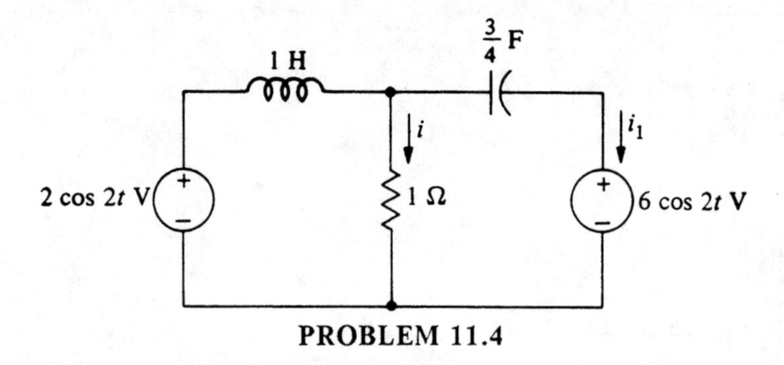

PROBLEM 11.4

11.5 Find the steady-state voltage v.

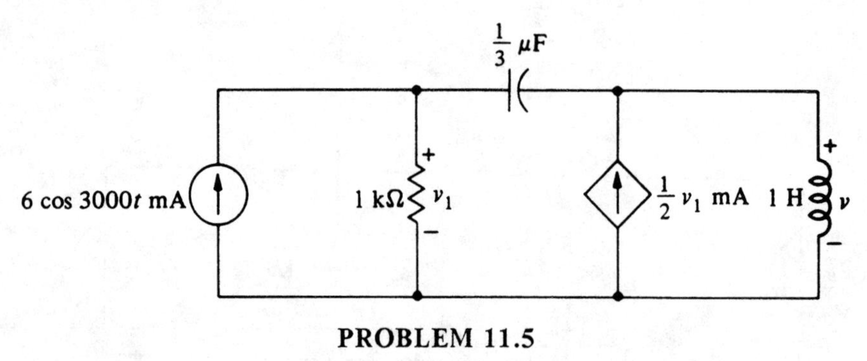

PROBLEM 11.5

11.6 Find the steady-state voltage v_1 in Prob. 11.5.

11.7 Find the steady-state voltage v if $v_g = 6\cos(5t)$ V.

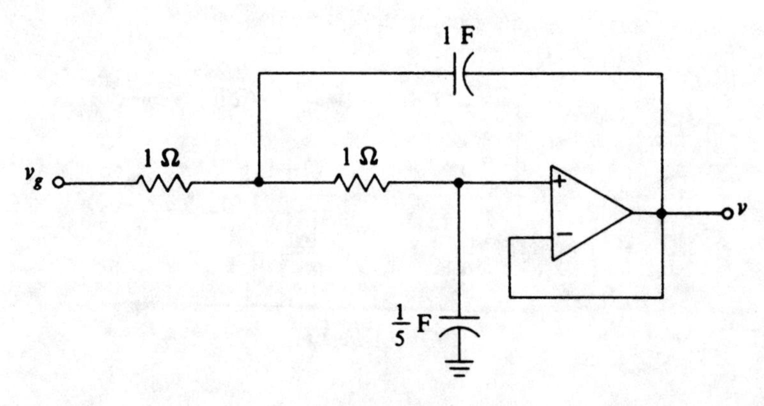

PROBLEM 11.7

11.8 Find the steady-state current i.

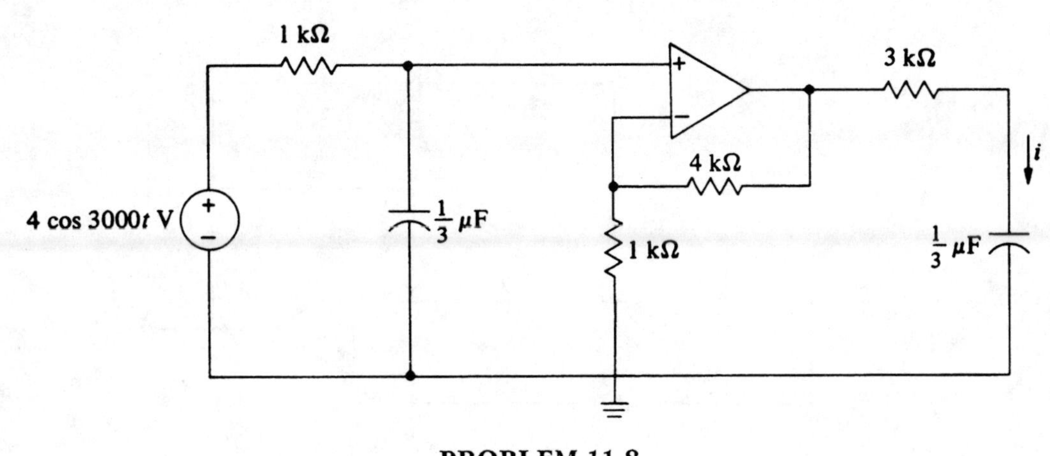

PROBLEM 11.8

11.9 Find the steady-state voltage v if $v_g = 5\cos(2t)$ V.

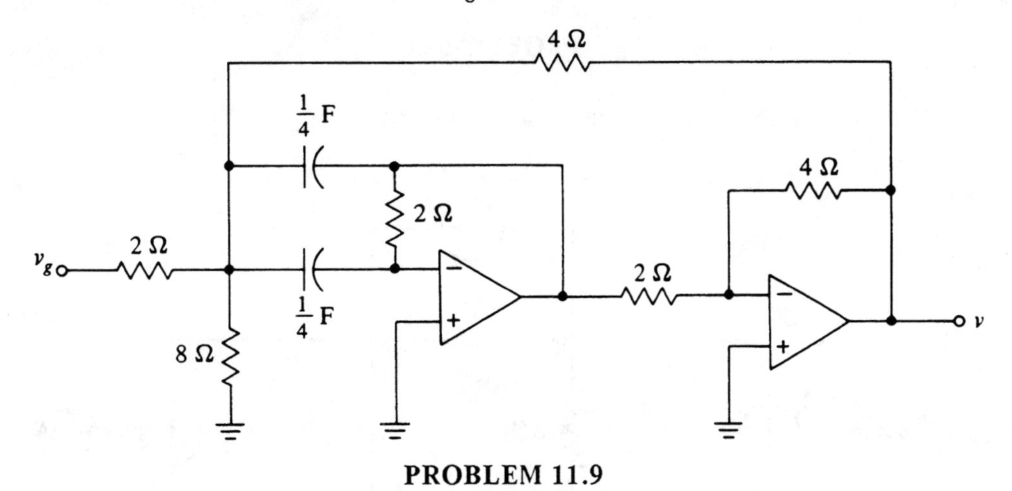

PROBLEM 11.9

11.10 Find the steady-state voltage v if $v_g = 6\cos(t)$ V.

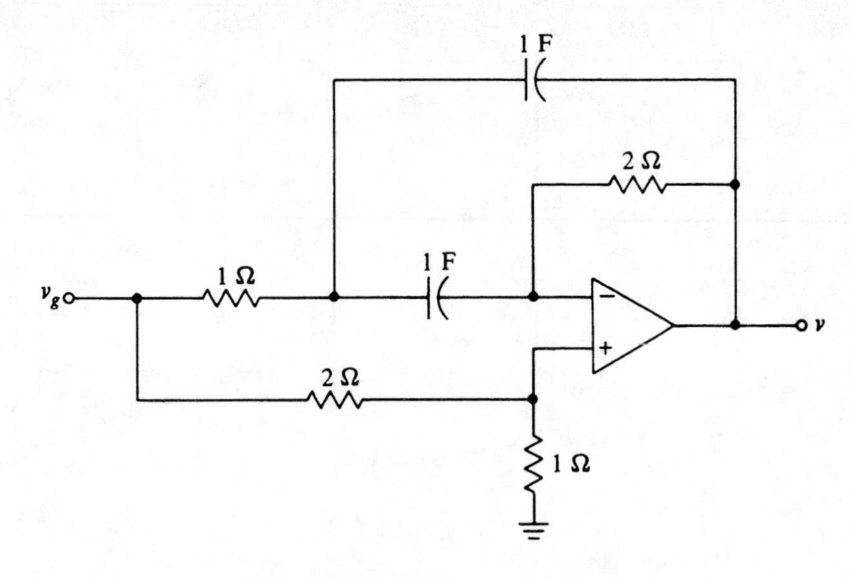

PROBLEM 11.10

11.11 Find the steady-state voltage v using nodal analysis.

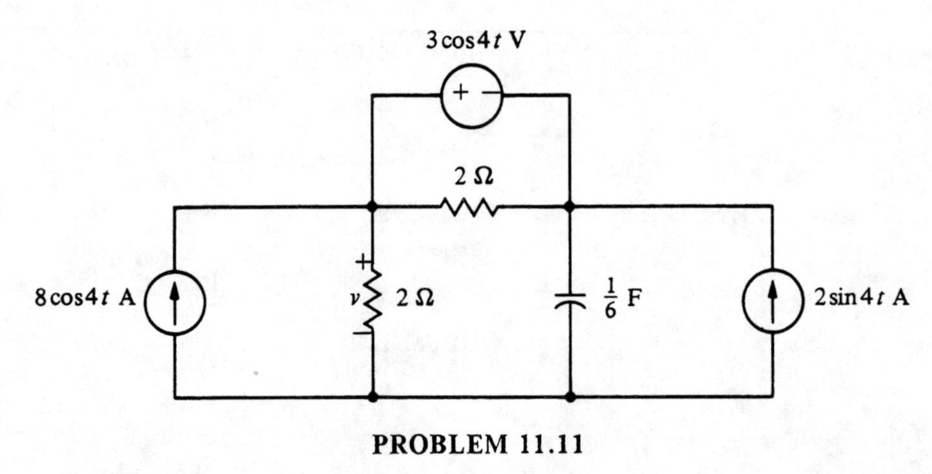

PROBLEM 11.11

11.12 Find the steady-state current i_1 using nodal analysis.

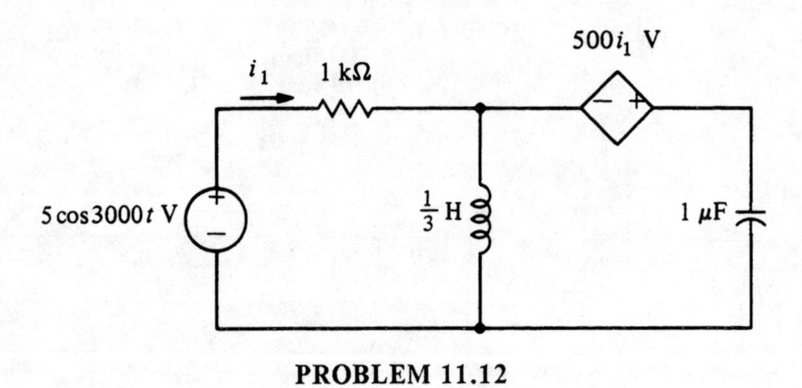

PROBLEM 11.12

11.2 Mesh Analysis

11.13 Find the steady-state voltage v using loop analysis.

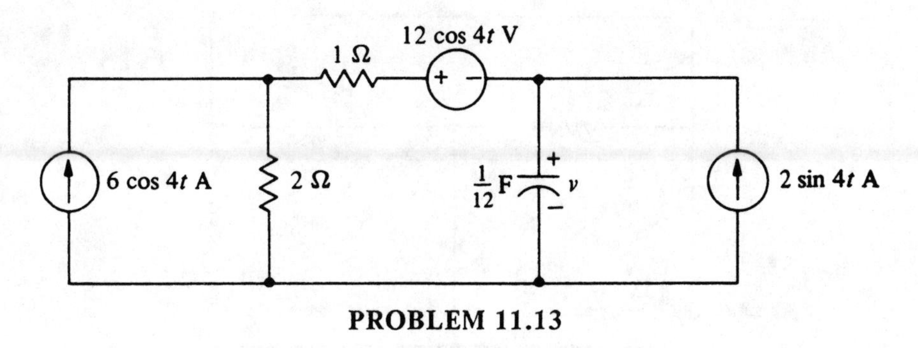

PROBLEM 11.13

11.14 Find the steady-state current i.

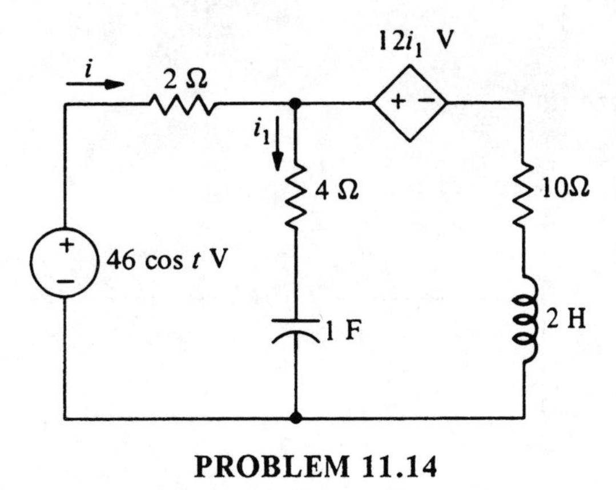

PROBLEM 11.14

11.15 Using mesh analysis, find the steady-state values of i_1 and i in Prob. 11.4.

11.16 Find the steady-state current i in Prob. 10.31 using mesh analysis.

11.17 Find the steady-state voltage v using mesh analysis if $i_{g1} = 6\cos(4t)$ A and $i_{g2} = 2\cos(4t)$ A.

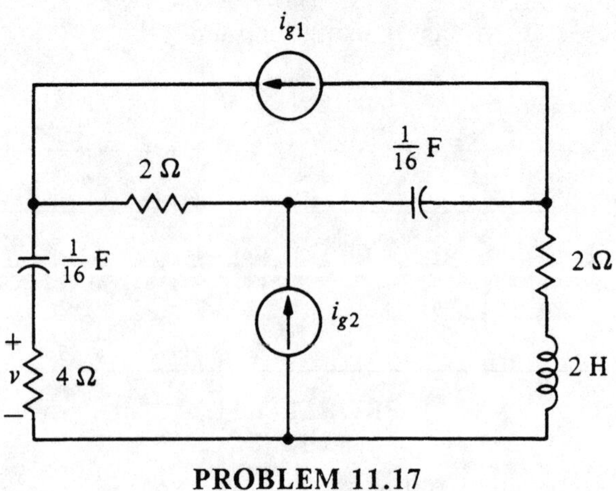

PROBLEM 11.17

11.3 Network Theorems

11.18 Find the steady-state response v.

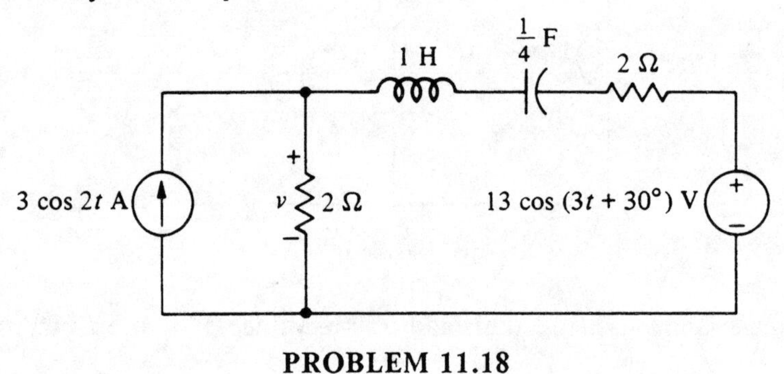

PROBLEM 11.18

11.19 In the corresponding phasor circuit, replace everything except the 1Ω resistor by its Norton equivalent, and use the result to find the steady-state current i.

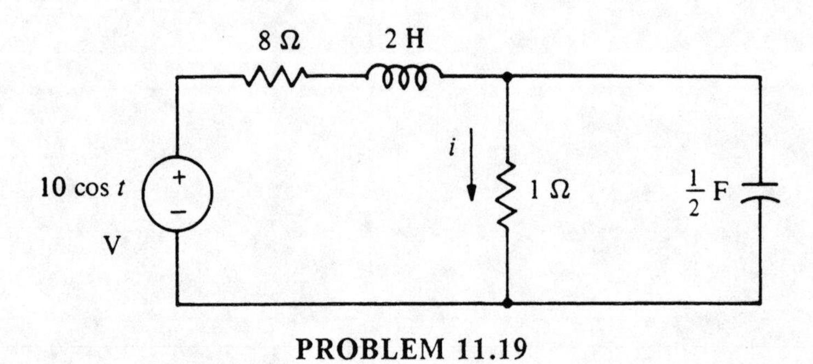

PROBLEM 11.19

11.20 Find the steady-state voltage v.

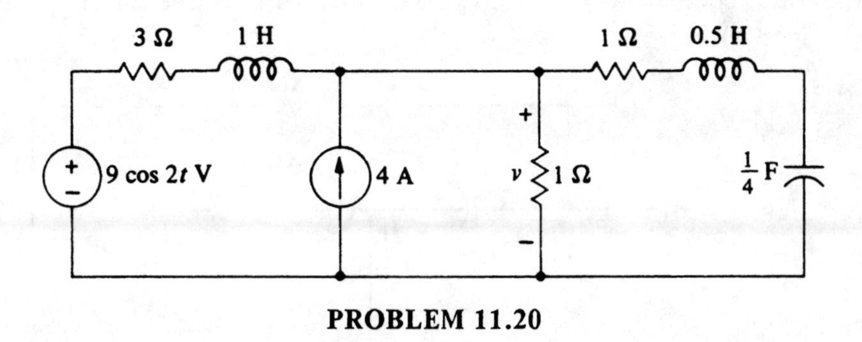

PROBLEM 11.20

11.21 Find the steady-state current i if $i_g = 20\cos(t) - 39\cos(2t) + 18\cos(3t)$ A.

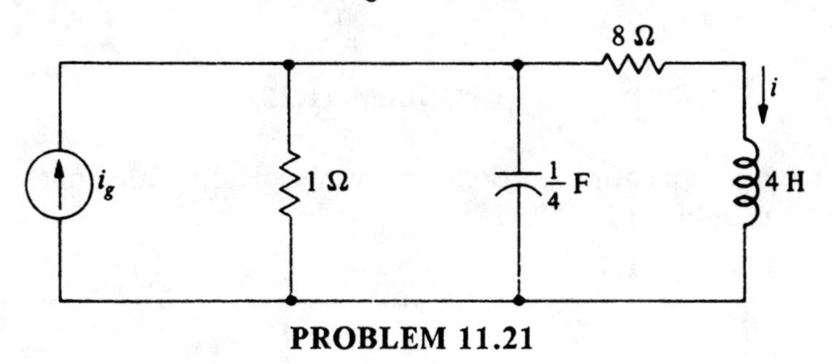

PROBLEM 11.21

11.22 Find the steady-state current i.

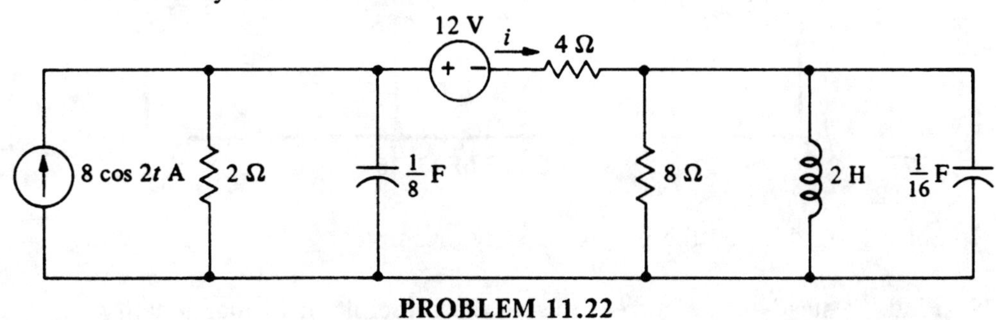

PROBLEM 11.22

11.23 For the phasor circuit corresponding to Prob. 10.29, replace the part to the left of terminals a-b by its Thevenin equivalent and find the steady-state value of v.

11.24 For the phasor circuit corresponding to Prob. 10.31, replace the part to the left of terminals a-b by its Thevenin equivalent and find the steady-state value of i.

11.25 Replace the phasor circuit except for the 1Ω resistor between terminals a-b by its Thevenin equivalent and find the steady-state value of i. The source is $v_g = 6\cos(2t)$ V.

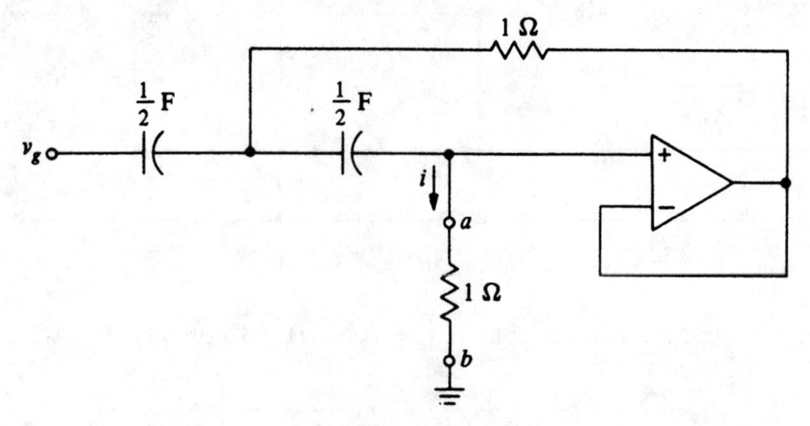

PROBLEM 11.25

11.26 Using the principle of proportionality on the corresponding phasor circuit to find the steady-state value of v.

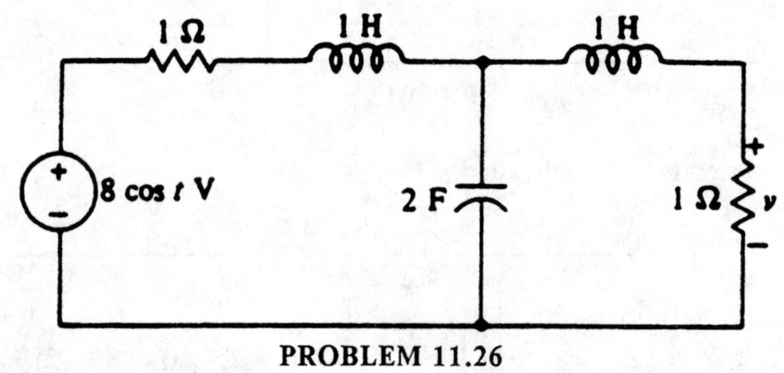

PROBLEM 11.26

11.27 Solve Prob. 11.7 by the proportionality principle.

11.28 Find the steady-state current i using the principle of proportionality.

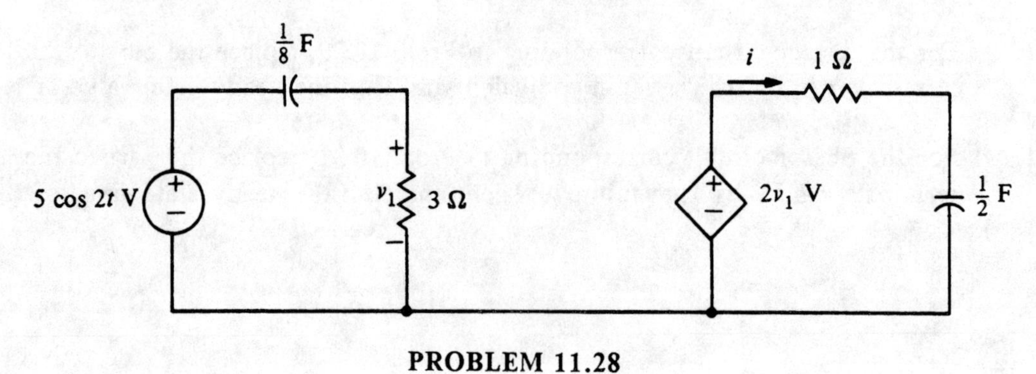

PROBLEM 11.28

11.4 Phasor Diagrams

11.29 Draw the phasor diagram for the circuit.

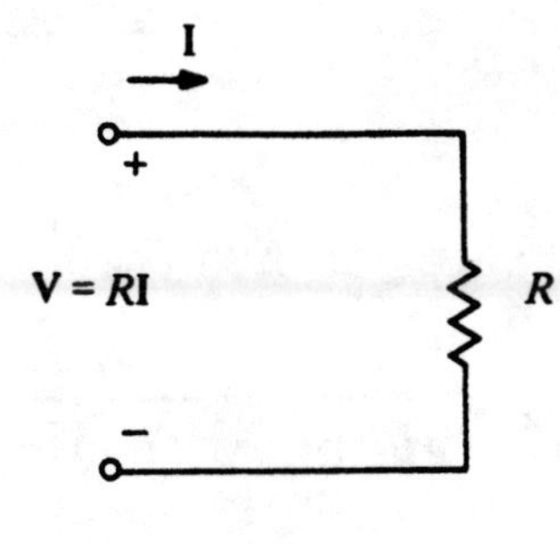

PROBLEM 11.29

11.30 Draw the phasor diagram for the circuit.

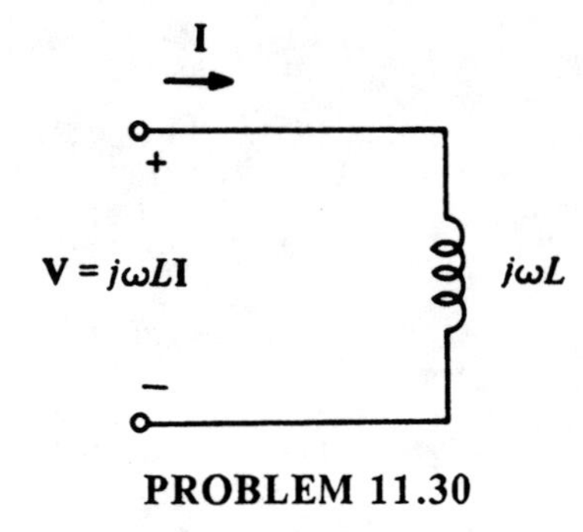

PROBLEM 11.30

11.31 Draw the phasor diagram for the circuit.

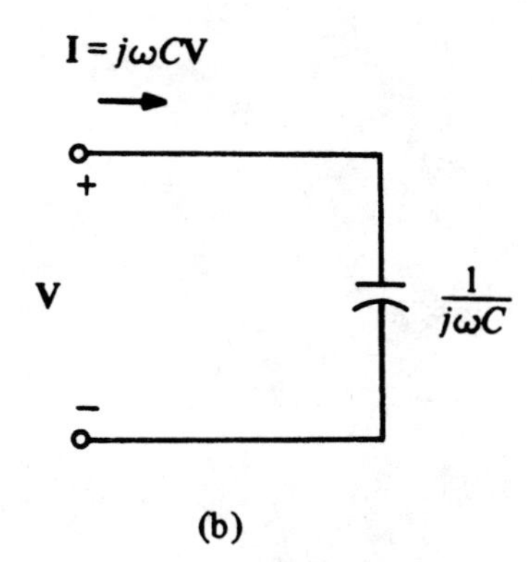

PROBLEM 11.31

11.32 Solve for V_m using a phasor diagram if $R = 2\Omega$, $L = 3$H and $\omega = 1$ rad/s, $\mathbf{I} = 1\angle 0°$A.

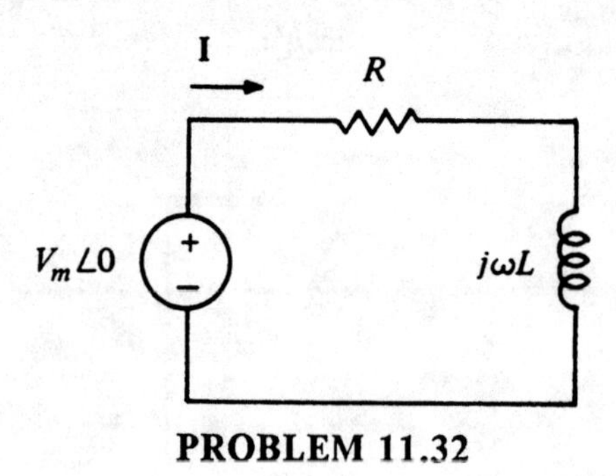

PROBLEM 11.32

11.1

8A, $\underline{V}$, 4Ω, $-j\frac{1}{4}$℧, 8Ω, $\underline{V}_1$, $j\frac{1}{6}$℧, $-j4$A. By KCL:

$\underline{V}(\frac{1}{4}+\frac{1}{8}-j\frac{1}{4})-\underline{V}_1(\frac{1}{8}-j\frac{1}{4})=8$ or

$\underline{V}(3-j2)+\underline{V}_1(-1+j2)=64$

$-\underline{V}(\frac{1}{8}-j\frac{1}{4})+\underline{V}_1(\frac{1}{8}+j\frac{1}{6}-j\frac{1}{4})=-j4$ or

$\underline{V}(-3+j6)+\underline{V}_1(3-j2)=-j96$

$$\underline{V}=\begin{vmatrix}8 & -1+j2\\ -j96 & 3-j2\end{vmatrix}\Big/\begin{vmatrix}3-j2 & -1+j2\\ -3+j6 & 3-j2\end{vmatrix}$$

$$=\frac{-j224}{14}=-j16$$

$\therefore v = \underline{16 \sin 8t \text{ V}}$

11.2 By KCL

$$\frac{\underline{V}-18}{6+j4}+\frac{\underline{V}}{2}+\frac{\underline{V}}{2+j2-j4}=0 \text{ or}$$

$$\underline{V}\left(\frac{1}{6+j4}+\frac{1}{2}+\frac{1}{2-j2}\right)=\frac{18}{6+j4} \text{ or}$$

$\underline{V}(18)=36-j36 \Rightarrow \underline{V}=2-j2$
$=2\sqrt{2}\angle-45°$

$\therefore v = \underline{2\sqrt{2}\cos(2t-45°)\text{ V}}$

11.3

4V, j100Ω, j200Ω, $\underline{V}_1$, j400Ω, −j200Ω, −j800Ω, $\underline{V}$. By KCL

$$\underline{V}_1\left(\frac{1}{j100}+\frac{1}{j400-j200}+\frac{1}{j200}\right)-\underline{V}\frac{1}{j200}=\frac{4}{j100} \text{ or}$$

$\underline{V}_1 4-\underline{V}=8,$

$\underline{V}(\frac{1}{j200}-\frac{1}{j800})-\underline{V}_1(\frac{1}{j200})=0$ or

$-\underline{V}_1 4+\underline{V}3=0$, Adding,

$\underline{V}=4 \therefore v = \underline{4\cos 1000t \text{ V}}$

11.4 By KCL

$$\underline{V}\left(1+\frac{1}{j2}-\frac{1}{j2/3}\right)=\frac{2}{j2}-\frac{6}{j2/3} \text{ or}$$

$\underline{V}(1+j)=j8$ or $\underline{V}=\frac{j8}{1+j}$

$\therefore \underline{V}=4+j4=4\sqrt{2}\angle45°$

$v=\underline{4\sqrt{2}\cos(2t+45°)\text{V}}$

11.5 By KCL using mA

$\underline{V}_1(1+j)-\underline{V}(j)=6$

$-\underline{V}_1(j)+\underline{V}(j-j\frac{1}{3})=\underline{V}_1(\frac{1}{2})$ or

$\underline{V}_1(-\frac{1}{2}-j)+\underline{V}(j\frac{2}{3})=0$

$$\underline{V}=\begin{vmatrix}1+j & 6\\ -\frac{1}{2}-j & 0\end{vmatrix}\Big/\begin{vmatrix}1+j & -j\\ -\frac{1}{2}-j & j\frac{2}{3}\end{vmatrix}$$

$$=\frac{3+j6}{\frac{1}{3}+j\frac{1}{6}}=18\angle36.87$$

$v=\underline{18\cos(3000t+36.87°)}$ V

11.6 From Prob. 11.5

$$\underline{V}_1=\frac{\begin{vmatrix}6 & -j\\ 0 & j2/3\end{vmatrix}}{\frac{1}{3}+j\frac{1}{6}}=\frac{j4}{\frac{1}{3}+j\frac{1}{6}}=10.73\angle63.4°$$

$v_1=\underline{10.73\cos(3000t+63.43°)}$ V

11.7 Let $\underline{V}_1$ be the phasor node voltage at the junction of the 1Ω resistors. and the $\underline{V}$ at the input terminals of opamp.

KCL gives:

$\underline{V}_1(1+1+j5)-\underline{V}(1+j5)=\underline{V}_g$ or

$\underline{V}_1(2+j5)-\underline{V}(1+j5)=6$

KCL at the noninverting terimınal of opamp gives

$-\underline{V}_1+\underline{V}(1+j1)=0$ or $\underline{V}_1=\underline{V}(1+j1)$

$$\therefore \underline{V}=\frac{6}{(1+j1)(2+j5)-(1+j5)}=1.34\angle-153°$$

$v=\underline{1.342\cos(5t-153.4°)}$ V

11.8 Let v = opamp output voltage and v_1 be the voltage at the input terminals.

$\underline{V}_1=\frac{1(4)}{1+j}$ (voltage division)

since $\underline{V}=(1+\frac{4}{1})\underline{V}_1=5\underline{V}_1$

then $\underline{V}=\frac{5(4)}{1+j}=10-j10$

$$\underline{I}=\frac{\underline{V}}{3000-j1000}=4.47\angle-26.57°\text{mA}$$

$i=\underline{4.47\cos(3000t-26.57°)\text{mA}}$

11.9 Let v_1 = node voltage at junction of 2Ω and 8Ω resistor, v_2 = output of 1st opamp. Then KCL at v_1 gives.

$$\underline{V}_1\left(\frac{1}{2}+\frac{1}{8}+\frac{1}{4}+j\frac{1}{2}+j\frac{1}{2}\right)-\underline{V}_2\left(j\frac{1}{2}\right)-\underline{V}\frac{1}{4}=\frac{5}{2}$$

or $\underline{V}_1(7+j8)-\underline{V}_2(j4)-\underline{V}2=20$

KCL at inverting terminal of first op amp gives $V_1 = jV_2$ and KCL at inverting terminal of 2nd op amp gives $\underline{V}_2=-\underline{V}/2$ therefore $V_1=-j\frac{V}{2}$ and

$$\underline{V}\left[(7+j8)(-j/2)+j\frac{4}{2}-2\right]=20 \text{ or}$$

$$\underline{V}=\frac{20}{2-j3/2}=8\angle 36.87^\circ \text{ V}$$

$$v=\underline{8\cos(2t+36.87^\circ)\text{ V}}$$

11.10 Let v_1 be the voltage at the input terminals of the op amp. then at the noninverting terminal KCL gives $\underline{V}_1\left(\frac{1}{2}+1\right)=\underline{V}_g\frac{1}{2} \Rightarrow \underline{V}_1=2V$

KCL at inverting terminal gives

$$\underline{V}_1\left(\frac{1}{2}+j\right)-\underline{V}_2(j)-\underline{V}\left(\frac{1}{2}\right)=0 \text{ or}$$

$\underline{V}_2 j+\underline{V}\frac{1}{2}=1+j2$ where v_2 is the node voltage to the right of the 1Ω resistor. KCL at $\underline{V}_2$ yields.

$$\underline{V}_2(1+j+j)-\underline{V}_1(j)-\underline{V}(j)=\underline{V}_g \text{ or}$$

$$\underline{V}_2(1+j2)-\underline{V}j=6+j2$$

$$\underline{V}=\begin{vmatrix} j & 1+j2 \\ 1+j2 & 6+j2\end{vmatrix} \Big/ \begin{vmatrix} j & \frac{1}{2} \\ 1+j2 & -j\end{vmatrix}$$

$$=\frac{1+j2}{\frac{1}{2}-j}=2\angle 126.9^\circ \text{ V}$$

$$v=\underline{2\cos(t+126.9^\circ)}\text{ V}$$

11.11

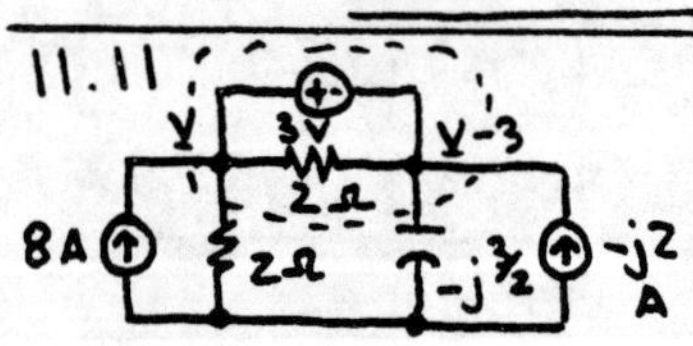

KCL for the generalized node yields:

$$\frac{\underline{V}}{2}+\frac{\underline{V}-3}{-j\,3/2}=8-j2 \text{ or } \underline{V}=\frac{48}{3+j4}=9.6\angle -53.1^\circ$$

$$v=\underline{9.6\cos(4t-53.1^\circ)\text{V}}$$

11.12 The node voltages are 5, $5-1000i_1$, $5-500i_1$. KCL for the supernode containing dependent source yields

$$-\underline{I}_1+\frac{5-1000\underline{I}_1}{j1000}+(5-500\underline{I}_1)(j3\times10^{-3})=0$$

$$\underline{I}_1=(4\sqrt{5}\times10^{-3})\angle 63.4^\circ \text{A}$$

$$i_1=\underline{4\sqrt{5}\cos(3000t+63.4^\circ)}\text{mA}$$

11.13 Let $\underline{I}$ be the phasor current in the center mesh. KVL yields

$$\underline{I}(1+2-j3)-6(2)+j3(-j2)+12=0$$

$$\underline{I}=\frac{6}{3-j3}=1+j \text{ A}$$

$$\underline{V}=(\underline{I}-j2)(-j3)=3\sqrt{2}\angle -135^\circ \text{ V}$$

$$v=\underline{3\sqrt{2}\cos(4t-135^\circ)}\text{ V}$$

11.14 KVL around the left mesh and the outside loop yields

$$2\underline{I}+\underline{I}_1(4-j)=46$$

$$\underline{I}2+(\underline{I}-\underline{I}_1)(10+j2)+12\underline{I}_1=46 \text{ or}$$

$$\underline{I}(12+j2)+\underline{I}_1(2-j2)=46$$

$$\underline{I}=\begin{vmatrix} 46 & 4-j \\ 46 & 2-j2\end{vmatrix} \Big/ \begin{vmatrix} 2 & 4-j \\ 12+j2 & 2-j2\end{vmatrix}$$

$$=\frac{-92-j46}{-46}=2+j=\sqrt{5}\angle 26.57^\circ$$

$$i=\underline{\sqrt{5}\cos(t+26.57^\circ)}\text{ A}$$

11.15 KVL around the right mesh and the outside loop yields

$$\underline{I}-\underline{I}_1(-j2/3)=6 \text{ or } \underline{I}+\underline{I}_1(j2/3)=6$$

$$(\underline{I}+\underline{I}_1)(j2)+\underline{I}_1(-j2/3)=2-6$$

$$\underline{I}(j2)+\underline{I}_1(j4/3)=-4$$

$$\underline{I}=\begin{vmatrix} 6 & j2/3 \\ -4 & j4/3\end{vmatrix} \Big/ \begin{vmatrix} 1 & j2/3 \\ j2 & j4/3\end{vmatrix}$$

$$=\frac{j32}{4+j4}=4\sqrt{2}\angle 45^\circ$$

$$\therefore i=\underline{4\sqrt{2}\cos(2t+45^\circ)}\text{ A}$$

$$\underline{I}_1=\frac{\begin{vmatrix} 1 & 6 \\ j2 & -4\end{vmatrix}}{4/3+j4/3}=\frac{-4-j12}{4/3+j4/3}=6.7\angle -153^\circ$$

$$i_1=\underline{6.7\cos(2t-153.4^\circ)}\text{ A}$$

11.16 Let i_1 be the mesh current in the left side (clockwise). KVL gives:

$$\mathbf{I}_1(6+j4)+(\mathbf{I}_1-\mathbf{I})2=18 \text{ or}$$

$$\mathbf{I}_1(8+j4)+\mathbf{I}(-2)=18$$

$$\mathbf{I}(2+j2-j4)+(\mathbf{I}-\mathbf{I}_1)2=0 \text{ or}$$

$$\mathbf{I}_1(-2)+\mathbf{I}(4-j2)=0$$

$$\mathbf{I}=\begin{vmatrix}8+j4 & 18\\ -2 & 0\end{vmatrix} \Big/ \begin{vmatrix}8+j4 & -2\\ -2 & 4-j2\end{vmatrix}$$

$$=\frac{36}{36}=1;\ \therefore i=\underline{\cos 2t \text{ A}}$$

11.17

KVL for loop with current $\mathbf{I}$:

$$(2-j4+2+j8+4-j4)\mathbf{I}+6(2-j4\Omega)+(-j4+2+j8)(2)=0$$

$$\mathbf{I}=2(-1+j)=2\sqrt{2}\angle 135^\circ \text{ A}$$

$$\mathbf{V}=-4\mathbf{I}=8\sqrt{2}\angle -45^\circ \text{ V}$$

$$v=\underline{8\sqrt{2}\cos(4t-45^\circ) \text{ V}}$$

11.18 With the current source dead

Voltage division yields

$$\mathbf{V}_1=\frac{2(13)\angle 30^\circ}{2+2+j3-j4/3}=\frac{26\angle 30^\circ}{4+j5/3}=6\angle 7.38^\circ$$

$$\therefore v_1=6\cos(3t-7.38^\circ) \text{ V}$$

With the voltage source dead

KCL gives:

$$\mathbf{V}_2\left(\frac{1}{2}+\frac{1}{2+2j-2j}\right)=3$$

$$\mathbf{V}_2=3 \text{ V}$$

$$\therefore v_2=3\cos(2t)$$

$$v=v_1+v_2=\underline{3\cos(2t)+6\ \text{os}(3t-7.38^\circ) \text{ V}}$$

11.19

$$\mathbf{Z}_{Th}=\frac{(-j2)(8+j2)}{8+j2-j2}=\frac{1}{2}-j2\ \Omega$$

$$\mathbf{I}_{sc}=\frac{10}{8+j2}=\underline{1.213\angle -14.04^\circ} \text{ A}$$

11.19

$1.2\cos(t-14^\circ)$ A

$$\mathbf{I}=\frac{(\frac{1}{2}-j2)(1.2\angle -14^\circ)}{1+\frac{1}{2}-j2}\quad \text{(By current division)}$$

$$=1\angle -36.87^\circ$$

$$i=\underline{\cos(t-36.87^\circ) \text{ A}}$$

11.20 With the current source dead, KCL gives

$$\mathbf{V}_1\left(1+\frac{1}{3+j2}+\frac{1}{1+j-j2}\right)=\frac{9}{3+j2} \text{ or}$$

$$\mathbf{V}_1=1-j=\sqrt{2}\angle -45^\circ$$

$$v_1=\sqrt{2}\cos(2t-45^\circ) \text{ V}$$

With the voltage source dead and the inductor a short circuit and all capacitors acting as open circuits. KCL gives:

$$\therefore v=v_1+v_2=\underline{3+\sqrt{2}\cos(2t-45^\circ)} \text{ V}$$

11.21 $i_{g0}=20\cos t$ produces i_1;

$i_{g1}=-39\cos(2t)$ produces i_2;

$i_{g2}=18\cos(3t)$ produces i_3;

By current division

$$\mathbf{I}_i=\frac{\frac{1}{8+j4\omega}\mathbf{I}_{gi}}{1+j\frac{\omega}{4}+\frac{1}{8+j4\omega}}=\frac{\mathbf{I}_{gi}}{9-\omega^2+j6\omega}$$

$$\omega=1 \text{ rad/s}: \mathbf{I}_{g1}=20\text{A}\Rightarrow \mathbf{I}_1=\frac{20}{8+j6}=2\angle -36.9^\circ \text{ A}$$

$$\omega=2 \text{ rad/s}: \mathbf{I}_{g2}=-39\text{A}\Rightarrow \mathbf{I}_2=\frac{-39}{5+j12}=3\angle 112.6^\circ \text{ A}$$

$$\omega=3 \text{ rad/s}: \mathbf{I}_{g3}=18\text{A}\Rightarrow \mathbf{I}_3=\frac{18}{j18}=1\angle -90^\circ \text{A}$$

$$i=i_1+i_2+i_3$$

$$=\underline{2\cos(t-36.9^\circ)+3\text{co}}$$

$$\underline{+\sin 3t \text{ A}}$$

11.22 With the current source dead

$$i_1=\frac{-12}{4+2}=-2 \text{ A}$$

With the voltage source dead

11.22 cont. By current division

$$\mathbf{I}_2 = \frac{\frac{1}{\frac{1}{2}+j\frac{1}{4}}(8)}{\frac{1}{\frac{1}{2}+j\frac{1}{4}} + 4 + \frac{1}{\frac{1}{8}-j\frac{1}{4}+j\frac{1}{8}}} = 1 - j = \sqrt{2}\angle -45^\circ$$

$$i_2 = \sqrt{2}\cos(2t - 45^\circ)\text{ A}$$

$$i = i_1 + i_2 = \underline{\sqrt{2}\cos(2t-45^\circ) - 2\text{ A}}$$

11.23 By voltage division

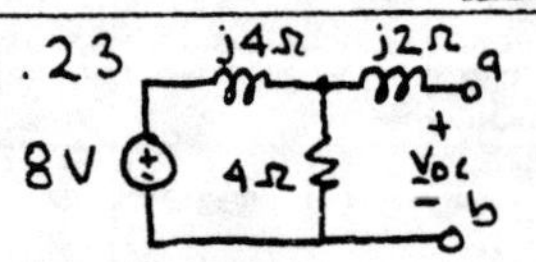

$$\mathbf{V}_{oc} = \frac{4}{4+j4}(8) = \frac{8}{1+j} = \underline{4\sqrt{2}\angle -45^\circ\text{ V}}$$

$$Z_{Th} = j2 + \frac{(4)(j4)}{4+j4} = \underline{2+j4\ \Omega}$$

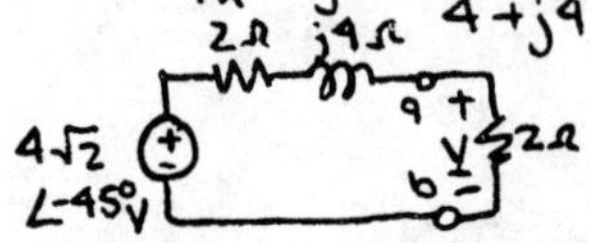

By voltage division

$$\mathbf{V} = \frac{2\,(\mathbf{V}_{oc})}{2+2+j4} = -j2$$

$$\therefore\ v = \underline{2\sin(2t)\text{ V}}$$

11.24 By voltage div.

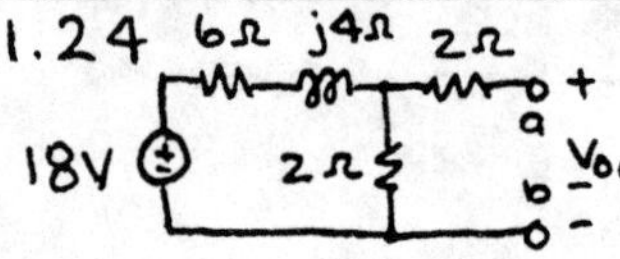

$$\mathbf{V}_{oc} = \frac{2(18)}{2+6+j4} = \underline{3.6 - j1.8\text{ V}}$$

$$Z_{Th} = 2 + \frac{2(6+j4)}{2+6+j4} = \underline{3.6 + j0.2\ \Omega}$$

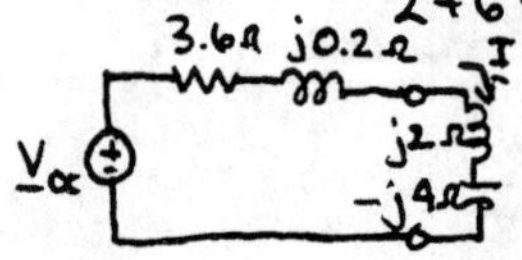

$$\mathbf{I} = \frac{\mathbf{V}_{oc}}{3.6+j0.2+j2-j4} = 1\text{ A}$$

$$\therefore\ i = \underline{\cos(2t)\text{ A}}$$

11.25 With a-b open $\mathbf{V}_{oc} = \mathbf{V}$. Current in right capacitor is zero so that $\mathbf{V}_1 = \mathbf{V}$. Hence the current in $1\,\Omega$ resistor and the other capacitor are zero. $\therefore\ \mathbf{V}_{oc} = \mathbf{V}_g = \underline{6\text{V}}$

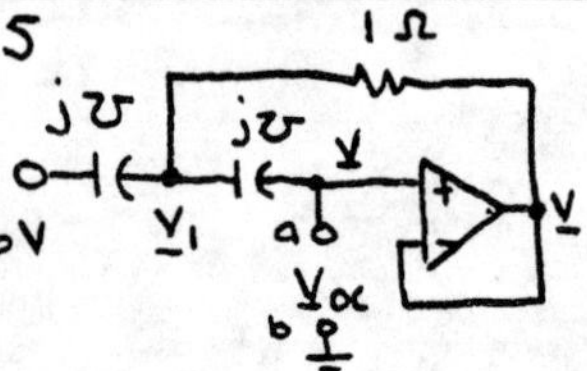

With a-b shorted $\mathbf{V} = 0$. KCL yields

$$\mathbf{V}_1(j+j+1) = j\mathbf{V}_g = j6 \Rightarrow \mathbf{V}_1 = \tfrac{6}{5}(2+j)\text{ V}$$

$$\mathbf{I}_{sc} = (j)\mathbf{V}_1 = j\tfrac{6}{5}(2+j)\text{ A}$$

$$Z_{th} = \frac{\mathbf{V}_{oc}}{\mathbf{I}_{sc}} = \frac{6}{j\frac{6}{5}(2+j)} = \underline{-1-j2\ \Omega}$$

The curent $\mathbf{I}$ from a to b is

$$\mathbf{I} = \frac{\mathbf{V}_{oc}}{Z_{th}+1} = \frac{6}{-1-j2+1} = j3\text{ A}$$

$$i = \underline{3\cos(2t+90^\circ)}\text{ A}$$

11.26 Assume $\mathbf{V} = 1\text{V}$ then

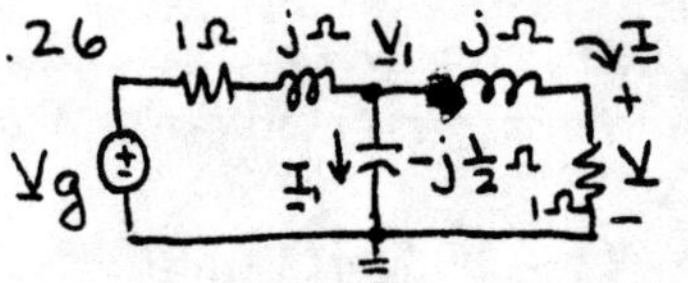

$$\mathbf{I} = \frac{\mathbf{V}}{1} = 1\text{A},\ \mathbf{V}_1 = \mathbf{I}(1+j) = 1+j\text{ V},$$

$$\mathbf{I}_1 = \mathbf{V}_1(j2) = -2+j2\text{ A}.$$

$$\frac{\mathbf{V}_g - \mathbf{V}_1}{1+j} = \mathbf{I} + \mathbf{I}_1 \Rightarrow \mathbf{V}_g = -2+j2\text{ V}$$

By Proportionality,

$$\mathbf{V} = \frac{8}{-2+j2} = 2\sqrt{2}\angle -135^\circ\text{ V}$$

$$v = \underline{2\sqrt{2}\cos(t - 135^\circ)\text{ V}}$$

11.27

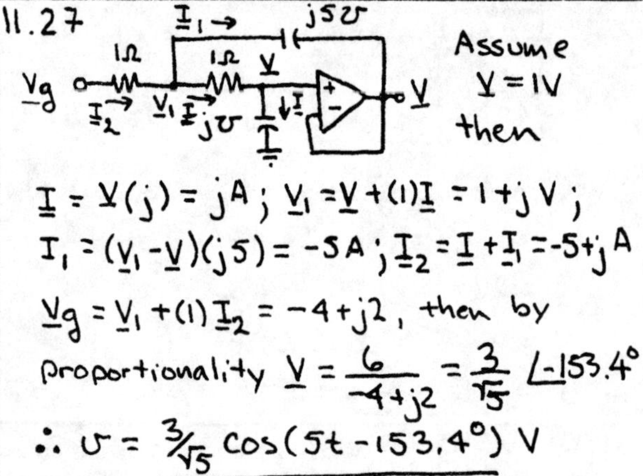

Assume $\mathbf{V} = 1\text{V}$ then

$$\mathbf{I} = \mathbf{V}(j) = j\text{A};\ \mathbf{V}_1 = \mathbf{V} + (1)\mathbf{I} = 1+j\text{ V};$$

$$\mathbf{I}_1 = (\mathbf{V}_1 - \mathbf{V})(j5) = -5\text{A};\ \mathbf{I}_2 = \mathbf{I} + \mathbf{I}_1 = -5+j\text{ A}$$

$\mathbf{V}_g = \mathbf{V}_1 + (1)\mathbf{I}_2 = -4+j2$, then by proportionality $\mathbf{V} = \frac{6}{-4+j2} = \frac{3}{\sqrt{5}}\angle -153.4^\circ$

$$\therefore\ v = \underline{\tfrac{3}{\sqrt{5}}\cos(5t - 153.4^\circ)\text{ V}}$$

11.28 Assume $\mathbf{I} = 1\text{A}$. Then

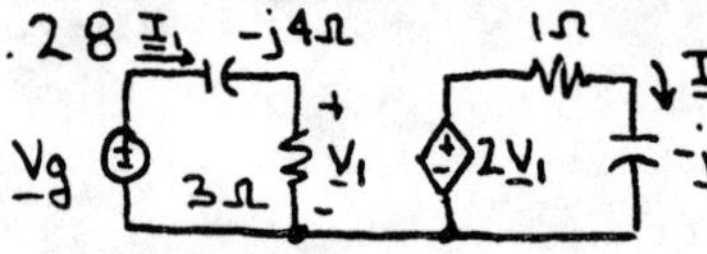

$$2\mathbf{V}_1 = (1-j1)\mathbf{V} \Rightarrow \mathbf{V}_1 = \frac{1-j1}{2}\text{ V}$$

$$\mathbf{I}_1 = \frac{\mathbf{V}_1}{3} = (1-j1)/6\text{ A}$$

$\mathbf{V}_g = (3-j4)(1-j1)/6\text{ V}$, By Proportionality,

$$\mathbf{I} = \frac{5}{(3-j4)(1-j1)/6} = 3\sqrt{2}\angle 98.1^\circ\text{ A}$$

$$i = \underline{3\sqrt{2}\cos(2t + 98.1^\circ)\text{ A}}$$

11.29

$\mathbf{V}$ $\mathbf{I}$

11.30

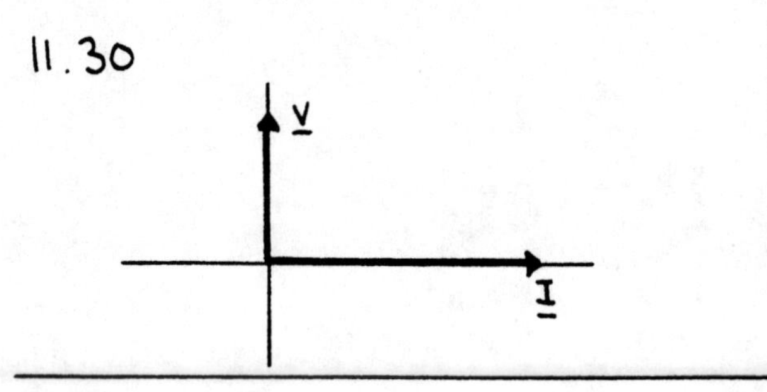

11.31

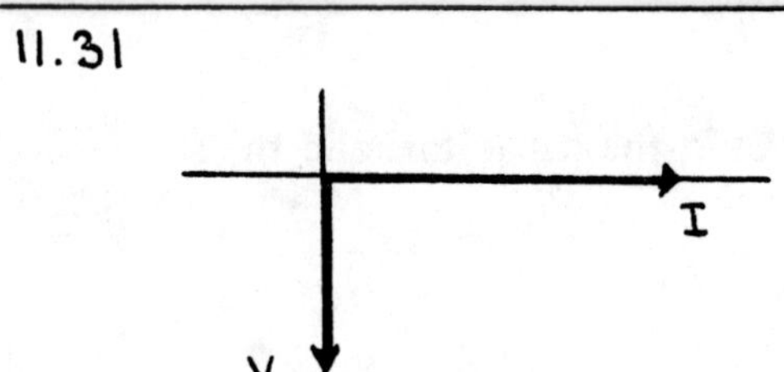

11.32

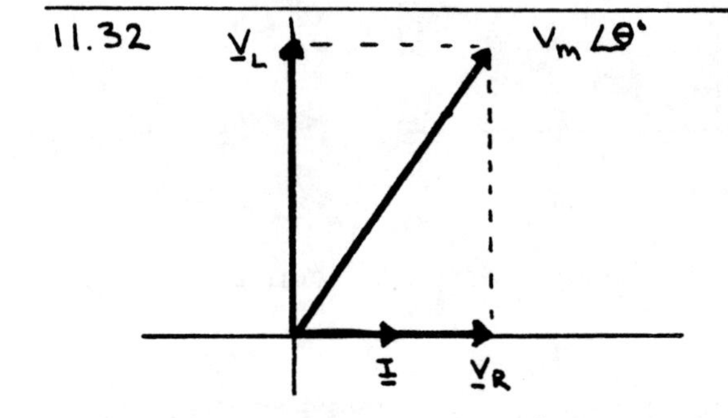

$\underline{V}_L = \omega L |\underline{I}| \angle 90° = 3 \angle 90°$

$\underline{V}_R = R|\underline{I}| = 2$

$\underline{V}_m = \underline{V}_R + \underline{V}_L = \sqrt{3^2+2^2} \angle \tan^{-1}(\frac{3}{2})$

$= \underline{\sqrt{13} \angle 56.31°}$

Chapter 12

AC Steady-State Power

12.1 Average Power

12.1 Find the average power absorbed by each resistor, the capacitor, and the source.

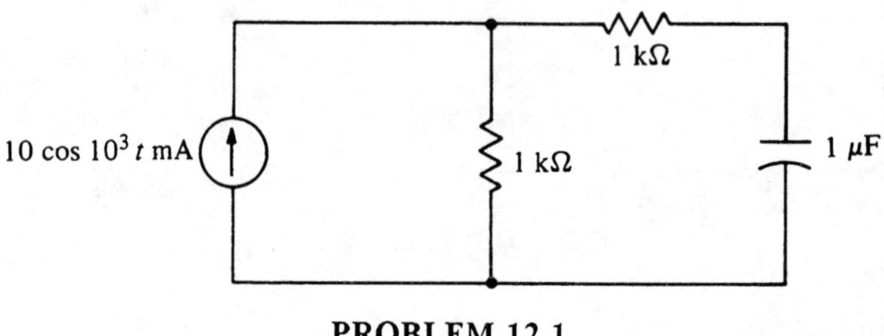

PROBLEM 12.1

12.2 Find the average power delivered to a 1Ω resistor carrying a current given by $5\cos(2\pi t)$.

12.3 Find the average power delivered to a 1Ω resistor carrying a current given by a square wave for which one cycle consists of 3A for 10ms followed by -1A for 20ms.

12.4 One cycle of periodic current is given by $i = 0.1(1 - e^{-t})$ A, $0 < t < 2$ if the current flows in a 10Ω resistor, find the average power.

12.5 Find the average power absorbed by the 2Ω resistor.

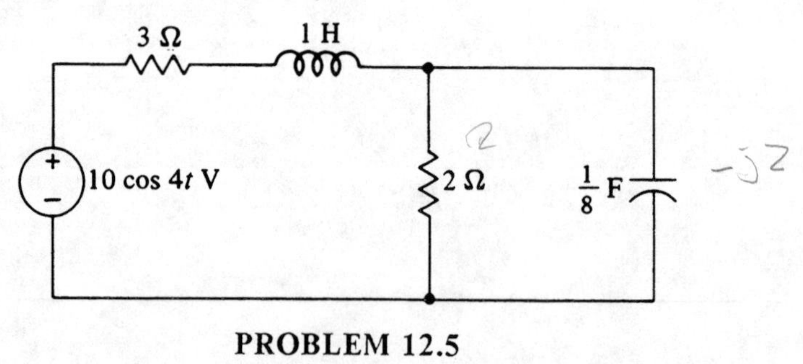

PROBLEM 12.5

12.6 Determine the average power for $p(t) = RI_m^2[1 + \cos(\omega t)]^2$.

12.7 Find the average power delivered the $\frac{1}{2}\Omega$ resistor.

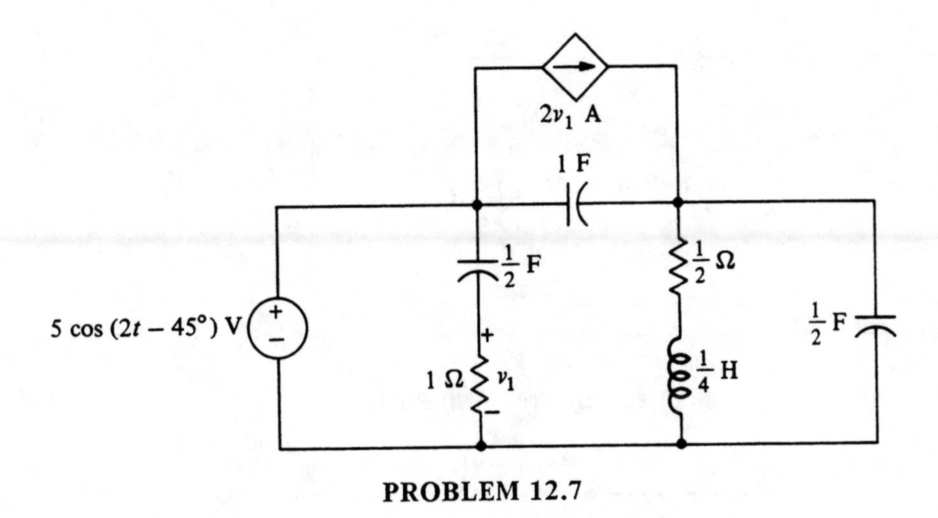

PROBLEM 12.7

12.2 Superposition and Power

12.8 Find the average power delivered to the resistor if R = 1kΩ and $v_{g1} = 5\cos(10t)$ V and $v_{g2} = 5\sin(10t)$ V.

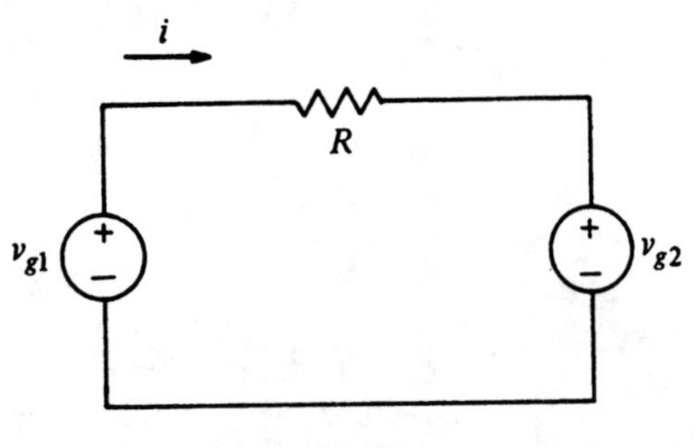

PROBLEM 12.8

12.9 Find the average power delivered to the resistor in Prob. 12.8 if R = 1kΩ and $v_{g1} = 5\cos(10t)$ V and $v_{g2} = 5\sin(20t)$ V.

12.10 Find the average power delivered to the resistor in Prob. 12.8 if R = 1kΩ and $v_{g1} = 5\cos(10t)$ V and $v_{g2} = 5$ V.

12.11 Find the average power delivered to R.

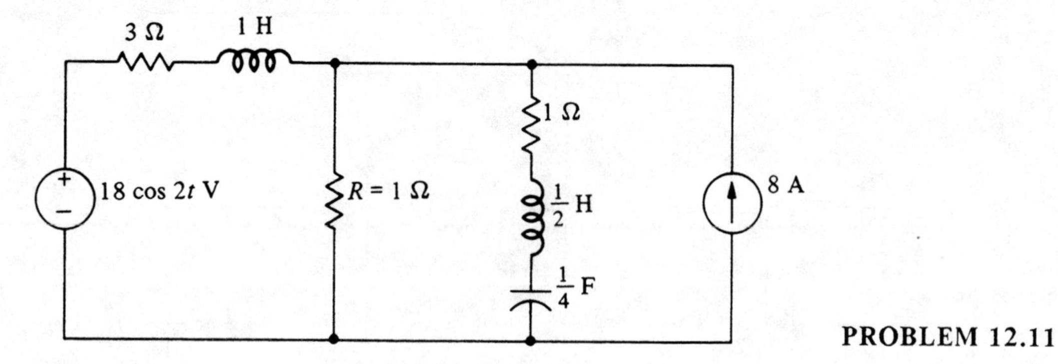

PROBLEM 12.11

12.12 Find the average power delivered to the 1Ω resistor.

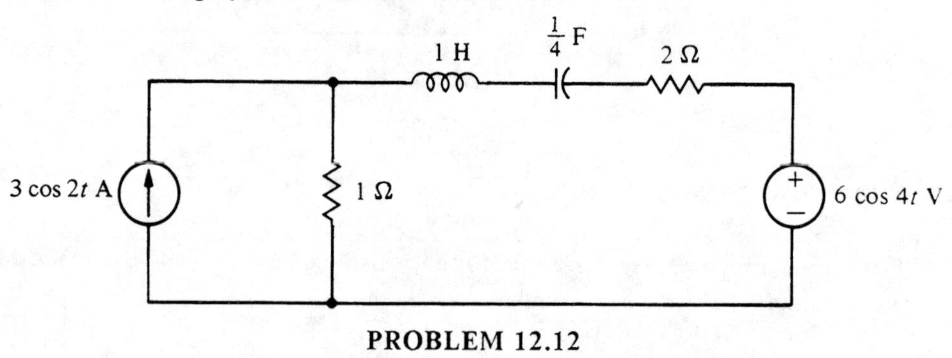

PROBLEM 12.12

12.3 RMS Values

12.13 Find I_{rms}.

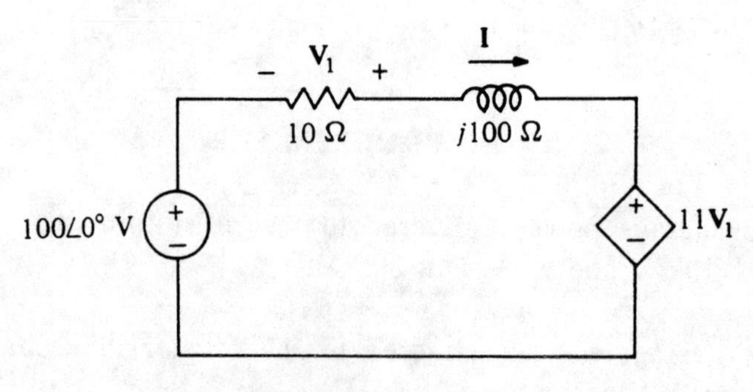

PROBLEM 12.13

12.14 Find rms value of $i = 5\sin(\omega t) + 8$ A.

12.15 Find the rms value of $i = 6\cos(3\omega t + 30°) + 2\cos(2\omega t + 20°)$ A.

12.16 Find the rms value of $i = 4\cos(\omega t) + 5\cos(\omega t + 15°)$ A.

12.17 Find the rms value of the voltage $v = 12 + 5\sqrt{2}\sin(t)$ V.

12.18 Find the rms value of the voltage $v = 9\sqrt{2}\cos(5t) + 3\sqrt{2}\cos(3t) + 2\sqrt{5}\cos(4t - 30°)$ V.

12.19 Find the rms value of the voltage $v = \cos(2t) - 6\cos(t) + 3\sqrt{2}\cos(3t)$ V.

12.20 Find the rms value of a periodic current for which one cycle is given by

$$i = \sqrt{t}\ \text{A}, \qquad 0 < t < 1s$$
$$= 0\ \text{A}, \qquad 1 < t < 3s.$$

12.21 Find the rms value of a periodic current for which one cycle is given by

$$i = t^2\ \text{A}, \qquad 0 < t < 3s$$
$$= 0\ \text{A}, \qquad 3 < t < 4s.$$

12.4 Power Factor

12.22 A load consists of a 200Ω resistor in a series with a 0.1H inductor. Find the parallel capacitance necessary to adjust to a *pf* of 0.95 lagging if ω = 1000 rad/s.

12.23 A load consists of a 200Ω resistor in a series with a 0.1H inductor. Find the parallel capacitance necessary to adjust to a *pf* of 0.95 leading if ω = 1000 rad/s.

12.24 A load consists of a 200Ω resistor in a series with a 0.1H inductor. Find the parallel capacitance necessary to adjust to a unity *pf* if ω = 1000 rad/s.

12.25 Find the power factor of the parallel connection of a 16Ω resistor, a 2H inductor, and a 1/32 F capacitor at a frequency of ω = 2 rad/s. At what frequency does a unity power factor occur?

12.26 Repeat Prob. 12.25 for the elements connected in series.

12.27 Find the power factor seen from the terminals for the source in Prob. 10.30. What reactance connected in parallel with the source will change the power factor to unity?

12.5 Complex Power

12.28 Find the complex power delivered to a load which has 0.707 lagging power factor and absorbs 40W.

12.29 Find the complex power delivered to a load which has 0.707 lagging power factor and absorbs 15 var.

12.30 Find the complex power delivered to a load which has 0.707 lagging power factor and absorbs 2 VA.

12.31 Find the impedance of the load in Prob. 12.28 if $V_{rms} = 10V$.

12.32 Find the impedance of the load in Prob. 12.29 if $V_{rms} = 10V$.

12.33 Find the impedance of the load in Prob. 12.30 if $V_{rms} = 10V$.

12.34 Three passive loads, $\mathbf{Z}_1$, $\mathbf{Z}_2$, and $\mathbf{Z}_3$, are receiving complex power values of $2 + j3$, $3 - j1$, and $1 + j6$ VA, respectively. If these loads and a voltage source of $20\angle 0°$ V rms are connected in series, find the rms value of the current that flows and the power factor seen by the source.

12.35 If $\mathbf{S} = 8 + j6$ VA, is the complex power delivered to the element N, find C so that the power factor seen by the source is unity.

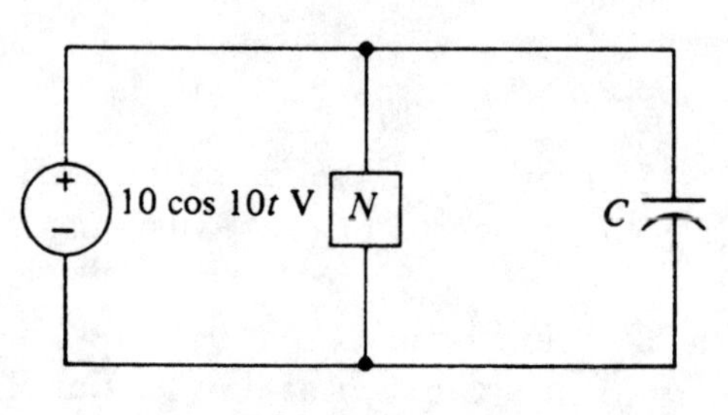

PROBLEM 12.35

12.36 Two loads $\mathbf{Z}_1$ and $\mathbf{Z}_2$ are in parallel across a source of $200\angle 0°$ V rms. If $\mathbf{Z}_1$ draws 8 kW at a power factor of 0.8 lagging and $\mathbf{Z}_2 = 4 - j3\Omega$, find the complex power delivered by the source.

12.37 For the circuit shown, $\mathbf{Z}_1$ draws $2 + j0$ VA and $\mathbf{Z}_2$ draws 10 VA at a power factor of 0.6 lagging. If $\mathbf{I}_g = 2\sqrt{2}\angle 15°$ A rms, find $\mathbf{V}_{rms}$.

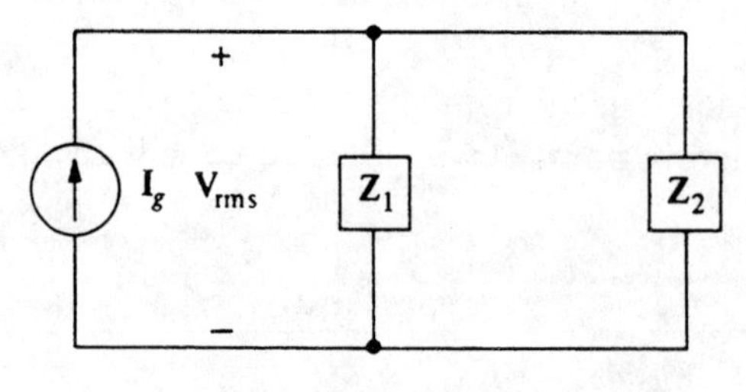

PROBLEM 12.37

12.38 Find the complex power delivered by the source and the power factor seen by the source.

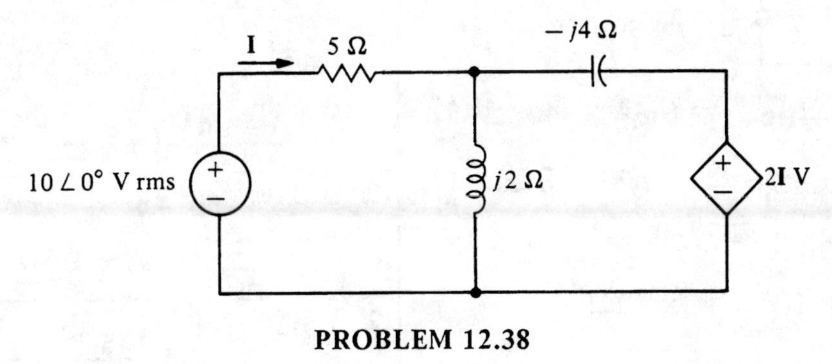

PROBLEM 12.38

12.6 Power Measurement

12.39 Find the wattmeter reading.

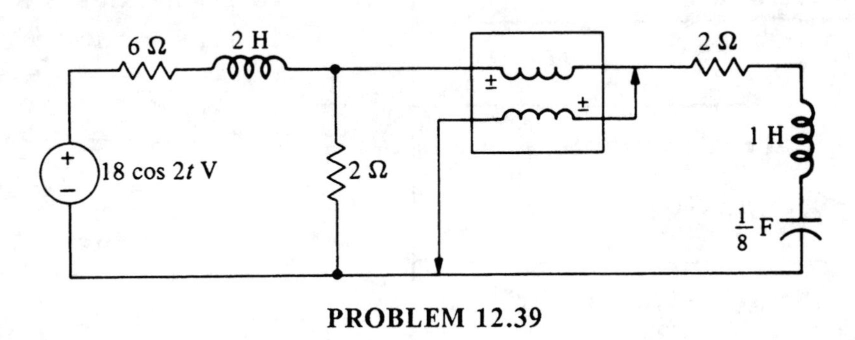

PROBLEM 12.39

12.1 (Circuit: 10 mA source; R_1 = 1 kΩ carrying I_1; R_2 = 1 kΩ; $-j1$ kΩ capacitor carrying I_2; node voltage V)

By current division:

$$I_1 = \frac{1-j}{1+1-j}(10\text{mA}) = 6-j2 = \sqrt{40}\angle -18.43^\circ \text{ mA}$$

$$I_2 = 10 - I_1 = 10-6+j2 = \sqrt{20}\angle 26.57^\circ \text{ mA}$$

$$V = I_1 R_1 = \sqrt{40}\angle -18.43^\circ \text{ V}$$

$$P_{R1} = \frac{I_{1m}^2}{2}(1\text{k}\Omega) = \underline{20\text{ mW}}$$

$$P_{R2} = \frac{I_{2m}^2}{2}(1\text{k}\Omega) = \underline{10\text{ mW}}$$

$$P_{\text{capacitor}} = \frac{I_{m2}^2}{2}\cdot \text{Re}(Z) = \underline{0\text{ W}}$$

$$P_{\text{source}} = -\frac{V_m}{2}(10\text{mA})\cos(-18.43^\circ - 0^\circ) = \underline{-30\text{ mW}}$$

12.2 $P = \frac{I_m^2}{2}(1) = \frac{25}{2} = \underline{12.5\text{ W}}$

12.3

$$P = \frac{1}{T}\int_0^t i^2 R\,dt = \frac{100}{3}\left[\int_0^{10^{-2}}(3)^2(1)dt + \int_{10^{-2}}^{3\times 10^{-2}}(-1)^2(1)dt\right]$$

$$= \frac{100}{3}\left[9+3-1\right](10^{-2}) = \underline{\frac{11}{3}\text{ W}}$$

12.4

$$P = \frac{1}{2}\int_0^2 i^2 R\,dt = \frac{1}{2}\int_0^2 (10)(.1)^2(1-e^{-t})^2 dt$$

$$= (0.05)\int_0^2 (1-2e^{-t}+e^{-2t})dt$$

$$= (0.05)\left[2+2e^{-2}-2-\frac{e^{-4}}{2}+\frac{1}{2}\right]$$

$$= (0.05)\left[\frac{1}{2}+2e^{-2}-\frac{e^{-4}}{2}\right] = \underline{38.08\text{ mW}}$$

12.5 V = voltage across 2Ω resistor

KCL: $V\left(\quad +j\frac{1}{2}+\frac{1}{3+j4}\right) = \frac{10}{3+j4}$ or

$$V = \frac{10}{(0.5+j3.5)} = 2.828\angle -81.87^\circ$$

$$P = \frac{V_m^2}{2(2\Omega)} = \underline{2\text{ W}}$$

12.6

$$P = \frac{\omega}{2\pi}\int_0^{2\pi/\omega} R I_m^2(1+2\cos\omega t + \cos^2\omega t)dt$$

$$= \frac{\omega}{2\pi}RI_m^2\left[\frac{2\pi}{\omega}+\frac{\pi}{\omega}\right] = \underline{\frac{3RI_m^2}{2}} \text{ (By Table 12.1)}$$

12.7 v = capacitor voltage of right ½F capacitor KCL gives

$$j2(V - 5\angle -45^\circ) + \frac{V}{\frac{1}{2}+j\frac{1}{2}} + jV = 2V_1,$$

where $V_1 = \frac{(1)(5\angle -45^\circ)}{1-j1} = \frac{5}{\sqrt{2}}$ V.

12.7 cont.

Then $V = 5\sqrt{2}\angle -36.9^\circ$ V

$$|I_{1/2\Omega}| = \frac{|V|}{\left|\frac{1}{2}+j\frac{1}{2}\right|} = 10\text{A};$$

$$P = \frac{|I_{1/2\Omega}|^2}{2}\left(\frac{1}{2}\right) = \underline{25\text{ W}}$$

12.8 $I = (V_{g1}-V_{g2})/R = \frac{5+j5}{1k} = 5\sqrt{2}\angle 45^\circ$ mA

$$P = \frac{I_m^2}{2}(1\text{k}\Omega) = \underline{25\text{ mW}}$$

12.9

$$P = \frac{|V_{g1}|^2}{2R} + \frac{|V_{g2}|^2}{2R} = \frac{5^2+5^2}{2(1\text{k}\Omega)} = \underline{25\text{ mW}}$$

12.10

$$P = \frac{|V_{g1}|^2}{2R} + \frac{v_{g2}^2}{R} = \frac{25}{2\times 10^3} + \frac{25}{10^3}$$

$$= \underline{37.5\text{ mW}}$$

12.11 v = voltage across R; $i = v/R = v$

with only the voltage source active, $v = v_1$ and KCL yields

$$\frac{V_1-18}{3+j2} + \frac{V_1}{1} + \frac{V_1}{1+j-j2} = 0 \Rightarrow V_1 = 2\sqrt{2}\angle -45^\circ \text{ V}$$

$\therefore I_1 = \frac{V_1}{1} = 2\sqrt{2}\angle -45^\circ$ A

with only the current source active $i = i_2$ and by current division

$$i_2 = \frac{3}{3+1}8 = 6\text{A (dc)}$$

$$P = i_2^2 R + \frac{1}{2}|I_1|^2 R = (6)^2 + \frac{1}{2}(2\sqrt{2})^2 = \underline{40\text{ W}}$$

12.12 i = current down in the 1Ω; with only the current source active, $i = i_1$ and by current div.

$$I_1 = \frac{j2-j2+2}{j2-j2+2+1}(3) = 2\angle 0^\circ \text{ A}$$

with only the voltage source active $i = i_2$ and KVL yields

$$I_2 = \frac{6}{1+j4-j+2} = \sqrt{2}\angle -45^\circ \text{ A}$$

$$P = P_1 + P_2 = \frac{1}{2}|I_1|^2(1) + \frac{1}{2}|I_2|^2(1)$$

$$= \frac{1}{2}\left[(2)^2+(\sqrt{2})^2\right] = \underline{3\text{ W}}.$$

12.13 $V_1 = -I(10)$ then KVL gives

$$I(10+j100) + 11V_1 = 100 \text{ or}$$

$$I = \frac{100}{-100+j100} = \frac{1}{\sqrt{2}}\angle -135^\circ \text{ A}$$

$$I_{rms} = \frac{1/\sqrt{2}}{\sqrt{2}} = \underline{\frac{1}{2}\text{ A}}$$

12.14 $I^2_{rms} = \frac{1}{2}(5)^2 + (8)^2 = 76.5$

$\therefore I_{rms} = \underline{8.746}$ A

12.15 $I^2_{rms} = \frac{1}{2}(6)^2 + \frac{1}{2}(2)^2 = 20$

$\therefore I_{rms} = \underline{4.472}$ A

12.16 $I = 4 + 5\angle 15^\circ = 8.924\angle 8.34^\circ$ A

$I_{rms} = \frac{8.924}{\sqrt{2}} = \underline{6.31}$ A

12.17 $V^2_{rms} = (12)^2 + \frac{1}{2}(5\sqrt{2})^2 = 169$

$\therefore V_{rms} = \underline{13\ V}$

12.18 $V^2_{rms} = \frac{1}{2}(9\sqrt{2})^2 + \frac{1}{2}(3\sqrt{2})^2 + \frac{1}{2}(2\sqrt{3})^2$

$= 100$

$\therefore V_{rms} = \underline{10\ V}$

12.19 $V^2_{rms} = \frac{1}{2}(1)^2 + \frac{1}{2}(-6)^2 + (3\sqrt{2})^2\frac{1}{2}$

$= 27.5$

$V_{rms} = \underline{5.244}$ V

12.20 $I^2_{rms} = \frac{1}{3}\left[\int_0^1 (\sqrt{t})^2 dt + \int_1^3 0 \cdot dt\right]$

$= \frac{1}{3}\left(\frac{1}{2}\right) = 1/6$

$I_{rms} = 1/\sqrt{6} = \underline{0.354\ A}$

12.21 $I^2_{rms} = \frac{1}{4}\left[\int_0^3 (t^2)^2 dt + \int_3^4 0 \cdot dt\right]$

$= \frac{1}{4}\left(\frac{243}{5}\right) = \frac{243}{20} = 12.15$

$I_{rms} = \underline{3.486}$ V

12.22 $X_1 = \frac{R^2 + X^2}{R\tan(\cos^{-1} PF) - X}$

$= \frac{(200)^2 + (100)^2}{200\tan(\cos^{-1} 0.95) - 100}$

$= -1459$

since negative $C = -1/\omega X_1$

$= \underline{0.685\ \mu F}$

12.23 $X_1 = \frac{R^2 + X^2}{-R\tan(\cos^{-1} PF) - X}$

$= -301.7$ since negative

$C = -1/\omega X_1 = \underline{3.315}\ \mu F$

12.24 $X_1 = \frac{(200)^2 + (100)^2}{-(100)} = -500$

$C = -1/\omega X_1 = \underline{2.0\ \mu F}$

12.25 $Y_T = \frac{1}{16} + j\frac{\omega}{32} - j\frac{1}{2\omega}$, at $\omega = 2$ rad/s

$Z_T = \frac{1}{Y} = 1.6 + j4.8 = 5.06\angle 71.57^\circ$

$PF = \cos(71.57^\circ) = \underline{0.316}$ (lagging)

for PF = 1, $\frac{\omega}{32} = \frac{1}{2\omega}$ or $\omega^2 = 16$

$\therefore \omega = \underline{4 \text{ rad/s}}$

12.26 $Z_T = 16 + j2\omega - j\frac{32}{\omega}$, at $\omega = 2$ rad/s

$Z_T = 16 - j12 = 20\angle -36.87$

$PF = \cos(-36.87^\circ) = \underline{0.80}$ (leading)

for PF = 1, $2\omega = \frac{32}{\omega}$ or $\omega^2 = 16$

$\therefore \omega = \underline{4 \text{ rad/s}}$

12.27 $Z_T = j4 + \frac{(4 + j4 - j4)(12)}{12 + 4} = 3 + j4$

$= 5\angle 53.13^\circ$

$PF = \cos(53.13^\circ) = \underline{0.60}$ (lagging)

for PF = 1 then

$X_1 = \frac{3^2 + 4^2}{-4} = \underline{-\frac{25}{4}\ \Omega}$

12.28 $\theta = \cos^{-1}(.707) = 45^\circ$; $Q = P\tan\theta$

$Q = 40\tan 45^\circ = 40$ VAR

$S = P + jQ = 40 + j40 = \underline{40\sqrt{2}\angle 45^\circ}$ VA

12.29 $\theta = \cos^{-1}(0.707) = 45^\circ$; $P = \frac{Q}{\tan\theta}$

$P = \frac{15}{\tan 45^\circ} = 15$ W

$S = P + jQ = 15 + j15 = \underline{15\sqrt{2}\angle 45^\circ}$ VA

12.30 $\theta = \cos^{-1}(0.707) = 45^\circ$

$S = \underline{2\angle 45^\circ \text{ VA}} = \sqrt{2} + j\sqrt{2}$ VA

12.31 $|S| = V_{rms} I_{rms} = 10 I_{rms}$

$|Z| = \frac{V_{rms}}{I_{rms}} = \frac{V_{rms}}{|S|/V_{rms}} = \frac{V^2_{rms}}{|S|} = \frac{(10)^2}{|S|}$

$\theta = \cos^{-1}(0.707) = 45^\circ$

$= \frac{100}{40\sqrt{2}}\angle 45^\circ = \underline{1.768\angle 45^\circ}\ \Omega$

12.32 From Prob. 12.31

$Z = \frac{100}{15\sqrt{2}}\angle 45^\circ = \underline{4.714\angle 45^\circ}\ \Omega$

12.33 From Prob. 12.31

$Z = \frac{100}{2}\angle 45^\circ = \underline{50\angle 45^\circ}\ \Omega$

12.34 $\underline{S} = \underline{S}_1 + \underline{S}_2 + \underline{S}_3$

$= 2 + j3 + 3 - j + 1 + j6$

$= 6 + j8 = 10\angle 53.13°$

$PF = \cos(53.13°) = \underline{0.6 \text{ (lagging)}}$

$\underline{S} = \underline{V}_{rms}\underline{I}^*_{rms} \Rightarrow 10\angle 53° = 20\,\underline{I}^*_{rms}$

$\underline{I}_{rms} = \underline{0.5\angle -53.13°} \text{ A rms}$

12.35 $Q_c = -6 = -\omega C V^2$

$C = \dfrac{6}{10(10/\sqrt{2})^2} = \underline{12 \text{ mF}}$

12.36 $\underline{S}_1 = \underline{P}_1 + jQ_1$

$\underline{P}_1 = 8000\text{W} : Q_1 = P_1 \tan\theta$

$= 8000 \tan(\cos^{-1} 0.8)$

$= 6000 \text{ VAR}$

$\underline{S}_1 = 8 + j6 \text{ KVA}$

$\underline{S}_2 = \dfrac{V^2_{rms}}{\underline{Z}^*} = \dfrac{(200)^2}{4 + j3} = 6.4 - j4.8 \text{ KVA}$

$\underline{S} = \underline{S}_1 + \underline{S}_2 = 14.4 + j1.2 = \underline{14.45\angle 4.76° \text{ kVA}}$

12.37 $\underline{S}_1 = 2 + j0 \text{ VA};$

$\theta_2 = \cos^{-1}(0.6) = 53.13°$

$\underline{S}_2 = 10\angle 53.13° = 6 + j8 \text{ VA}$

$\underline{S} = \underline{S}_1 + \underline{S}_2 = 2 + 6 + j8 = 8\sqrt{2}\angle 45° \text{ VA}$

$\underline{S} = \underline{V}_{rms}\underline{I}^*_{rms} \Rightarrow \underline{V}_{rms} = \dfrac{8\sqrt{2}\angle 45°}{2\sqrt{2}\angle -15°}$

$= \underline{4\angle 60° \text{ V}}$

12.38 KCL yields

$\dfrac{10 - 5\underline{I}}{j2} - \underline{I} + \dfrac{10 - 5\underline{I} - 2\underline{I}}{-j4} = 0$

$\underline{I} = 2\angle -53.1° \text{ A rms}$

$\underline{S} = \underline{V}_{rms}\underline{I}^*_{rms} = 10(2\angle 53.1°$

$= 20\angle 53.1° \text{ VA}$

$pf = \cos(53.1°) = \underline{0.6 \text{(lagging)}}$

$\underline{S} = 20\angle 53.1° = \underline{12 + j16 \text{ VA}}$

12.39 KCL yields

$\underline{V}\left(\dfrac{1}{2} + \dfrac{1}{6+j4} + \dfrac{1}{2+j2-j4}\right) = \dfrac{18}{6+j4}$ or

$\underline{V} = 2 - j2 = 2\sqrt{2}\angle -45° \text{ V}$

$\underline{I} = \dfrac{\underline{V}}{2+j2-j4} = 1\angle 0° \text{ A}$

$P = \dfrac{|V||I|}{2}\cos(45°) = \underline{1\text{W}}$.

Chapter 13

Three-Phase Circuits

13.1 Single-Phase, Three-Wire Systems

13.1 If $\mathbf{V}_{an} = \mathbf{V}_{nb} = 120\angle 0°$ V rms, the impedance between terminals A-N is $8\angle 15°\Omega$, and that between terminals N-B is $8\angle 0°\Omega$, find the neutral current $\mathbf{I}_{nN}$.

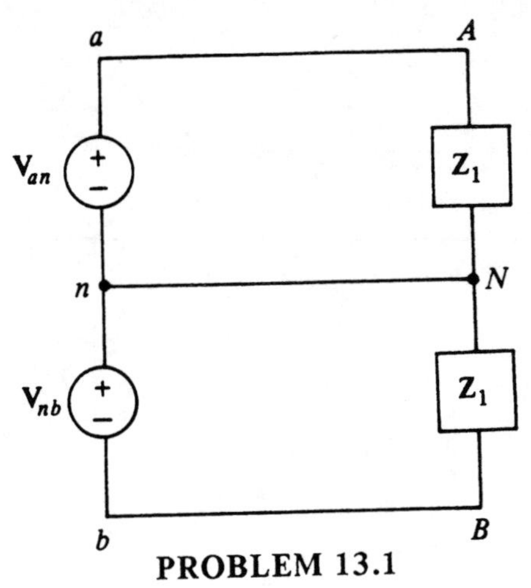

PROBLEM 13.1

13.2 Let $\mathbf{V}_1 = 15\angle 0°$ V rms, $\mathbf{Z}_1 = 2 + j1\Omega$, $\mathbf{Z}_2 = 5\Omega$, $\mathbf{Z}_3 = 2\Omega$, and $\mathbf{Z}_4 = 8\Omega$. Find the average power absorbed by the loads, lost in the lines, and delivered by the sources.

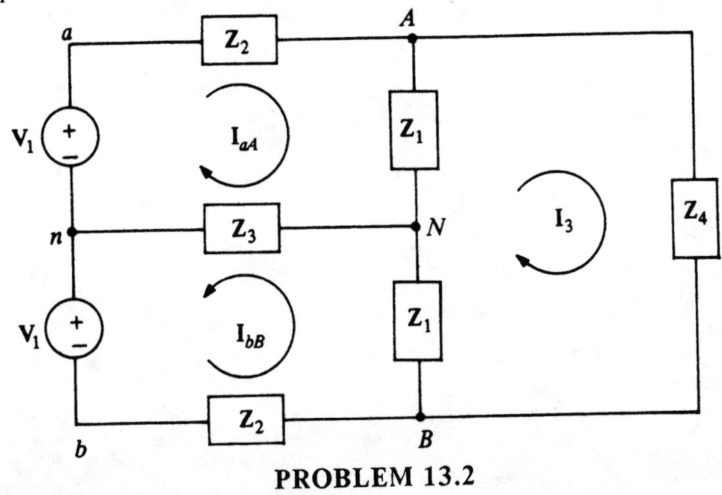

PROBLEM 13.2

13.2 Three-Phase Y-Y Systems

13.3 Given: $V_{ab} = 100\angle 30°$ V rms is a line voltage of a balanced Y-connected three-phase source. If the phase sequence is *abc*, find the phase voltages.

13.4 The source voltages are determined by Prob. 13.3, and the load in each phase is a series combination of a 10Ω resistor, a 20μF capacitor to the load. The frequency is $\omega = 500$ rad/s. Find the line currents and the power delivered to the load.

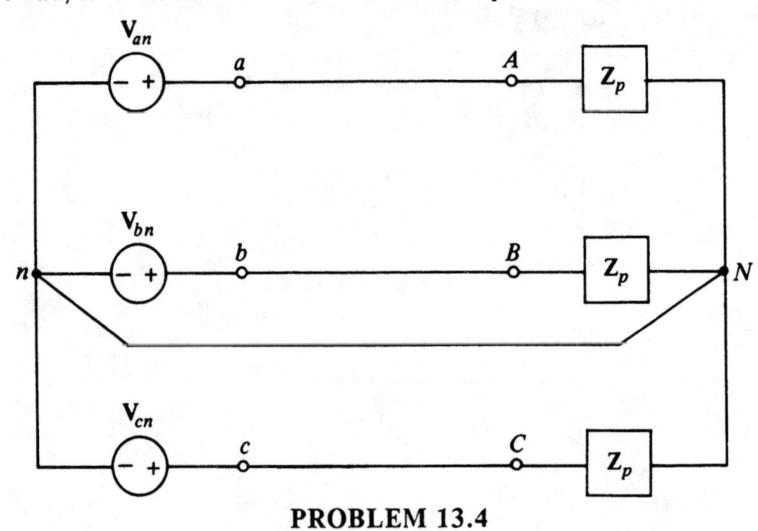

PROBLEM 13.4

13.5 If $\mathbf{Z}_1 = 3 + j3\Omega$, $\mathbf{Z}_2 = 3 - j3\Omega$, and the line voltage is $V_L = 300$ V rms, find the current I_L in each line.

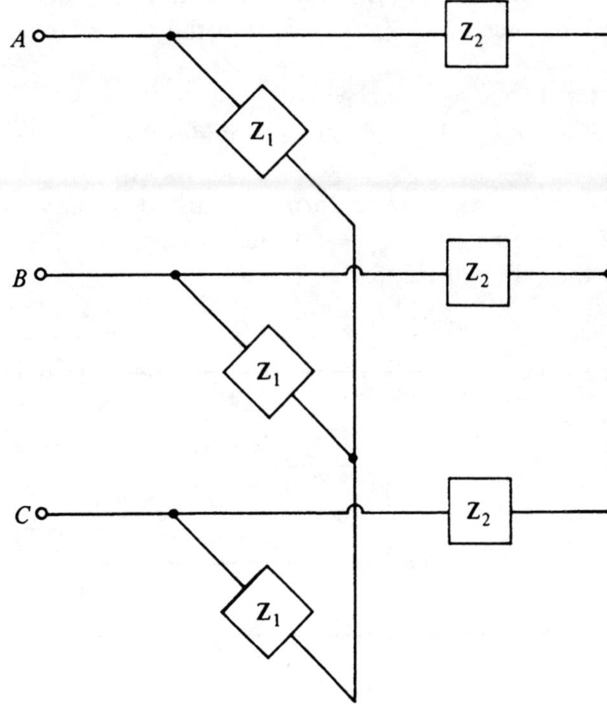

PROBLEM 13.5

13.6 In Prob. 3.4 the line currents form a balanced, positive sequence set with $\mathbf{I}_{aA} = 20\angle 0°$ A rms and $\mathbf{V}_{ab} = 60\angle 60°$ V rms. Find $\mathbf{Z}_p$ and the power delivered to the three-phase load.

13.7 A balanced Y-connected load is present on a 240 V rms (line-to-line) three phase system. If the phase impedance is $3\angle 60°\Omega$, find the total average power delivered to the load.

13.8 A balanced, positive-sequence Y-connected load has $\mathbf{V}_{ab} = 240\angle 0°$ V rms and $\mathbf{Z}_p = 8\sqrt{3}\angle -30°\Omega$. Find the power delivered to the three-phase load.

13.9 A balanced Y-Y system has a positive sequence source with $\mathbf{V}_{an} = 200\angle 0°$ V rms and $\mathbf{Z}_p = 5\angle 60°\Omega$. Find the line voltage, the line current, and the power delivered to the load.

13.10 A balanced Y-Y three-wire, positive sequence system has $\mathbf{V}_{an} = 80\angle 0°$ V rms and $\mathbf{Z}_p = 3 + j4\Omega$. If the lines have an impedance of 1Ω, find the line current and the total power delivered to the load.

13.11 A balance Y-connected source, $\mathbf{V}_{an} = 100\angle 0°$ V rms, negative sequence, is connected by four perfect conductors (having zero impedance) to an unbalanced Y-connected load, $\mathbf{Z}_{AN} = 5\Omega$, $\mathbf{Z}_{BN} = j15\Omega$, and $\mathbf{Z}_{CN} = -j10\Omega$. Find the four line currents.

13.12 A balanced three-phase Y-connected load draws 1 kW at a power factor of 0.707 leading. A balanced Y of capacitors is to be placed in parallel with the load so that the power factor of the combination is 1.0. If the frequency is 60 Hz and the line voltages are a balanced 200 V rms set, find the capacitancies required.

13.13 In Prob. 13.4 if the source is balanced, with positive phase sequence, and $\mathbf{V}_{an} = 200\angle 0°$ V rms, and if the source provides 9 kW at $pf = 0.707$ lagging, find $\mathbf{Z}_p$.

13.14 In the diagram, the line currents form a balanced, positive sequence set with $\mathbf{I}_{aA} = 20\angle -30°$ A rms and $\mathbf{V}_{ab} = 60\angle 30°$ V rms. Find $\mathbf{Z}_p$ and the power delivered to the three-phase load.

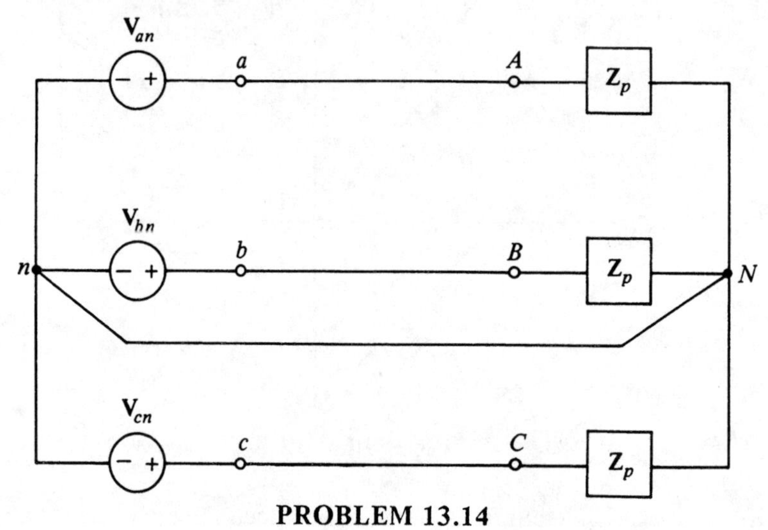

PROBLEM 13.14

13.15 A balanced three-phase Y-connected load draws 6 kW at a power factor of 0.8 lagging. If the line voltages are a balanced 200 V rms set, find the line current I_L.

13.3 The Delta Connection

13.16 Solve Prob. 13.4 if the source and load are unchanged except that the load is Δ-connected.

13.17 A balanced Δ-connected load has $\mathbf{Z}_p = 2 - j\Omega$, and the line voltage is $V_L = 100$ V rms at the load terminals. Find the total power delivered to the load.

13.18 A balanced Δ-connected load has a line voltage of $V_L = 100$ V rms at the load terminals and absorbs a total power of 4.8 kW. If the power factor of the load is 0.8 leading, find the phase impedance.

13.19 Repeat Prob. 13.13 if the load is a balanced Δ.

13.20 Repeat Prob. 13.7 if the load is Δ-connected and the phase impedance is $12\angle 60°\Omega$.

13.21 A balanced three-phase, positive-sequence source with $\mathbf{V}_{ab} = 200\angle 0°$ V rms is supplying a Δ-connected load, $\mathbf{Z}_{AB} = 10\angle 30°\Omega$, $\mathbf{Z}_{BC} = 10\angle 30°\Omega$, $\mathbf{Z}_{CA} = 20\angle 30°\Omega$. Find the line currents. (Assume perfectly conducting lines.)

13.22 In the Y-Δ system shown, the source is positive sequence with $\mathbf{V}_{an} = 200\angle 0°$ V rms and the phase impedance is $\mathbf{Z}_p = 4 + j3\Omega$. Find the line voltage V_L, the line current I_L, and the power delivered to the load.

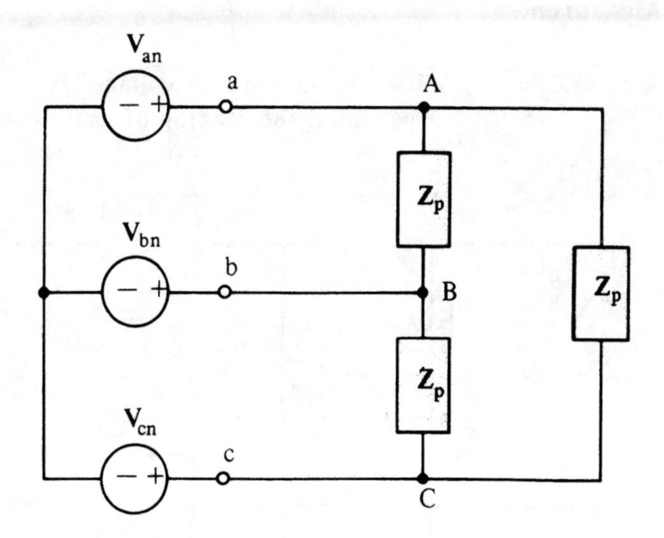

PROBLEM 13.22

13.23 A balanced three-phase, positive-sequence source with $\mathbf{V}_{ab} = 200\angle 0°$ V rms is supplying a Δ-connected lead, $\mathbf{Z}_{AB} = 3 - j4\Omega$, $\mathbf{Z}_{BC} = 20\angle 60°\Omega$, and $\mathbf{Z}_{CA} = 50\angle 30°\Omega$. Find the phasor line currents. (Assume perfectly conducting lines.)

13.4 Y - Δ Transformations

13.24 A balanced three-phase source with $V_L = 100$ V rms is delivering power to a balanced Y-connected load with phase impedance $\mathbf{Z}_1 = 5 + j12\Omega$ in parallel with a balanced Δ-connected load with phase impedance $\mathbf{Z}_2 = 15\Omega$. Find the power delivered by the source.

13.25 If the lines in Prob. 13.17 each have a resistance of 0.1Ω, find the power lost in the lines.

13.26 Repeat Prob. 13.19 if each line contains an impedance of 1Ω.

13.27 A balanced Y-Δ system with $\mathbf{V}_{an} = 100\angle 0°$ V rms, positive phase sequence, has $\mathbf{Z}_p = 6 - j12\Omega$ and an impedance of 1Ω in the lines. Find the power delivered to the load.

13.28 A balanced three-phase positive-sequence source with $\mathbf{V}_{ab} = 100\angle 0°$ V rms is supplying a parallel combination of a Y-connected load and a Δ-connected load. If the Y and Δ loads are balanced with phase impedances of $3 - j3\Omega$ and $9 + j9\Omega$, respectively, find the line current $\mathbf{I}_L$ and the power supplied by the source, assuming perfectly conducting lines.

13.5 Power Measurement

13.29 Let $\mathbf{Z}_1 = \mathbf{Z}_2 = \mathbf{Z}_3 = 20\angle 60°\Omega$, and let the line voltages be a balanced *abc* sequence set, with $\mathbf{V}_{ab} = 115\angle 0°$ V rms. Find the reading of each meter.

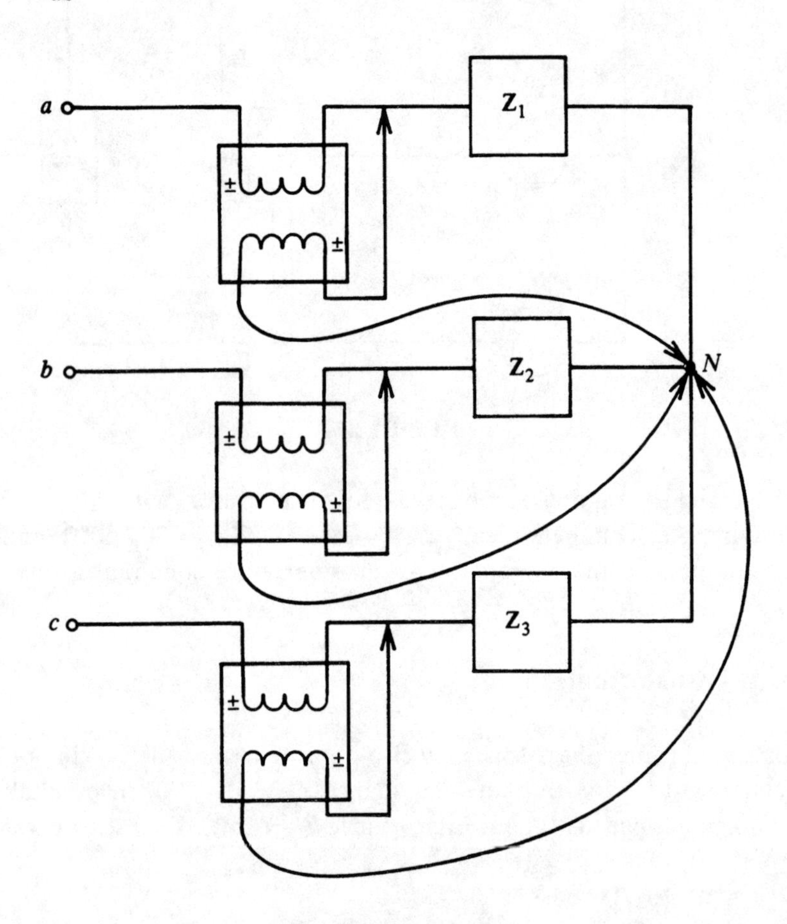

PROBLEM 13.29

13.30 If the power delivered to the load of Prob. 13.29 is measured by the two wattmeters A and C connected as shown, find the wattmeter readings. Check for consistency with the answer of Prob. 13.29.

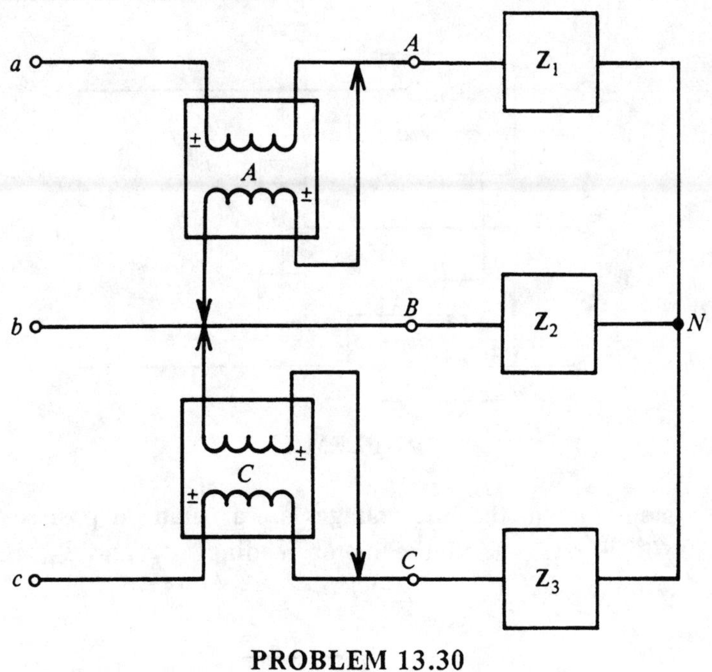

PROBLEM 13.30

13.31 Find the wattmeter readings P_A and P_C and the total power P if the line voltages are as given in Prob. 13.29 and $Z_1 = Z_2 = Z_3 = 10\angle 45°\Omega$.

13.32 The line voltages are a balanced, positive-sequence set, wit $V_{ab} = 200\angle 0°$ V rms, and $Z_1 = Z_2 = Z_3 = 15\angle 45°\Omega$. Find the power delivered to the load by finding the readings of the two wattmeters and by $P = 3P_p$.

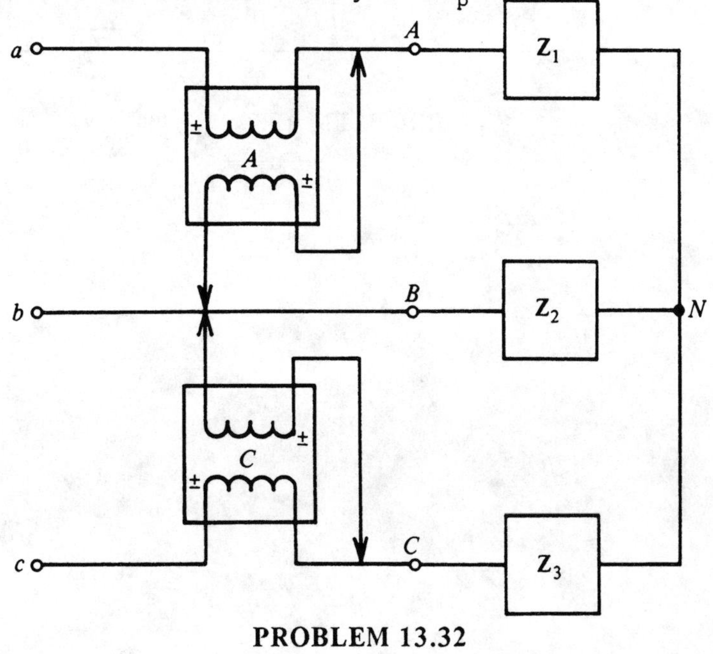

PROBLEM 13.32

13.33 Find the readings P_A and P_B of the wattmeters and the total power delivered to the load if the source is a balanced Y-connected *abc* sequence-source with $V_{an} = 115\angle 0°$ V rms, and the phase impedances are each $\mathbf{Z}_p = 15\angle -60°\Omega$.

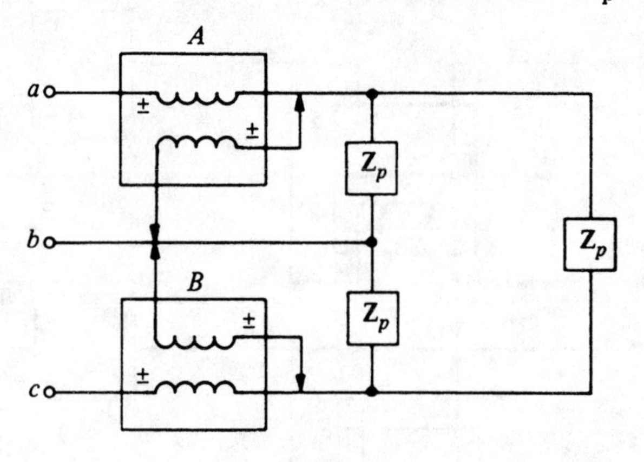

PROBLEM 13.33

13.34 For the system shown, the line voltages are a balanced positive sequence set with $V_{ab} = 120\angle 0°$ V rms. Find the meter readings P_A and P_B and the total power delivered to the load if $\mathbf{Z}_p = 20\angle -90°\Omega$.

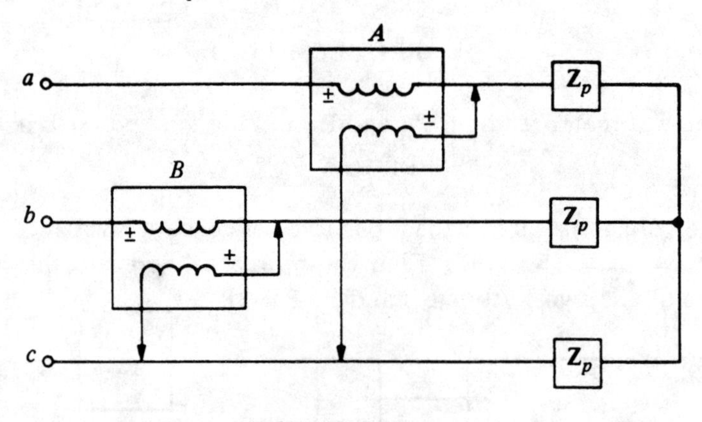

PROBLEM 13.34

13.1 $\mathbf{I}_{AN} = \dfrac{120\angle 0^\circ}{8\angle 15^\circ} = 15\angle -15^\circ$ A rms

$\mathbf{I}_{NB} = \dfrac{120\angle 0^\circ}{8\angle 0^\circ} = 15\angle 0^\circ$ A rms.

$\mathbf{I}_{nN} = -(\mathbf{I}_{AN} + \mathbf{I}_{BN}) = -(15\angle -15^\circ - 15\angle 0^\circ)$

$= \underline{3.916\angle 82.5^\circ \text{ A rms}}.$

13.2 Since $\mathbf{I}_{aA} + \mathbf{I}_{bB} = 0$, the first mesh equation can be written, using $\mathbf{I}_{aA} = \mathbf{I}_1$,

$(\mathbf{Z}_1 + \mathbf{Z}_2)\mathbf{I}_1 - \mathbf{Z}_1\mathbf{I}_3 = \mathbf{V}_1$

The equation for the right mesh can be written

$-2\mathbf{Z}_1\mathbf{I}_1 + (2\mathbf{Z}_1 + \mathbf{Z}_4)\mathbf{I}_3 = 0$

Then $\mathbf{I}_1 = \dfrac{(2\mathbf{Z}_1 + \mathbf{Z}_4)\mathbf{V}_1}{(2\mathbf{Z}_1 + \mathbf{Z}_4)\mathbf{Z}_2 + \mathbf{Z}_1\mathbf{Z}_4}$

For $\mathbf{Z}_1 = 2 + j$, $\mathbf{Z}_2 = 5$, $\mathbf{Z}_3 = 2$ and $\mathbf{Z}_4 = 8$

$\mathbf{I}_1 = \mathbf{I}_{aA} = -\mathbf{I}_{bB} = \dfrac{(4 + 2j + 8)(15)}{(12 + j2)5 + 16 + j8}$

$= 2.336\angle -3.86^\circ$ A rms

$\mathbf{I}_3 = \dfrac{2\mathbf{Z}_1}{2\mathbf{Z}_1 + \mathbf{Z}_4}\mathbf{I}_1 = \dfrac{2(2+j)}{12 + j2}(2.34\angle -3.7^\circ)$

$= 0.429\angle 13.24$ A rms

$P_{Z4} = 8|\mathbf{I}_3|^2 = 1.475$ W

$P_{Z1} = \text{RE}(2+j)|\mathbf{I}_3 - \mathbf{I}_1|^2$

$\mathbf{I}_3 - \mathbf{I}_1 = (0.429\angle 13.24^\circ - 2.336\angle -3.86^\circ)$

$= 1.93\angle 172.4^\circ$ A rms

$P_{Z1} = 2(1.93)^2 = 7.451$ W,

$P_{aA} = 5|\mathbf{I}_{aA}|^2 = 27.30$ W,

$P_{bB} = 5|\mathbf{I}_{aA}|^2 = 27.30$ W

$P_{\text{TOP source}} = |\mathbf{V}_1|\cdot|\mathbf{I}_{aA}|\cos[\text{ang}\,\mathbf{V}_1 - \text{ang}\,\mathbf{I}_{aA}]$

$= 15(2.34)\cos(3.86^\circ)$

$= 34.97$ W

$P_{\text{BOTTOM source}} = |\mathbf{V}_1|\cdot|\mathbf{I}_{bB}|\cos[-$

$= 15(2.34)\cos(3.86^\circ)$

$= 34.97$ W

$P_{\text{LOAD}} = 2P_{Z_1} + P_{Z4} = \underline{16.38 \text{ W}}$

$P_{\text{LOSS}} = P_{aA} + P_{bB} = \underline{54.59 \text{ W}}$

$P_{\text{delivered}} = P_{\text{LOAD}} + P_{\text{LOSS}} = \underline{70.97 \text{ W}}$

13.3 $\mathbf{V}_{ab} = 100\angle 30^\circ$ V rms

$\mathbf{V}_{an} = \dfrac{100}{\sqrt{3}}\angle 30^\circ - 30^\circ = \underline{57.74\angle 0^\circ} \text{ V rms}$

$\mathbf{V}_{bn} = \underline{57.74\angle -120^\circ} \text{ V rms}$

$\mathbf{V}_{cn} = \underline{57.74\angle 120^\circ} \text{ V rms}$

13.4 $\mathbf{Z}_p = 10 - j\dfrac{1}{(500)(20\times 10^{-6})} = 100.5\angle -84.3^\circ\ \Omega$

$\mathbf{I}_{aA} = \dfrac{\mathbf{V}_{an}}{\mathbf{Z}_p} = \dfrac{57.74\angle 0^\circ}{100.5\angle -84.3^\circ} = \underline{0.574\angle 84.3^\circ \text{ A rms}}$

$\mathbf{I}_{bB} = 0.574\angle 84.3 - 120^\circ = \underline{0.574\angle -35.71^\circ \text{ A rms}}$

$\mathbf{I}_{cC} = \underline{0.574\angle -155.7^\circ} \text{ A rms}$

$P = 3V_p I_p \cos\theta$

$= 3(57.74)(0.574)\cos(-84.3)^\circ$

$= \underline{9.90 \text{ W}}$

13.5 $\mathbf{Z}_p = \dfrac{\mathbf{Z}_1\mathbf{Z}_2}{\mathbf{Z}_1 + \mathbf{Z}_2} = \dfrac{(3+j3)(3-j3)}{3+j3+3-j3} = 3\Omega$

$V_p = \dfrac{V_L}{\sqrt{3}} = \dfrac{300}{\sqrt{3}} = 173.2$ V rms

$I_L = I_p = \dfrac{V_p}{|\mathbf{Z}_p|} = \dfrac{173.2}{3} = \underline{57.74 \text{ A rms}}$

13.6 $\mathbf{V}_{aN} = \dfrac{60\angle 30^\circ}{\sqrt{3}}$ V rms

$\mathbf{I}_{aA} = 20\angle 0^\circ$ A rms

$\mathbf{Z}_p = \mathbf{V}_{an}/\mathbf{I}_{aA} = \dfrac{60}{20\sqrt{3}}\angle 30^\circ$

$= \sqrt{3}\angle 30^\circ\ \Omega$

$P = 3V_p I_p \cos\theta = 3\left(\dfrac{60}{\sqrt{3}}\right)20\cos 30^\circ$

$= \underline{1{,}800 \text{ W}}$

13.7 $V_p = \dfrac{240}{\sqrt{3}} = 80\sqrt{3}$ V rms

$I_p = \dfrac{V_p}{|\mathbf{Z}_p|} = \dfrac{80\sqrt{3}}{3} = \dfrac{80}{\sqrt{3}}$ A rms

$P = 3V_p I_p \cos\theta$

$= 3(80\sqrt{3})(80/\sqrt{3})\cos 60^\circ$

$= \underline{9.6 \text{ kW}}$

13.8 $V_p = \dfrac{V_L}{\sqrt{3}} = 80\sqrt{3}$ V rms

$I_p = \dfrac{V_p}{|\mathbf{Z}_p|} = \dfrac{80\sqrt{3}}{8\sqrt{3}} = 10$ A rms

$P = 3V_p I_p \cos\theta$

$= 3(80\sqrt{3})(10)\cos(-30^\circ) = \underline{3.6 \text{ kW}}$

13.9 $V_L = \sqrt{3}V_p = \underline{200\sqrt{3} \text{ V rms}}$

$I_p = I_L = \dfrac{V_p}{|\mathbf{Z}_p|} = \dfrac{200}{5} = \underline{40 \text{ A rms}}$

$P = 3V_p I_p \cos\theta = 3(200)(40)\cos 60^\circ$

$= \underline{12 \text{ kW}}$

13.10 $\mathbf{Z}'_p = \mathbf{Z}_p + 1 = 4 + j4 = 4\sqrt{2}\angle 45^\circ\ \Omega$

$V_p = 80\text{ V rms},\ I_p = \dfrac{V_p}{|\mathbf{Z}'_p|} = \dfrac{80}{4\sqrt{2}} = 10\sqrt{2}\text{ A rms}$

$P = 3(I_p^2)\text{Re}\,\mathbf{Z}_p = 3(10\sqrt{2})^2(3)$

$= \underline{1.8\text{ KW}}$

13.11 $\mathbf{I}_{AN} = \dfrac{\mathbf{V}_{an}}{\mathbf{Z}_{AN}} = \dfrac{100}{5} = \underline{20\text{ A rms}}$

$\mathbf{I}_{BN} = \dfrac{\mathbf{V}_{bn}}{\mathbf{Z}_{BN}} = \dfrac{100\angle 120^\circ}{j15} = \underline{6.67\angle 30^\circ\text{ A rms}}$

$\mathbf{I}_{CN} = \dfrac{\mathbf{V}_{cn}}{\mathbf{Z}_{CN}} = \dfrac{100\angle -120^\circ}{-j10} = \underline{10\angle -30^\circ\text{ A rms}}$

$\mathbf{I}_{nN} = -(\mathbf{I}_{aN} + \mathbf{I}_{bN} + \mathbf{I}_{cN})$

$= -(20 + 5.77 + j3.33 + 8.66 - j5)$

$= \underline{34.47\angle 177.2^\circ\text{ A rms}}$

13.12 $P_1 = 1/3\text{ KW},\ Q_1 = P_1\tan(\cos^{-1}0.707)$

$= P_1 = 1/3\text{ KW}$

$Q_T = Q_1 + Q_2 = P_T\tan(\cos^{-1}1) = 0$

$Q_2 = -Q_1 = -1/3\text{ KVAR}$

$C = \dfrac{-Q_2}{2\pi f V_p^2} = \dfrac{1/3\times 10^3}{377(200/\sqrt{3})^2} = \underline{66.3\ \mu\text{F}}$

13.13 $P = 3V_pI_p\cos\theta = 3\dfrac{V_p^2}{|\mathbf{Z}_p|}\cos\theta$

$|\mathbf{Z}_p| = \dfrac{3V_p^2\cos\theta}{P} = \dfrac{3(200)^2(0.707)}{9000}$

$= 9.43\ \Omega$

$\mathbf{Z}_p = 9.43\angle\cos^{-1}(0.707) = \underline{9.43\angle 45^\circ\ \Omega}$

13.14 $\mathbf{V}_{an} = 60/\sqrt{3}\angle 0^\circ\text{ Vrms}$,

$\mathbf{I}_{aA} = 20\angle -30^\circ\text{ A rms}$,

$\mathbf{Z}_p = \mathbf{V}_{an}/\mathbf{I}_{aA} = \left(\dfrac{60}{\sqrt{3}}/20\right)\angle 30^\circ = \sqrt{3}\angle 30^\circ\ \Omega$

$P = 3V_pI_p\cos\theta = 3\left(\dfrac{60}{\sqrt{3}}\right)(20)\cos 30^\circ$

$= \underline{1.8\text{ KW}}$.

13.15 $I_p = I_L,\ V_p = V_L/\sqrt{3} = 200/\sqrt{3}$

$I_L = \dfrac{P}{3V_p\cos\theta} = \dfrac{6000}{3\left(\frac{200}{\sqrt{3}}\right)(0.8)} = \underline{21.65\text{ A rms}}$

13.16 $\mathbf{V}_{ab} = 100\angle 30^\circ\text{ Vrms}$

$\mathbf{Z}_p = 10 - j100\ \Omega = 100.5\angle -84.3^\circ\ \Omega$

$\mathbf{I}_{AB} = \dfrac{\mathbf{V}_{ab}}{\mathbf{Z}_p} = \dfrac{100\angle 30^\circ}{100.5\angle -84.3^\circ} = 0.995\angle 114.3^\circ\text{ A rms}$

$\mathbf{I}_{aA} = 0.995\sqrt{3}\angle 114.3^\circ - 30^\circ$

$= \underline{1.723\angle 84.3^\circ\text{ A rms}}$

$\mathbf{I}_{bB} = \underline{1.723\angle -35.7^\circ\text{ A rms}}$

$\mathbf{I}_{cC} = \underline{1.723\angle -155.7^\circ\text{ A rms}}$

13.17 $\mathbf{Z}_p = 2 - j\ \Omega = \sqrt{5}\angle -26.57^\circ\ \Omega$

$I_p = V_L/|\mathbf{Z}_p| = \dfrac{100}{\sqrt{5}}\text{ A rms}$

$P = 3(100)\left(\dfrac{100}{\sqrt{5}}\right)\cos(-26.57^\circ)$

$= \underline{12\text{ KW}}$

13.18 $|\mathbf{Z}_p| = \dfrac{3V_p^2\cos\theta}{P} = \dfrac{3(100)^2(0.8)}{4800} = 5\ \Omega$

$\mathbf{Z}_p = 5\angle -\cos^{-1}(0.8) = 5\angle -36.9^\circ\ \Omega$

$= \underline{4 - j3\ \Omega}$

13.19 $\mathbf{V}_{ab} = 200\sqrt{3}\angle 0^\circ\text{ V rms}$

$|\mathbf{Z}_p| = \dfrac{3V_p^2\cos\theta}{P} = \dfrac{3(200\sqrt{3})^2(0.707)}{9000}$

$= 28.28\ \Omega$

$\mathbf{Z}_p = \underline{28.28\angle 45^\circ\ \Omega}$

13.20 $V_p = 240\text{ V rms}$

$P = \dfrac{3V_p^2\cos\theta}{|\mathbf{Z}_p|} = \dfrac{3(240)^2\cos 60^\circ}{12}$

$= \underline{7.2\text{ W}}$

13.21 $\mathbf{I}_{AB} = \dfrac{\mathbf{V}_{ab}}{\mathbf{Z}_{AB}} = \dfrac{200\angle 0^\circ}{10\angle 30^\circ} = 20\angle -30^\circ\text{ A rms}$

$\mathbf{I}_{BC} = \dfrac{\mathbf{V}_{bc}}{\mathbf{Z}_{BC}} = \dfrac{200\angle -120^\circ}{10\angle 30^\circ} = 20\angle -150^\circ\text{ A rms}$

$\mathbf{I}_{CA} = \dfrac{\mathbf{V}_{ca}}{\mathbf{Z}_{CA}} = \dfrac{200\angle 120^\circ}{20\angle 30^\circ} = 10\angle 90^\circ\text{ A rms}$

$\mathbf{I}_{aA} = \mathbf{I}_{AB} - \mathbf{I}_{CA} = (17.32 - j10) - (j10)$

$= 17.32 - j20 = \underline{26.46\angle -49.1^\circ\text{ A rms}}$

$\mathbf{I}_{bB} = \mathbf{I}_{BC} - \mathbf{I}_{AB} = (-17.32 - j10) - (17.32 - j10)$

$= \underline{-34.64\text{ A rms}}$

$\mathbf{I}_{cC} = \mathbf{I}_{CA} - \mathbf{I}_{BC} = (j10) - (-17.32 - j10)$

$= 17.32 + j10 = \underline{26.46\angle 49.1^\circ\text{ A rms}}$

13.22 $V_L = \underline{200\sqrt{3}\text{ Vrms}},\ \mathbf{Z}_p = 5\angle 36.9^\circ\ \Omega$

$I_p = \dfrac{200\sqrt{3}}{5} = 40\sqrt{3}\text{ A rms}$

$I_L = \sqrt{3}I_p = \underline{120\text{ A rms}}$

$P = 3(200\sqrt{3})(40\sqrt{3})\cos 36.9^\circ$

$= \underline{57.6\text{ W}}$

13.23 $\mathbf{I}_{AB} = \dfrac{\mathbf{V}_{ab}}{\mathbf{Z}_{AB}} = \dfrac{200}{3 - j4} = 20\angle 53.1^\circ\text{ A rms}$

$\mathbf{I}_{BC} = \dfrac{\mathbf{V}_{bc}}{\mathbf{Z}_{BC}} = \dfrac{200\angle -120^\circ}{20\angle 60^\circ} = -10\text{ A rms}$

$\mathbf{I}_{CA} = \dfrac{\mathbf{V}_{ca}}{\mathbf{Z}_{CA}} = \dfrac{200\angle 120^\circ}{50\angle 30^\circ} = j4\text{ A rms}$

13.23 cont.

$I_{aA} = I_{AB} - I_{CA} = \underline{12 + j12 \text{ A rms}}$

$I_{bB} = I_{BC} - I_{AB} = \underline{-22 - j16 \text{ A rms}}$

$I_{cC} = I_{CA} - I_{BC} = \underline{10 + j4 \text{ A rms}}$

13.24 $Z_{2Y} = \frac{1}{3}Z_2 = \frac{1}{3}(15) = 5\Omega$

$Z_p = \frac{Z_1 Z_{2Y}}{Z_1 + Z_{2Y}} = \frac{(5+j12)(5)}{5+5+j12} = 4.16\angle 17.19^\circ \Omega$

$I_p = \frac{V_P}{|Z_p|} = \frac{100/\sqrt{3}}{4.16} = 13.87 \text{ A rms}$

$P = \sqrt{3} V_L I_L \cos\theta = \sqrt{3}(100)(13.87)\cos 17.19^\circ$

$= \underline{2{,}296 \text{ W}}$

13.25 $P_L = 3(|I|^2 R) = 3(I_P\sqrt{3})^2 R$

$= 3\left(\frac{100}{\sqrt{3}}\right.$

13.26 Z_{TY} = Total Impedance of Y load

$= 9.43\angle 45^\circ = \frac{20}{3} + j\frac{20}{3}\Omega$

from Prob. 13.13

$Z_{PY} = Z_{TY} - 1 = \frac{17}{3} + j\frac{20}{3}\Omega$

$Z_{P\Delta} = 3Z_{PY} = 17 + j20\Omega$

$= \underline{26.25\angle 49.64^\circ \Omega}$

13.27 $Z_Y = \frac{1}{3}Z_P = \frac{6-j12}{3} = 2 - j4\Omega$

$Z'_Y = Z_Y + 1 = 3 - j4 = 5\angle -53.13^\circ$

$I_P = I_L = \frac{V_P}{|Z'_P|} = \frac{100}{5} = 20 \text{ A rms}$

$P_L = 3I_P^2 \text{ Re } Z_Y = 3(20)^2 2 = \underline{2.4 \text{ KW}}$

13.28 Δ-Y: $Z_Y = \frac{1}{3}Z_d = \frac{1}{3}(9+j9) = 3+j3\Omega$

$Z_{eq} = \frac{(3-j3)(3+j3)}{3-j3+3+j3} = 3\Omega$

$I_L = \frac{V_P}{|Z_{eq}|} = \frac{100/\sqrt{3}}{3} = 19.25 \text{ A rms}$

$P = 3I_L^2 \text{ Re } Z_{eq} = 3(19.25)^2(3) = \underline{\frac{10}{3} \text{ kW}}$

13.29 Each meter reads phase power which is the same for each phase.

$V_L = 115 \text{ Vrms} \Rightarrow V_P = 115/\sqrt{3} \text{ Vrms}$

$I_P = \frac{115}{20\sqrt{3}} = 3.32 \text{ A rms}$

$V_{an} = 115/\sqrt{3}\angle -30^\circ;\ I_{an} = 3.32\angle -30^\circ - 60^\circ = 3.32\angle -90^\circ \text{ V rms}$

$P = |V_{an}|\cdot|I_{an}|\cos(\text{ang } V_{an} - \text{ang } I_{an})$

$= (115/\sqrt{3})(3.32)\cos(-30+90) = \underline{110.2 \text{ W}}$

13.30 $P_A = |V_{ab}||I_{aN}|\cos(\text{ang } V_{ab} - \text{ang } I_{aN})$

$= (115)(3.32)\cos(0+90^\circ) = \underline{0 \text{ W}}$

$P_C = |V_{cb}|\cdot|I_{cN}|\cos(\text{ang } V_{cb} - \text{ang } I_{cN})$

$V_{bc} = 115\angle -120^\circ \Rightarrow V_{cb} = 115\angle 60^\circ \text{ Vrms}$

$I_{cN} = 3.32\angle -90+120 = 3.32\angle 30^\circ \text{ A rms}$

$P_C = (115)(3.32)\cos(60-30)$

$= \underline{330.6 \text{ W}}$

$P_C = 3P = 3(110.2) = 330.6$ (checks)

13.31 $V_{ab} = 115\angle 0^\circ$ Vrms; $I_{an} = \frac{V_{an}}{Z_P} = \frac{115/\sqrt{3}\angle -30^\circ}{10\angle 45^\circ}$

$= 6.64\angle -75^\circ \text{ A rms}$

$P_A = |V_{ab}|\cdot|I_{an}|\cos(\text{ang } V_{ab} - \text{ang } I_{an})$

$= (115)(6.64)\cos(0+75^\circ)$

$= \underline{197.6 \text{ W}}$

$V_{bc} = 115\angle -120^\circ \Rightarrow V_{cb} = 115\angle 60^\circ \text{ V rms}$

$I_{cn} = 6.64\angle -75+120^\circ = 6.64\angle 45^\circ \text{ A rms}$

$P_C = (115)(6.64)\cos(60^\circ - 45^\circ)$

$= \underline{737.5 \text{ W}}$

$P_T = P_A + P_C = \underline{935.1 \text{ W}}$

13.32 $V_{ab} = 200\angle 0^\circ \text{V};\ V_{cb} = 200\angle 60^\circ \text{V}$

$V_{an} = \frac{200}{\sqrt{3}}\angle -30^\circ \text{ V rms};\ I_{an} = \frac{200\angle -30}{15\sqrt{3}\angle 45^\circ}$

$= 7.698\angle -75^\circ \text{ A rms}$

$I_{cn} = 7.7\angle -75+120^\circ = 7.7\angle 45^\circ \text{ A rms}$

$P_A = |V_{ab}||I_{an}|\cos(\text{ang } V_{ab} - \text{ang } I_{an})$

$= (200)(7.7)\cos(0+75^\circ)$

$= \underline{398.5 \text{ W}}$

$P_C = |V_{cb}||I_{cn}|\cos(\text{ang } V_{cb} - \text{ang } I_{cn})$

$= (200)(7.7)\cos(60 - 45^\circ)$

$= \underline{1{,}487 \text{ W}}$

$P_L = P_A + P_C = \underline{1886 \text{ W}}$

$P = 3V_P I_P \cos\theta = 3\left(\frac{200}{\sqrt{3}}\right)(7.7)\cos 45^\circ$

$= \underline{1886 \text{ W}}$ (checks)

13.33 $P_A = |V_{ab}||I_{aA}|\cos(\text{ang } V_{ab} - \text{ang } I_{aA})$

$V_{ab} = 115\sqrt{3}\angle 30^\circ \text{ V rms}$

$I_{AB} = \frac{V_{AB}}{Z_P} = \frac{115\sqrt{3}\angle 30^\circ}{15\angle -60^\circ} = \frac{23}{\sqrt{3}}\angle 90^\circ \text{ A rms}$

$I_{aA} = \frac{23}{\sqrt{3}}\sqrt{3}\angle 90^\circ - 30^\circ = 23\angle 60^\circ \text{ A rms}$

$P_A = (115\sqrt{3})(23)\cos(30^\circ - 60^\circ)$

$= \underline{3968 \text{ W}}$

13.33 cont.

$P_B = |\underline{V}_{cb}||I_{cC}|\cos(\text{ang } V_{cb} - \text{ang } I_{cC})$

$\underline{V}_{cb} = 115\sqrt{3}\angle 30^\circ - 120^\circ + 180^\circ$

$= 115\sqrt{3}\angle 90^\circ$ Vrms

$I_{cC} = 23\angle 60^\circ + 120^\circ = 23\angle 180^\circ$ Arms

$P_B = (115\sqrt{3})(23)\cos(90 - 180) = \underline{0\text{ W}}$

13.34 $P_A = |\underline{V}_{ac}||I_{aA}|\cos(\text{ang } \underline{V}_{ac} - \text{ang } I_{aA})$

$\underline{V}_{ac} = 120\angle 0 + 120^\circ - 180^\circ = 120\angle -60^\circ\ V_{rms}$

$I_{aA} = \frac{\underline{V}_{an}}{\underline{Z}_p} = \frac{\frac{120\angle -30}{\sqrt{3}}}{20\angle -90^\circ} = 2\sqrt{3}\angle 60^\circ\ A_{rms}$

$P_A = (120)(2\sqrt{3})\cos(-60^\circ - 60^\circ)$

$= \underline{-120\sqrt{3}\text{ W}}$

$P_B = |\underline{V}_{bc}||I_{bB}|\cos(\text{ang } \underline{V}_{bc} - \text{ang } I_{bB})$

$V_{bc} = 120\angle -120^\circ\ V_{rms}$; $I_{bB} = 2\sqrt{3}\angle -60^\circ\ A_{rms}$

$P_B = (120)(2\sqrt{3})\cos(-120^\circ + 60^\circ)$

$= 120\sqrt{3}\text{ W}$

$P_T = P_A + P_B = \underline{0\text{ W}}$

Chapter 14

Complex Frequency and Network Functions

14.1 The Damped Sinusoid

14.1 Excite the circuit by the complex function $v_1 = 15e^{-3t}e^{jt}$ V and find the forced response.

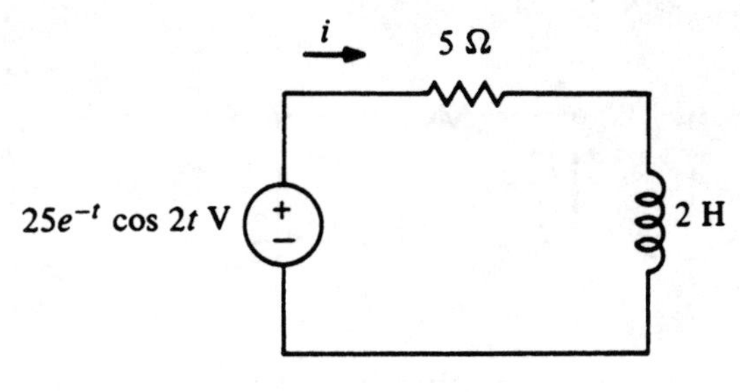

PROBLEM 14.1

14.2 Excite the circuit of Prob. 14.1 by the complex function $v_1 = 6e^{-j3t}$ V and find the forced response.

14.2 Complex Frequency and Generalized Phasors

14.3 Find the complex frequencies associated with (a) $3 + e^{-5t}$, (b) $\cos(3\omega t + 15°)$, (c) $4e^{-2t}\cos(\omega t)$.

14.4 Find s and $\mathbf{V}(s)$ if $v(t)$ is given by (a) $3e^{-t}$ V, (b) $4e^{-t}\cos(2t + 10°)$ V.

14.5 Find $v(t)$ if $\mathbf{V} = 6\angle{-30°}$ V rms, $s = -2 + j4$.

14.6 Find the phasor representation $\mathbf{V}(s)$ of (a) $v(t) = 8e^{-t}\cos(5t + 30°)$ V, (b) $v(t) = 5e^{-9t}$ V, and (c) $v(t) = e^{-2t}[4\cos(3t) + 3\sin(3t)]$ V.

14.3 Impedance and Admittance

14.7 Find the forced response i if $v_{g1} = 8e^{-t}\cos(t)$ V and $i_{g2} = 2e^{-2t}$ A.

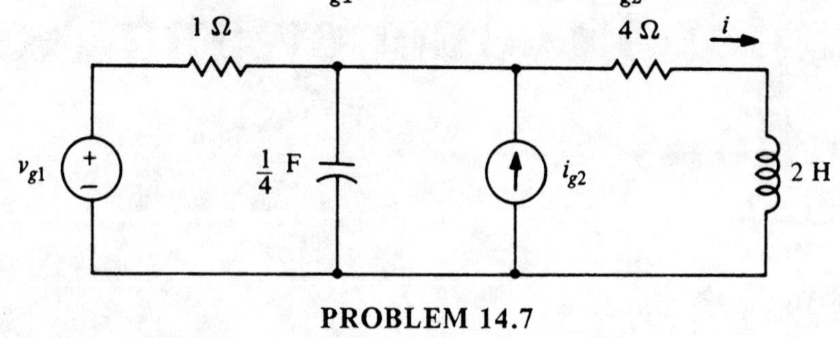

PROBLEM 14.7

14.8 Replace all the circuit except the inductor, by its Thevenin equivalent, and find $\mathbf{I}_1$.

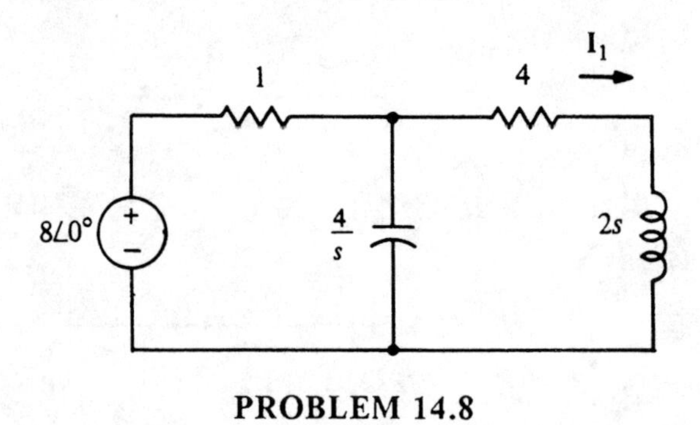

PROBLEM 14.8

14.9 Assume $\mathbf{I}_1 = 1$A, in Prob. 14.8 and use the method of proportionality to find the true $\mathbf{I}_1$.

14.10 Find the impedance $\mathbf{Z}(s)$ seen by the source. If the source is $v_g = 10e^{-t}\cos(2t)$ V, find the forced component of the current it delivers.

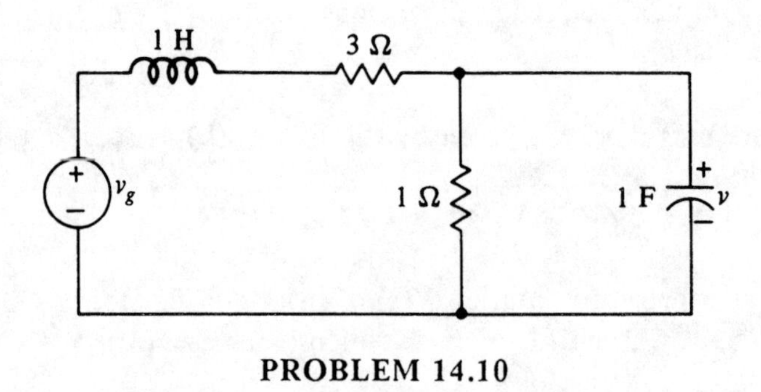

PROBLEM 14.10

14.11 Find the forced component of v in Prob. 14.10 if $v_g = 16e^{-4t}\cos(2t)$ V.

14.12 Find the forced component of i.

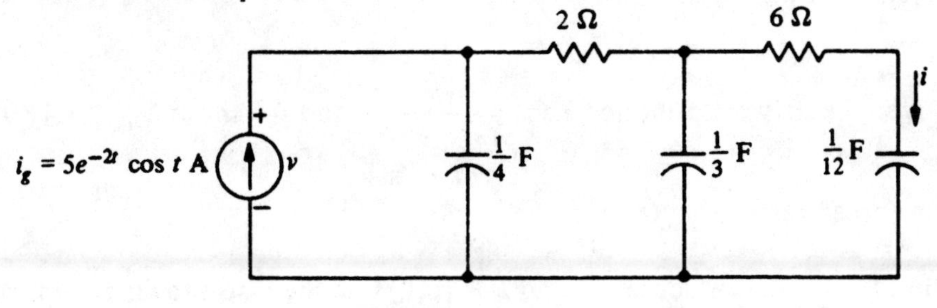

PROBLEM 14.12

14.13 Replace in the corresponding phasor circuit everything except the capacitor by its Thevenin equivalent circuit and use the result to find the forced response i if $v_g = 4e^{-2t}\cos(2t)$ V.

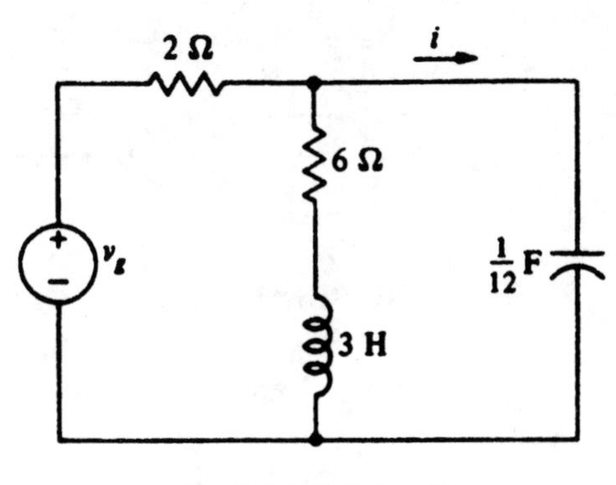

PROBLEM 14.13

14.14 Find the complex frequencies which characterize (a) $v = e^{-t}[\cos(2t) + \sin(2t)]$ V, (b) $v = e^{-3t} + e^{-4t} + 2\cos(t)$ V, and (c) $v = 10$V.

14.15 Find the impedance $\mathbf{Z}(s)$ seen by the source and locate its poles and zeros. If the source is $v_g = 10e^{-t}\cos(2t)$ V, find the forced component of current it delivers.

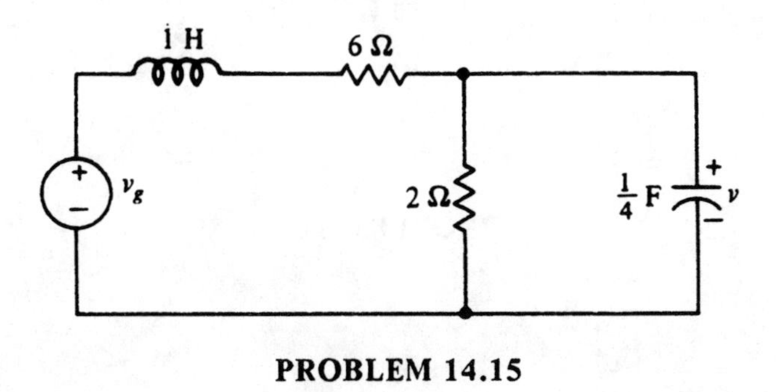

PROBLEM 14.15

14.4 Network Functions

14.16 Given the network function $\mathbf{H}(s) = \dfrac{2(s+4)}{s^2+2s+2}$ and the input $\mathbf{V}_1(s) = 4\angle 0°$ V rms find the forced response $v_0(t)$ if $s = -1 + j3$.

14.17 Find $\mathbf{H}(s)$ if the response is v and $R = 16\Omega$. Use the result to find the forced response if $v_g = 3e^t$ V.

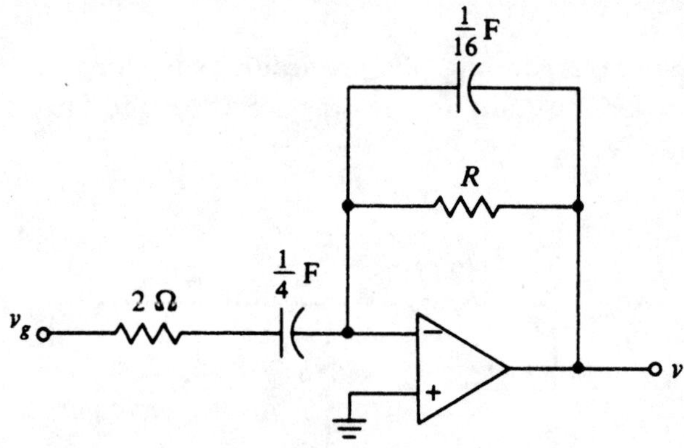

PROBLEM 14.17

14.18 Find the network function $\mathbf{V}(s)/\mathbf{V}_g(s)$ and use the result to find the forced response v if $v_g = 3e^{-2t}\cos(4t)$ V.

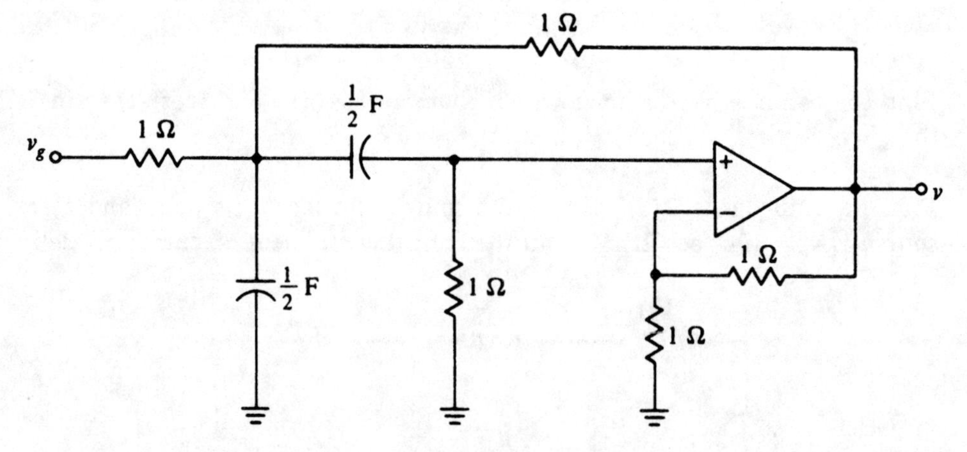

PROBLEM 14.18

14.19 Find $\mathbf{H}(s) = \mathbf{V}(s)/\mathbf{V}_g(s)$ and use the result to find the forced response v if $v_g = e^{-t}\cos(t)$ V.

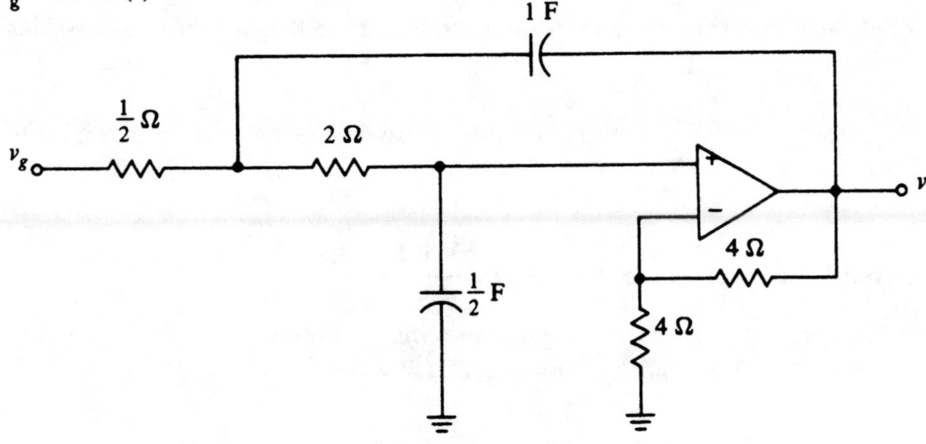

PROBLEM 14.19

14.20 For the corresponding phasor circuit, replace everything to the left of terminals a-b by its Thevenin equivalent and use the result to find the forced response i.

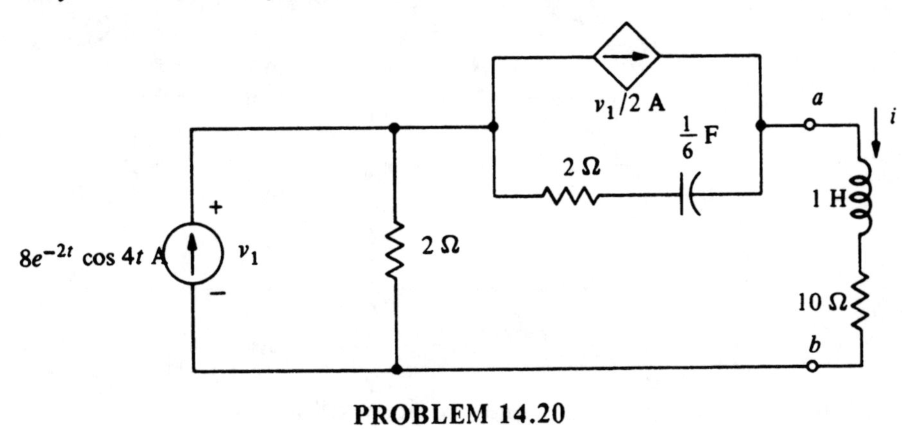

PROBLEM 14.20

14.5 Poles and Zeros

14.21 If the zeros of $\mathbf{H}(s)$ are $s = -4 \pm j3$, the poles are $s = -3, -2 \pm j$, and $\mathbf{H}(0) = 5$, find $\mathbf{H}(s)$.

14.22 Find the pole-zero plot of the network function of Prob. 14.21.

14.23 Find the poles and zeros of Prob. 14.10.

14.24 Find the poles and zeros of Prob. 14.12.

14.6 The Natural Response from the Network Function

14.25 Find the natural response in Prob. 14.16 assuming that there is no cancellation in the network function.

14.26 Find the complete response in Prob. 14.16 if the natural response is as given in Prob. 14.25 and $v_o(0^+) = dv_o(0^+)/dt = 0$.

14.27 Given the network function, $\mathbf{H}(s) = \dfrac{\mathbf{V}_o(s)}{\mathbf{V}_i(s)} = \dfrac{4s(s+2)}{s^2 + 2s + 2}$,

If no cancellation has occurred, find the complete response $v_o(t)$, for $t > 0$, if $v_i(t) = 6e^{-t}\cos(2t)\text{V}$ and $v_o(0^+) = dv_o(0^+)/dt = 0$.

14.7 Natural Frequencies

14.28 Find the natural frequencies by killing the source and using (a) a pliers entry in series with the 3H inductor and (b) a soldering entry across the 3H inductor.

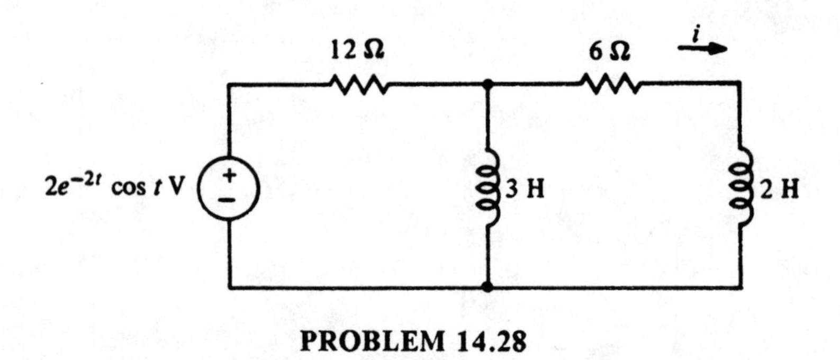

PROBLEM 14.28

14.29 Find the natural frequencies of the circuit by killing the source and using (a) a pliers entry in series with the inductor and (b) a soldering entry across the inductor.

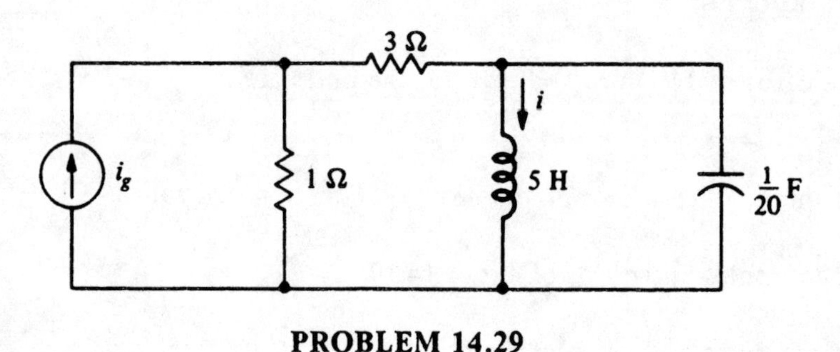

PROBLEM 14.29

14.8 Two-Port Networks

14.30 Draw an equivalent circuit for a two-port network with parameters $\mathbf{Z}_{11} = 8\Omega$, $\mathbf{Z}_{12} = 2\Omega$, and $\mathbf{Z}_{22} = 4\Omega$.

14.31 Two sets of hybrid parameters, h_{11}, h_{12}, h_{21}, h_{22}, and g_{11}, g_{12}, g_{21}, g_{22}, are defined by

$$V_1 = h_{11}I_1 + h_{12}V_2$$
$$I_2 = h_{21}I_1 + h_{22}V_2$$

and

$$I_1 = g_{11}V_1 + g_{12}I_2$$
$$V_2 = g_{21}V_1 + g_{22}I_2$$

Find the h-parameters and the g-parameters of the network of Prob. 14.30.

14.32 Considering the figure as a two-port network with terminals after the voltage source and to the left of the 1Ω resistor, find the z-parameters as functions of s.

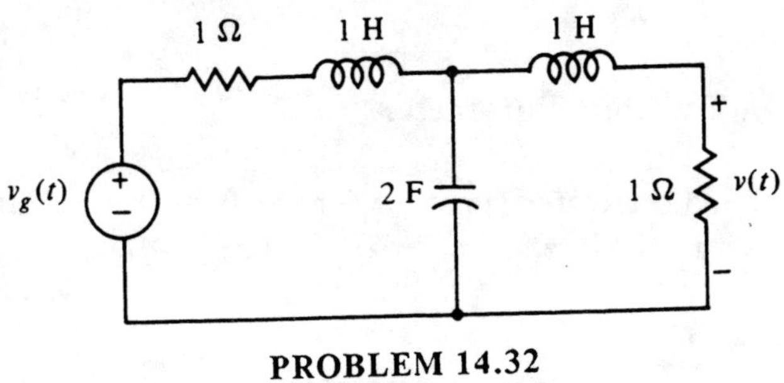

PROBLEM 14.32

14.33 Find the h-parameters of the two-port network of Prob. 14.32.

14.34 Find the transmission parameters of the two-port network of Prob. 14.32.

14.35 Let $h_{11} = 3k\Omega$, $h_{12} = 10^{-3}$, $h_{21} = 200$, and $h_{22} = 10^{-3}$S, and find the network function V_2/V_1 if port 2 is open circuited.

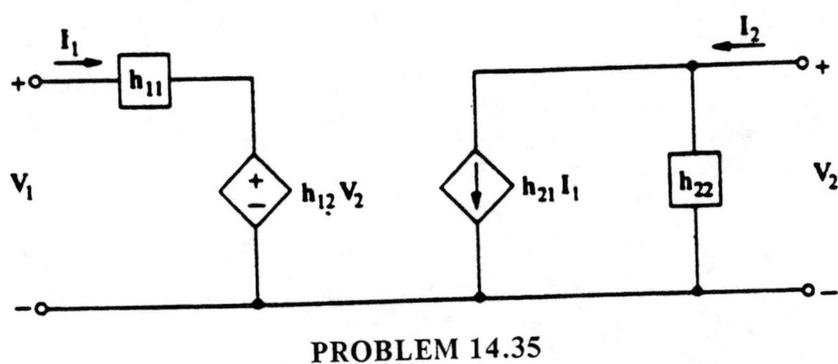

PROBLEM 14.35

14.36 Considering the figure as a two-port network with terminals as shown, find the z- and y-parameters as functions of s.

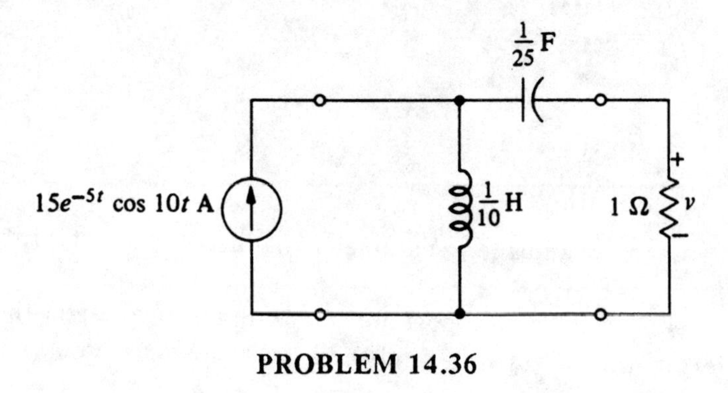

PROBLEM 14.36

14.9 Applications of Two-Port Parameters

14.37 Find the voltage radio transfer function for the two-port terminated in 1Ω, shown with z-parameters z_{11} = 8Ω, z_{12} = z_{21} = 2Ω, and z_{22} = 4Ω.

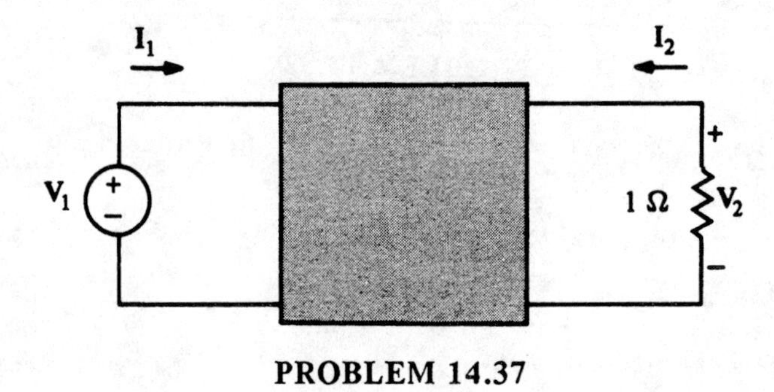

PROBLEM 14.37

14.1 $2\frac{di}{dt}+5i=15e^{-3t}e^{jt}$; $i=Ie^{(-3+j)t}$

$2(-3+j)Ie^{(-3+j)t}+5Ie^{(-3+j)t}=15e^{(-3+j)t}$

or $(-1+j2)I=15 \Rightarrow I=\frac{15}{-1+j2}$

$I=-3-j6=6.7\angle -116.6^\circ$ A

$\therefore i=\underline{(6.7\angle -116.6^\circ)e^{(-3+j)t}\text{ A}}$

14.2 $2\frac{di}{dt}+5i=6e^{-j3t}$; $i=Ie^{-j3t}$

$2(-j3)Ie^{-j3t}+5Ie^{-j3t}=6e^{-j3t}$ or

$(5-j6)I=6 \Rightarrow I=\frac{6}{5-j6}=0.77\angle 50.2^\circ$ A

$\therefore i=\underline{(0.77\angle 50.2^\circ)e^{-j3t}\text{ A}}$

14.3 (a) $3+e^{-5t}=3e^{0t}+e^{-5t}$; $s=\underline{0,5}$

(b) $\cos(3\omega t+15^\circ)$

$=\frac{1}{2}\left(e^{j15^\circ}e^{j3\omega t}+e^{-j15^\circ}e^{-j3\omega t}\right)$; $s=\underline{\pm j3\omega}$

(c) $4e^{-2t}\cos\omega t=\frac{4}{2}e^{-2t}\left(e^{j\omega t}+e^{-j\omega t}\right)$

$=2\left(e^{(-2+j\omega)t}+e^{(-2-j\omega)t}\right)$

$s=\underline{-2\pm j\omega}$

14.4 (a) $v=3e^{-t} \Rightarrow \underline{s=-1}$, $\underline{V(s)=3\angle 0^\circ\text{ V}}$

(b) $v=4e^{-t}\cos(2t+10^\circ)$

$=\frac{4}{2}(e^{-t})\left(e^{j(2t+10^\circ)}+e^{-j(2t+10^\circ)}\right)$

$=2\left(e^{j10^\circ}e^{(-1+j2)t}+e^{-j10^\circ}e^{(-1-j2)t}\right)$

$s=\underline{-1\pm j2}$, $V(s)=\underline{4\angle 10^\circ\text{ V}}$

14.5 $\mathbf{V}=6\angle -30^\circ$ V, $s=-2+j4$

$v=\underline{6e^{-2t}\cos(4t-30^\circ)\text{ V}}$

14.6 (a) $V(s)=\underline{8\angle 30^\circ}$ V where $s=-1+j5$

(b) $V(s)=\underline{5\angle 0^\circ\text{ V}}$ where $s=-9$

(c) $v=e^{-2t}(4\cos 3t+3\sin 3t)$ V

$=e^{-2t}\,5\cos(3t-36.9^\circ)$ V

$V(s)=\underline{5\angle -36.9^\circ}$ where $s=-2+j3$

14.7 $\mathbf{V}_{g1}=8\angle 0^\circ$, $\mathbf{I}_{g2}=2\angle 0^\circ$ when $s=-1+j$

with Node voltage $\mathbf{V}$, KCL gives

$\mathbf{V}\left(1+\frac{s}{4}+\frac{1}{2s+4}\right)=\frac{\mathbf{V}_{g1}}{1}+\mathbf{I}_{g2}$ or

$\mathbf{V}=\frac{(2+8)2(2s+4)}{s^2+6s+10}$

14.7 cont

$\mathbf{I}=\frac{\mathbf{V}}{2s+4}=\frac{20}{s^2+6s+10}$, @ $s=-1+j$

$\mathbf{I}=\frac{20}{4+j4}=\frac{5}{\sqrt{2}}\angle -45^\circ$ A

$i=\underline{\frac{5}{\sqrt{2}}e^{-t}\cos(t-45^\circ)\text{ A}}$

14.8 $\mathbf{V}_{oc}=\frac{4/s}{1+4/s}(8)=\frac{32}{s+4}$ V

$R_{th}=4+\frac{4/s}{1+4/s}=\frac{4(s+5)}{s+4}$

$\mathbf{I}_1=\frac{\mathbf{V}_{oc}}{R_{th}+2s}=\frac{32/s+4}{\frac{4(s+5)}{s+4}+2s}$

$=\underline{\frac{16}{s^2+6s+10}}$

14.9 $\mathbf{I}_1=1$A, $\mathbf{V}_1=\mathbf{I}_1(2s+4)=2s+4$

$\mathbf{V}_g=\mathbf{V}_1\left(1+\frac{s}{4}+\frac{1}{2s+4}\right)$

$=\frac{(2s+4)(s^2+6s+10)}{2(2s+4)}$

$=\frac{s^2+6s+10}{2}=8$

$\therefore \mathbf{I}_1=\frac{8}{\frac{s^2+6s+10}{2}}=\underline{\frac{16}{s^2+6s+10}}$

14.10 $Z(s)=s+3+\frac{1}{s+1}=\frac{s^2+4s+4}{s+1}$

$=\underline{\frac{(s+2)^2}{s+1}}$

$I(s)=\frac{V_g(s)}{Z(s)}=\frac{10\angle 0^\circ(s+1)}{(s+2)^2}$ when $s=-1+j2$

$\mathbf{I}=\frac{10(j2)}{(1+j2)^2}=4\angle -36.9^\circ$ A

$i=\underline{4e^{-t}\cos(2t-36.9^\circ)\text{ A}}$

14.11 KCL yields

$\mathbf{V}\left(1+s+\frac{1}{s+3}\right)=\frac{\mathbf{V}_g}{s+3}$ or

$\mathbf{V}=\frac{\mathbf{V}_g}{s^2+4s+4}=\frac{\mathbf{V}_g}{(s+2)^2}$

$\mathbf{V}=\frac{16}{(s+2)^2}$ when $s=-4+j2$

$=\frac{16}{(-2+j2)^2}=\frac{16}{-j8}=2\angle 90^\circ$

$v=\underline{2e^{-4t}\cos(2t+90^\circ)\text{ V}}$

$=-2e^{-4t}\sin(2t)$ V

14.12 v_1 = voltage across the $\frac{1}{3}$-F capacitor
i_1 = current through the 2-Ω

Assume $\mathbf{I} = 1$A. Then

$\mathbf{V}_1 = (6 + \frac{12}{s})\mathbf{I} = \frac{6s+12}{s}$,

$\mathbf{I}_1 = \frac{s}{3}\mathbf{V}_1 + \mathbf{I} = 2s + 5$,

$\mathbf{V} = 2\mathbf{I}_1 + \mathbf{V}_1 = \frac{4s^2+16s+12}{s}$,

$\mathbf{I}_g = \frac{s}{4}\mathbf{V} + \mathbf{I}_1 = s^2 + 6s + 8$

Therefore

$\frac{\mathbf{I}}{\mathbf{I}_g} = \frac{1}{s^2+6s+8}$; $s^2+6s+8 = (s+2)^2 + 2(s+2)$

$\mathbf{I} = \frac{5}{(-2+j)^2+6(-2+j)+8} = \frac{5}{(j)^2+2(j)}$

$= \frac{5}{-1+j2} = \sqrt{5}\angle{-116.6^\circ}$

$i = \sqrt{5}\,e^{-2t}\cos(t - 116.6^\circ)$ A

14.13 $\mathbf{V}_{oc} = \frac{6+3s}{8+3s}\mathbf{V}_g$ by voltage div.

$R_{th} = \frac{2(3s+6)}{3s+8}$

$\mathbf{I} = \frac{\mathbf{V}_{oc}}{R_{th} + 12/s} = \frac{\frac{4(3s+6)}{8+3s}}{\frac{2(3s+6)}{3s+8} + \frac{12}{s}}$

$= \frac{2s(s+2)}{s^2+8s+16} = \frac{2s(s+2)}{(s+4)^2}$

$\mathbf{I}(-2+j2) = \frac{2(-2+j2)(j2)}{(2+j2)^2} = \sqrt{2}\angle{135^\circ}$

$i = \sqrt{2}\,e^{-2t}\cos(2t + 135^\circ)$ A

14.14 (a) $v = e^{-t}\left[\frac{e^{j2t}+e^{-j2t}}{2} + \frac{e^{j2t}-e^{-j2t}}{j2}\right]$;

$s = -1 \pm j2$.

(b) $v = e^{-3t} + e^{-4t} + \cos t$

$= e^{-3t} + e^{-4t} + \frac{1}{2}[e^{jt} + e^{-jt}]$;

$s = -3, -4, \pm j$

(c) $v = 10e^{0}$; $s = 0$.

14.15 $Z(s) = s + 6 + \frac{2(4/s)}{2+(4/s)} = \frac{s^2+8s+16}{s+2}$

$s = -2+j4$: $Z = \frac{4-16-j16-16+j32+16}{j4}$

$= 4 + j3$ Ω

$\mathbf{I} = \frac{10\angle 0^\circ}{4+j3} = 2\angle{-36.9^\circ}$ A

$i = 2e^{-2t}\cos(4t - 36.9^\circ)$ A

14.16 $\mathbf{V}_o = \mathbf{H}\mathbf{V}_1 = \frac{2(s+4)4}{s^2+2s+2} = \frac{8(s+4)}{(s+1)^2+1}$

$\mathbf{V}_o(-1+j3) = \frac{8(3+j3)}{(j3)^2+1} = -3\sqrt{2}\angle{45^\circ}$

$= 3\sqrt{2}\angle{-135^\circ}$ V

$v_o(t) = 3\sqrt{2}\,e^{-t}\cos(3t - 135^\circ)$ V

14.17 KCL at inverting input yields

$\frac{\mathbf{V}_g}{2+(4/s)} + \mathbf{V}(\frac{1}{16} + \frac{1}{16}s) = 0 \Rightarrow$

$H(s) = \frac{\mathbf{V}}{\mathbf{V}_g} = \frac{1}{(2+4/s)(\frac{1}{16}+\frac{1}{16}s)}$

$= \frac{-8s}{(s+2)(s+1)}$

For $s = 1$ and $\mathbf{V}_g = 3\angle 0^\circ$

$\mathbf{V} = \frac{-8(1)(3\angle 0^\circ)}{(3)(2)} = -4\angle 0^\circ$ V

$v = -4e^{t}$ V

14.18 $\frac{v}{2}$ = input voltage for VCVS

v_1 = voltage at common node of two 1Ω resistor and two $\frac{1}{2}$F - capacitor.

KCL yields:

$\mathbf{V}_1(1+1+\frac{s}{2}+\frac{s}{2}) - \frac{\mathbf{V}}{2}(\frac{s}{2}) - \mathbf{V} = \mathbf{V}_g$ or

$\mathbf{V}_1(8+4s) - \mathbf{V}(s+4) = 4\mathbf{V}_g$

$-\mathbf{V}_1(\frac{s}{2}) + \frac{\mathbf{V}}{2}(1+\frac{s}{2}) = 0$ or $\mathbf{V}_1 = \mathbf{V}\left(\frac{2+s}{2s}\right)$

substituting for $\mathbf{V}_1$

$\mathbf{V}(\frac{2+s}{2s})(8+4s) - \mathbf{V}(s+4) = 4\mathbf{V}_g$

$H(s) = \frac{V(s)}{V_g(s)} = \frac{4s}{s^2+4s+8}$

For $s = -2+j4$ and $\mathbf{V}_g = 3\angle 0^\circ$

$\mathbf{V} = \frac{4(-2+j4)(3)}{4-16-j16-8+j16+8} = -2\sqrt{5}\angle{116.6^\circ}$

$= 2\sqrt{5}\angle{-63.4^\circ}$

$v = 2\sqrt{5}\,e^{-2t}\cos(4t - 63.4^\circ)$ V

14.19 $\frac{v}{2}$ = input voltage for the VCVS

v_1 = voltage at common node of $\frac{1}{2}$Ω and 2Ω resistors

KCL yields

$2(\mathbf{V}_1 - \mathbf{V}_g) + \frac{1}{2}(\mathbf{V}_1 - \frac{\mathbf{V}}{2}) + s(\mathbf{V}_1 - \mathbf{V}) = 6$

$\frac{1}{2}(\frac{\mathbf{V}}{2} - \mathbf{V}_1) + \frac{1}{2}s\frac{\mathbf{V}}{2} = 0$

14.19 cont.

$$\mathbf{V} = \frac{8\mathbf{V}_g}{2s^2+3s+4} \Rightarrow H(s) = \frac{\mathbf{V}}{\mathbf{V}_g} = \frac{8}{2s^2+3s+4}$$

$$\mathbf{V}(-1+j) = \frac{8(1)}{2(-1+j)^2+3(-1+j)+4} = 4\sqrt{2}\angle 45^\circ \text{ V}$$

$$v = 4\sqrt{2}\, e^{-t}\cos(t+45^\circ) \text{ V}$$

14.20 With a-b open,

$$V_{oc} = \mathbf{V}_1 + \left(\frac{6}{s}+2\right)\left(\frac{\mathbf{V}_1}{2}\right) = (2+3/s)\mathbf{V}_1$$

where $\mathbf{V}_1 = 2(8\angle 0^\circ) = 16$ V. Therefore

$$V_{oc} = \frac{16(2s+3)}{s}$$

With a-b shorted,

$$I_{sc} = 8 - \frac{\mathbf{V}_1}{2} = \frac{\mathbf{V}_1}{2} + \frac{\mathbf{V}_1}{2+6/s}.$$

Therefore $\mathbf{V}_1 = \dfrac{16(s+3)}{3s+6}$ and

$$I_{sc} = \frac{8(2s+3)}{3(s+2)}$$

$$\therefore Z_{th} = \frac{V_{oc}}{I_{sc}} = \frac{16(2s+3)/s}{8(2s+3)/3(s+2)} = \frac{6(s+2)}{s}$$

$$\mathbf{I} = \frac{V_{oc}}{Z_{th}+s+10} = \frac{16(2s+3)/s}{\frac{6(s+2)}{s}+s+10} = \frac{16(2s+3)}{s^2+16s+12}$$

$$\mathbf{I}(-2+j4) = \frac{16(-1+j18)}{16(-2+j3)} = 2-j = \sqrt{5}\angle -26.6^\circ \text{ A}$$

$$\therefore i = \sqrt{5}\, e^{-2t}\cos(4t-26.6^\circ) \text{ A}$$

14.21

$$H(s) = \frac{K(s+4+j3)(s+4-j3)}{(s+3)(s+2+j)(s+2-j)}$$

$$H(0) = \frac{K(25)}{3(5)} = 5 \Rightarrow K = 3,$$

$$H(s) = \frac{3(s^2+8s+25)}{(s+3)(s^2+4s+5)}$$

14.22

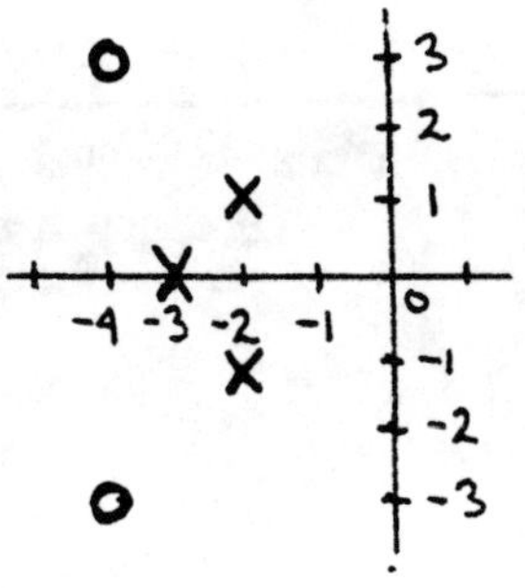

14.23 Using $\mathbf{Z}(s)$ from Prob 14.10

$$\frac{\mathbf{I}}{\mathbf{V}_g} = \frac{1}{\mathbf{Z}(s)} = \frac{s+1}{(s+2)^2}$$

zero at -1; poles at $-2, -2, \infty$.

14.24 From Prob 14.12

$$\frac{\mathbf{I}}{\mathbf{I}_g} = \frac{1}{s^2+6s+8} = \frac{1}{(s+4)(s+2)}$$

no zeros; poles at $-4, -2$.

14.25

$$H(s) = \frac{2(s+4)}{s^2+2s+2}$$

characteristic equation:

$$s^2+2s+2 = 0 \Rightarrow s_{1,2} = -1 \pm j$$

$$v_n = e^{-t}(A_1\cos t + A_2\sin t) \text{ V}$$

14.26 $v_{of} = 3\sqrt{2}\, e^{-t}\cos(3t-135^\circ)$

$$v_o = e^{-t}[A_1\cos t + A_2\sin t + 3\sqrt{2}\cos(3t-135^\circ)]$$

$$v_o(0^+) = 0 = A_1 + 3\sqrt{2}\cos(-135^\circ) = A_1 - 3 \Rightarrow A_1 = 3$$

$$\frac{dv_o}{dt}(0^+) = 0 = -A_1 - 3\sqrt{2}\cos(-135^\circ) + A_2 - 9\sqrt{2}\sin(-135^\circ) = -3+3+A_2+9 \Rightarrow A_2 = -9$$

$$v_o = e^{-t}[3\cos t - 9\sin t + 3\sqrt{2}\cos(3t-135^\circ)] \text{ V}$$

14.27 characteristic equation:

$$s^2+2s+2 = 0;\ s_{1,2} = -1 \pm j$$

$$v_{on} = e^{-t}(A_1\cos t + A_2\sin t)$$

For $s = -1+j2$ and $\mathbf{V}_i = 6\angle 0^\circ$

$$\mathbf{V}_{of} = \frac{4(-1+j2)(1+j2)(6)}{1-4-j4-2+j4+2} = 40\angle 0^\circ$$

$$v_{of} = 40\, e^{-t}\cos 2t$$

$$v_o = e^{-t}(A_1\cos t + A_2\sin t + 40\cos 2t)$$

$$v_o(0^+) = A_1 + 40 \Rightarrow A_1 = -40$$

$$\frac{dv_o}{dt}(0^+) = -A_1 - 40 + A_2 \Rightarrow A_2 = 0$$

$$v_o = e^{-t}(40\cos 2t - 40\cos t) \text{ V}$$

14.28 (a)

$$\mathbf{Z}(s) = 3s + \frac{(12)(6+2s)}{18+2s} = \frac{3(s^2+13s+12)}{s+9}$$

natural frequencies:

$$s^2+13s+12 = 0 \Rightarrow s = -1, -12$$

14.28 cont.

(b) $Y(s) = \frac{1}{12} + \frac{1}{3s} + \frac{1}{6+2s}$

$= \frac{s^2+13s+12}{6(s)(2s+6)}$

natural frequencies:

$s^2+13s+12=0 \Rightarrow \underline{s=-1,-12}$.

14.29 (a) $\underline{Z}(s) = 5s + \frac{4(20/s)}{4+(20/s)}$

$= \frac{5(s^2+5s+4)}{s+5}$

natural frequencies:

$s^2+5s+4=0 \Rightarrow \underline{s=-1,-4}$.

(b) $Y(s) = \frac{1}{4} + \frac{1}{5s} + \frac{s}{20}$

$= \frac{s^2+5s+4}{20s}$

natural frequencies:

$s^2+5s+4=0 \Rightarrow \underline{s=-1,-4}$.

14.30 $z_{11}-z_{12} = 6\Omega$, $z_{22}-z_{12} = 2\Omega$, $z_{12} = 2\Omega$

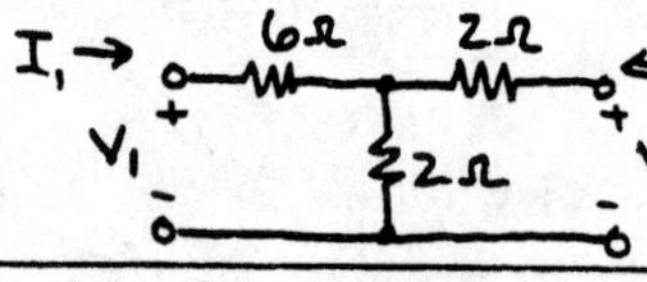

14.31 By (14.34),

$I_2 = -\frac{z_{21}}{z_{22}} I_1 + \frac{1}{z_{22}} V_2$,

$V_1 = z_{11} I_1 + z_{12}\left(-\frac{z_{21}}{z_{22}} I_1 + \frac{1}{z_{22}} V_2\right)$

$= \frac{\Delta}{z_{22}} I_1 + \frac{z_{12}}{z_{22}} V_2$ where

$\Delta = z_{11}z_{22} - z_{12}z_{21}$. Comparing these with the relations for the h-parameters gives

$h_{11} = \frac{\Delta}{z_{22}} = \frac{28}{4} = \underline{7\Omega}$, $h_{12} = \frac{z_{12}}{z_{22}} = \frac{2}{4} = \underline{1/2}$

$h_{21} = -\frac{z_{21}}{z_{22}} = \underline{-1/2}$, $h_{22} = \frac{1}{z_{22}} = \underline{\frac{1}{4}\,℧}$

Also by (14.34)

$I_1 = \frac{1}{z_{11}} V_1 - \frac{z_{12}}{z_{11}} I_2$,

$V_2 = z_{21}\left(\frac{1}{z_{11}} V_1 - \frac{z_{12}}{z_{11}} I_2\right) + z_{22} I_2$

$= \frac{z_{21}}{z_{11}} V_1 + \frac{\Delta}{z_{11}} I_2$, By compare

$g_{11} = \frac{1}{z_{11}} = \underline{\frac{1}{8}\,℧}$, $g_{12} = -\frac{z_{12}}{z_{11}} = -\frac{2}{8} = \underline{-\frac{1}{4}}$

$g_{21} = \frac{z_{21}}{z_{11}} = \underline{1/4}$, $g_{22} = \frac{\Delta}{z_{11}} = \frac{28}{8} = \underline{7/2\,\Omega}$

14.32 $z_{12} = 1/2s$

$z_{11} = (1+s) + z_{12} = \underline{\frac{2s^2+2s+1}{2s}}$

$z_{22} = s + z_{12} = \underline{\frac{2s^2+1}{2s}}$

14.33 $\Delta = z_{11}z_{22} - z_{12}z_{21}$

$= \frac{(2s^2+2s+1)}{2s}\frac{(2s^2+1)}{2s} - \left(\frac{1}{2s}\right)\left(\frac{1}{2s}\right)$

$= \frac{2s^3+2s^2+2s+1}{2s}$

$h_{11} = \frac{\Delta}{z_{22}} = \underline{\frac{2s^3+2s^2+2s+1}{2s^2+1}}\ \Omega$

$h_{12} = \frac{z_{12}}{z_{22}} = \underline{\frac{1}{2s^2+1}}$, $h_{21} = -\frac{z_{21}}{z_{22}} = \underline{-\frac{1}{2s^2+1}}$

$h_{22} = \frac{1}{z_{22}} = \underline{\frac{2s}{2s^2+1}}\ ℧$

14.34 From (14.34)

$I_1 = \frac{1}{z_{21}} V_2 - \frac{z_{22}}{z_{21}} I_2 = CV_2 - DI_2$

$V_1 = z_{11}\left(\frac{1}{z_{21}} V_2 - \frac{z_{22}}{z_{21}} I_2\right) + z_{12} I_2$

$= \frac{z_{11}}{z_{21}} V_2 - \frac{\Delta}{z_{21}} I_2 = AV_2 - BI_2$

$A = \frac{z_{11}}{z_{21}} = \underline{2s^2+2s+1}$

$B = \frac{\Delta}{z_{21}} = \underline{2s^3+2s^2+2s+1}\ \Omega$

$C = \frac{1}{z_{21}} = \underline{2s}\ ℧$

$D = \frac{z_{22}}{z_{21}} = \underline{2s^2+1}$

14.35 $I_2 = 0$ then from Prob 14.31

$I_2 = 0 = h_{21} I_1 + h_{22} V_2 \Rightarrow I_1 = -\frac{h_{22}}{h_{21}} V_2$

$V_1 = h_{11}\left(-\frac{h_{22}}{h_{21}} V_2\right) + h_{22} V_2$

$\frac{V_2}{V_1} = \frac{h_{21}}{h_{12}h_{21} - h_{11}h_{22}} = \frac{200}{(200)10^{-3} - 3(10^3)(10^{-3})}$

$= \underline{-71.43}$

14.36

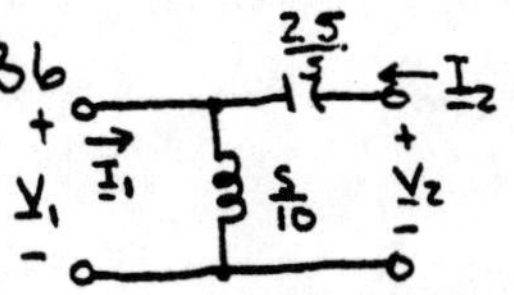

the loop equations are

$$\underline{V}_1 = \frac{s}{10}\underline{I}_1 + \frac{s}{10}\underline{I}_2,$$

$$\underline{V}_2 = \frac{s}{10}\underline{I}_1 + \left(\frac{s}{10} + \frac{25}{s}\right)\underline{I}_2,$$

$$\therefore z_{11} = \frac{s}{10},\ z_{12} = z_{21} = \frac{s}{10},\ z_{22} = \frac{s}{10} + \frac{25}{s}$$

The nodal equations are

$$\underline{I}_1 = \left(\frac{10}{s} + \frac{s}{25}\right)\underline{V}_1 - \frac{s}{25}\underline{V}_2,$$

$$\underline{I}_2 = -\frac{s}{25}\underline{V}_1 + \frac{s}{25}\underline{V}_2,$$

$$\therefore y_{11} = \frac{10}{s} + \frac{s}{25},\ y_{12} = y_{21} = -\frac{s}{25},\ y_{22} = \frac{s}{25}$$

14.37 $I_2 = -V_2$, substitute into (14.34)

$$V_2 = z_{21} I_1 - z_{22} V_2 \text{ or}$$

$$I_1 = \frac{(1 + z_{22})}{z_{21}} V_2$$

$$V_1 = (z_{11})\left(\frac{1 + z_{22}}{z_{21}}\right) V_2 - z_{12} V_2$$

$$\frac{V_2}{V_1} = \frac{z_{21}}{z_{11}(1 + z_{22}) - z_{12} z_{21}} = \frac{z_{21}}{z_{11} + \Delta}$$

$$= \frac{2}{8(1+4) - (2)(2)} = \frac{1}{18}$$

Chapter 15

Frequency Response

15.1 Amplitude and Phase Responses

15.1 Let $R = 1\text{k}\Omega$, $L = 1\text{mH}$, and $C = 10\mu\text{F}$, and find the maximum amplitude and the point where it occurs.

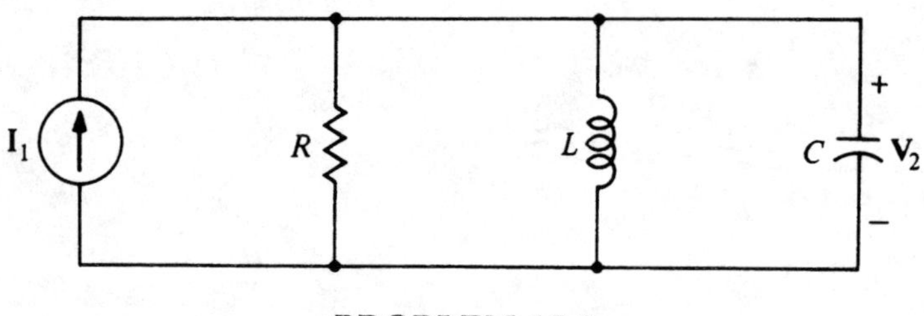

PROBLEM 15.1

15.2 For the circuit shown, $R = 20\Omega$, $L = 0.1\text{H}$, and $C = 0.001\text{F}$. If the input and output are V_1 and V_2, respectively, find the network function and show that the peak amplitude and zero phase occur at $\omega = 100$ rad/s.

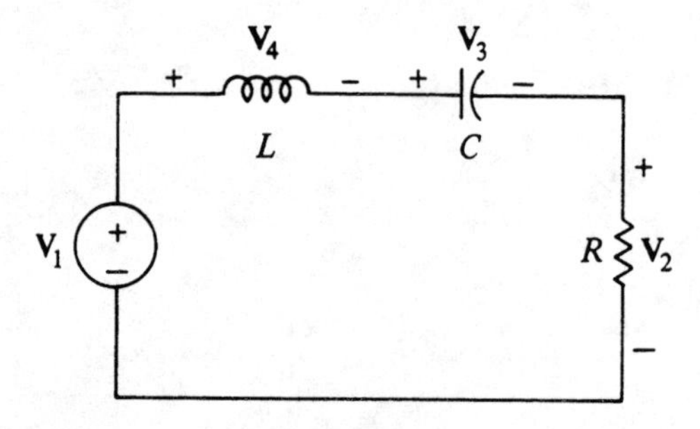

PROBLEM 15.2

15.3 Find $\mathbf{H}(s) = \mathbf{V}_2(s)/\mathbf{V}_1(s)$ and sketch the amplitude and phase responses. Show that the peak amplitude and zero phase occur at $\omega = 10$ rad/s.

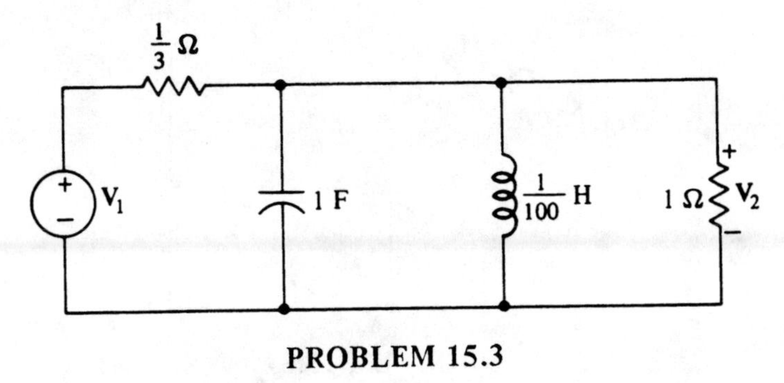

PROBLEM 15.3

15.4 For the circuit shown, $R_1 = R_2 = 0.1\Omega$ and $R_3 = 0.005\Omega$. If the input and output are $\mathbf{V}_1$ and $\mathbf{V}_2$, respectively, find the network function and sketch the amplitude and phase responses.

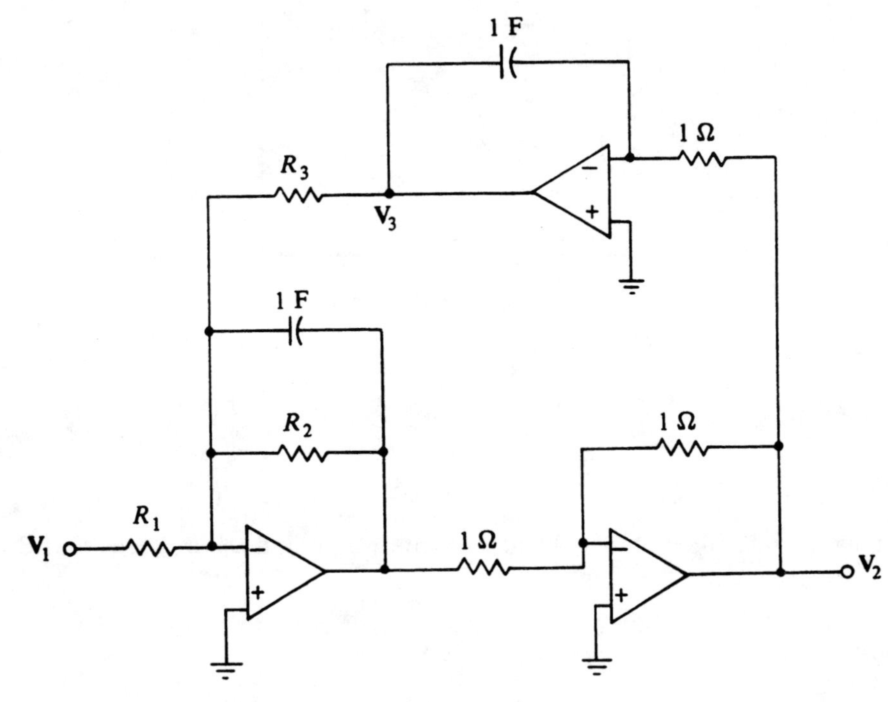

PROBLEM 15.4

15.5 For the circuit shown, find the network function, $\mathbf{H}(s) = \mathbf{V}_2(s)/\mathbf{V}_1(s)$, show that the peak amplitude and zero phase occur at $\omega = 0$.

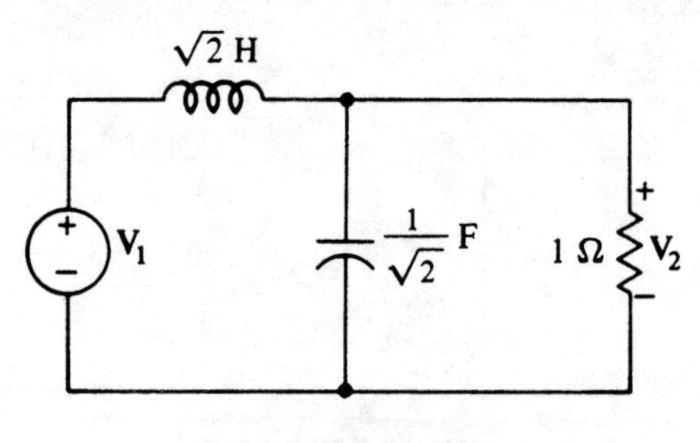

PROBLEM 15.5

15.6 For the circuit shown, $R = 1\Omega$, $L = \frac{1}{2}$H, and $C = 0.02$F. If the input and output are $\mathbf{V}_1$ and $\mathbf{V}_2$ respectively, find the network function and sketch the amplitude and phase responses. Show that the peak amplitude and zero phase occur at $\omega = 10$ rad/s.

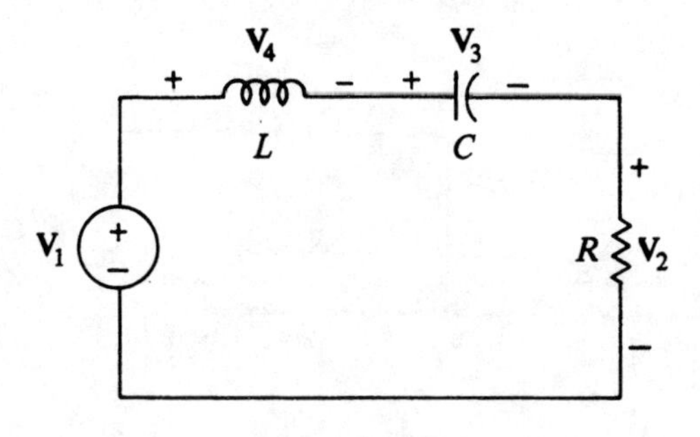

PROBLEM 15.6

15.2 Filters

15.7 Show that $$\mathbf{H}(s) = \frac{6s}{s^2 + s + 4}$$
is the network function of a bandpass filter, and find ω_0, ω_{c1}, ω_{c2}, and **B**.

15.8 Show that $$\mathbf{H}(s) = \frac{6s^2}{s^2 + s + 4}$$
is the network function of a high-pass filter, and find $|\mathbf{H}(j\omega)|_{max}$ and ω_c.

15.9 Show that $$\mathbf{H}(s) = \frac{6(s^2 + 4)}{s^2 + s + 4}$$
is the network function of a band-reject filter, and find $|\mathbf{H}(j\omega)|_{max}$, ω_0, ω_{c1}, and ω_{c2}.

15.10 Show that the given circuit is a low-pass filter with $\omega_c = 1$ rad/s by finding $\mathbf{H}(s) = \mathbf{V}_2(s)/\mathbf{V}_1(s)$ and the amplitude response.

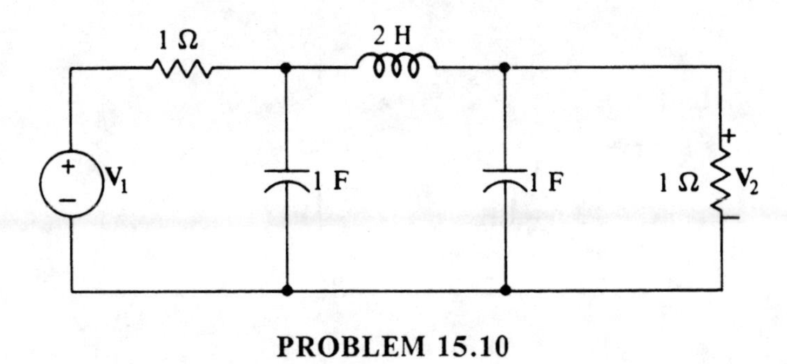

PROBLEM 15.10

15.11 Show by finding $\mathbf{V}_2/\mathbf{V}_1$ that the circuit is a bandpass filter, and find the gain, the bandwidth, and the center frequency.

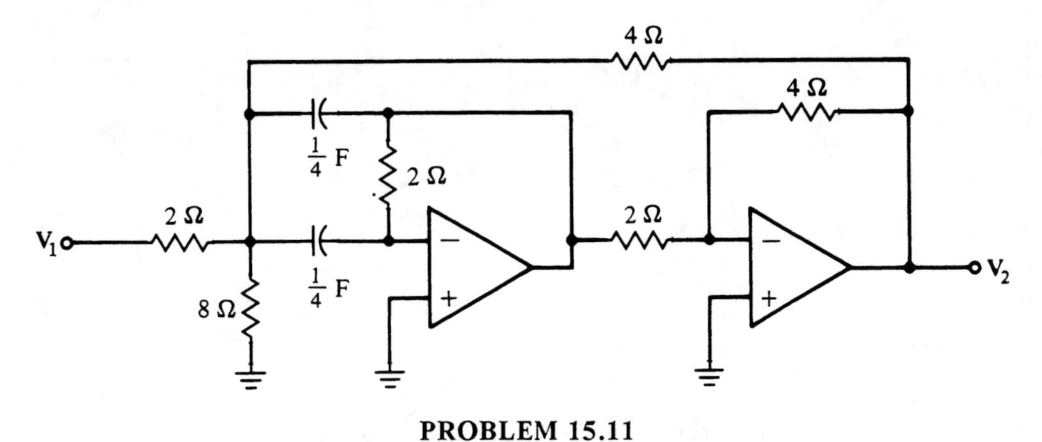

PROBLEM 15.11

15.3 Resonance

15.12 Show that the resonant frequency in the case of the function of Prob. 15.7 coincides with the frequency for which the function is real.

15.13 Find the resonant frequency for the parallel *RLC* circuit described by (a) $R = 5\Omega$, $L = 2$H, and $C = 2$F; (b) $\omega_d = 20$ rad/s and $\alpha = 10$Np/s; (c) $\alpha = 10$Np/s, $R = 20\Omega$, and $L = 2$H.

15.14 In Prob. 15.13 (a) find the amplitude of the voltage across the combination if the current source is $i_g = \cos(\omega t)$ mA and ω is (a) 0.05, (b) 0.5, and (c) 5 rad/s.

15.15 Find the resonance frequency and the frequency at which the network function $\mathbf{Y}(j\omega)$ is real.

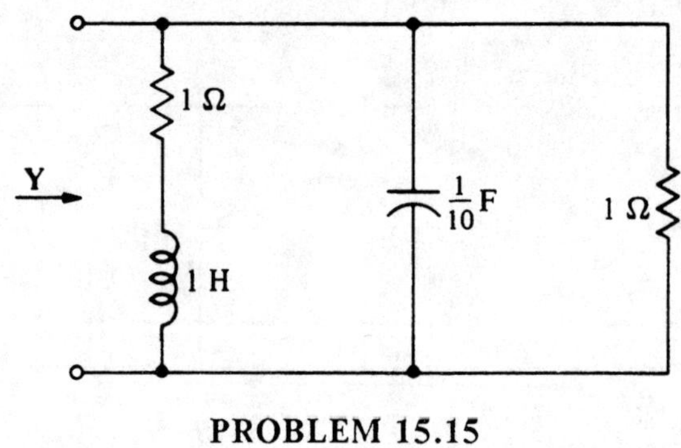

PROBLEM 15.15

15.4 Bandpass Functions and Quality Factor

15.16 Find Q in Prob. 15.1.

15.17 Find Q in Prob. 15.2.

15.5 Use of Pole-Zero Plots

15.18 Using pole zero plots find ω_0, ω_{c1}, ω_{c2}, and Q, B if $\mathbf{H}(s) = \dfrac{4s}{s^2 + 2s + 101}$.

15.19 Find the exact values for the answers given in Prob. 15.18.

15.6 Scaling the Network Function

15.20 Frequency- and impedance-scale the circuit to obtain $\omega_c = 400\pi$ rad/s, using capacitors of 2μF and 1μF.

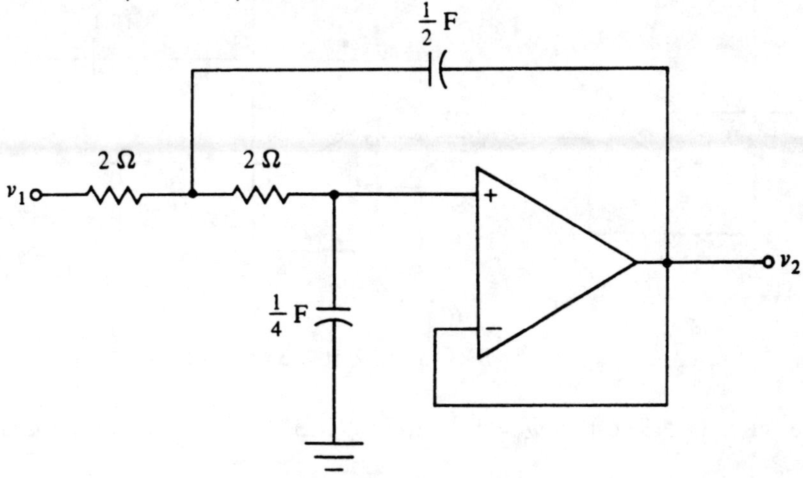

PROBLEM 15.20

15.21 Apply impedance and frequency scaling to find L and Q in Prob. 15.2 to obtain a bandpass filter with a center frequency $f_0 = 10^6 Hz$, and using a capacitor of 5 nF.

15.22 Determine R_1 and R_2 so that the circuit is a first-order low-pass filter ($\mathbf{H} = \mathbf{V}_2/\mathbf{V}_1$) with $\omega_c = 10^5$ rad/s and H(0) = -2.0. Use a 1 nF capacitor.

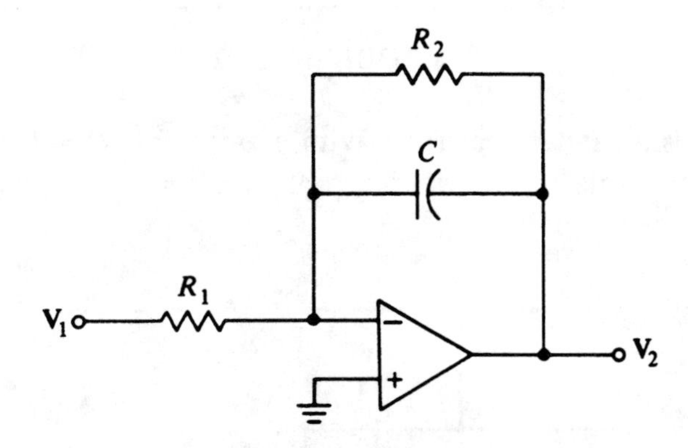

PROBLEM 15.22

15.23 The circuit is a third-order low-pass Butterworth filter with $\omega_c = 1$ rad/s and a gain of 1. Scale the circuit so that the capacitors are 1μF each and $\omega_c = 2\text{x}10^4$ rad/s.

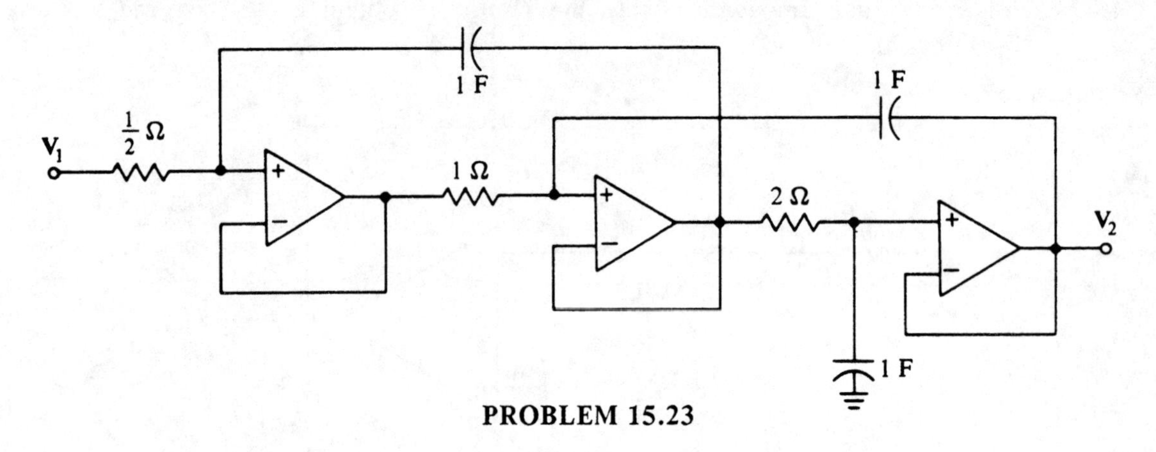

PROBLEM 15.23

15.24 Scale the network so that $\omega_0 = 10^6$ rad/s, $Q = 5$, $K = 0.5$, and the capacitance is 10 nF.

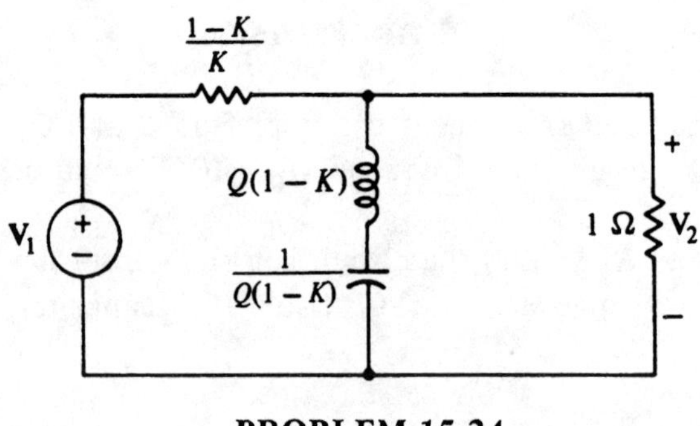

PROBLEM 15.24

15.25 The circuit is a band-reject filter with a gain of 1, $Q = 1$, and a center frequency $\omega_0 = 1$ rad/s. Scale the network to obtain a center frequency $f_0 = 1000$Hz using capacitances of 5 and 10 nF.

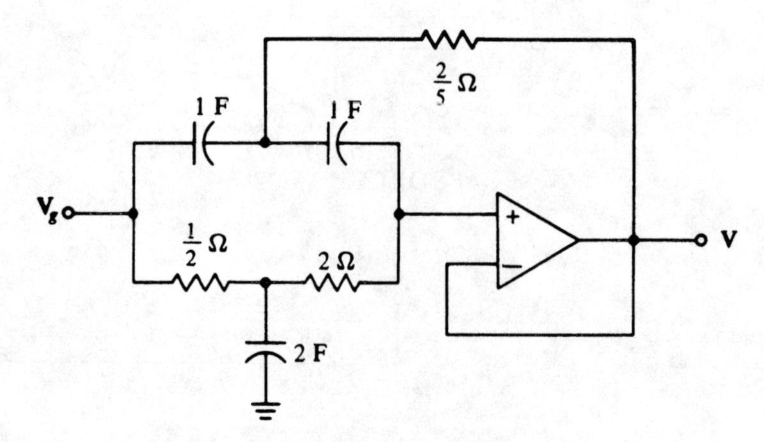

PROBLEM 15.25

15.26 Scale Prob. 15.1 to $\omega_{\text{peak}} = 1$ Krad/s, $C = 5$nF.

15.27 Scale Prob. 15.3 to $\omega_{\text{peak}} = 10^6$ rad/s, $L = 0.1$mH.

15.28 Scale Prob. 15.10 to $\omega_c = 2000$ rad/s, $R = 10\text{k}\Omega$.

15.29 Apply impedance and frequency scaling to the circuit of Prob. 15.6 to obtain a bandpass filter with a center frequency $f_0 = 10^5$ Hz, with a quality factor $Q = 5$, and using a capacitor of 1 nF.

15.7 The Decibel

15.30 Given the bandpass filter function $\mathbf{H}(s) = \dfrac{s}{s^2 + s + 4}$

find the loss in decibels at $\omega = 0.02$, 1.56, 2, 2.56, 20, and 200 rad/s.

15.31 Find dB of Prob. 15.2 at $\omega = 1$, 10, 100, 1000, and 10,000 rad/s.

15.32 Find dB of Prob. 15.27 at $\omega =$ 10k, 100k, 1M, 10M, and 100M rad/s.

15.33 Find dB of Prob. 15.10 at $\omega = 0$, 0.1, 1, 10, 100, and 1k rad/s.

15.1 $\omega_0 = 1/\sqrt{LC} = 1/\sqrt{(10^{-3})(10^{-5})} = \underline{10 \text{ Krad/s}}$

$|H(j\omega)|_{max} = |H(j\omega_0)| = R = \underline{1000}$.

15.2 $H(s) = \frac{V_2}{V_1} = \frac{R}{R + sL + 1/sC}$

$= \frac{s200}{s^2 + s200 + 10^4}$;

$|H(j\omega)| = \frac{R}{\sqrt{R^2 + (\omega L - \frac{1}{\omega C})^2}}$

$|H(j\omega)|_{max}$ occurs when

$\omega L - \frac{1}{\omega C} = 0 \therefore \omega_0 = \frac{1}{\sqrt{LC}} = 100 \text{ rad/s}$

$|H(j\omega_0)| = \underline{1} \Rightarrow \phi(\omega_0) = \underline{0^0}$

15.3 KCL yields

$V_2(1 + 3 + \frac{1}{s} + \frac{s}{100}) = V_1 3$

$H(s) = \frac{V_2}{V_1} = \frac{s300}{s^2 + s400 + 100}$.

$|H(j\omega)|^2 = \frac{9\times10^4 \omega^2}{(100-\omega^2)^2 + 16\times10^4\omega^2}$

$= \frac{9\times10^4}{(\frac{100-\omega^2}{\omega})^2 + 16\times10^4}$

$|H(j\omega)|_{max}$ occurs when

$100 - \omega^2 = 0 \Rightarrow \omega_0 = \underline{10 \text{ rad/s}}$

$H(j10) = \frac{j3000}{j4000} = \frac{3}{4} \Rightarrow \phi(10) = 0$

$|H(j\omega)|$, 3/4, 0, $10 = \omega_0$, ω

$\phi(\omega)$, $\pi/2$, 0, $\omega_0 = 10$, $-\pi/2$, ω

15.4 V_4 = output of lossy integrator

KCL at the inverting terminals gives:

$V_2 + V_4 = 0 \Rightarrow V_4 = -V_2$;

$V_2 + sV_3 = 0 \Rightarrow V_3 = -\frac{1}{s}V_2$;

$\frac{1}{R_1}V_1 + (s + \frac{1}{R_2})V_4 + \frac{1}{R_3}V_3 = 0$ or

$[(s + \frac{1}{R_2})(-1) + \frac{1}{R_3}(-\frac{1}{s})]V_2 = -\frac{1}{R_1}V_1$

$\therefore H(s) = \frac{V_2}{V_1} = \frac{1/R_1\, s}{s^2 + 1/R_2\, s + 1/R_3}$

$= \frac{10s}{s^2 + 10s + 200}$

$\omega_0 = 10\sqrt{2}$ rad/s, $H(j10\sqrt{2}) = 1$

Amplitude and phase respones are similiar to those in Prob. 15.3 with $\omega_0 = 10\sqrt{2}$ and $|H(j\omega)|_{max} = 1$.

15.5 KCL yields

$V_2(1 + s/\sqrt{2} + 1/s\sqrt{2}) = V_1/s\sqrt{2}$.

$H(s) = \frac{V_2}{V_1} = \frac{1}{s^2 + s\sqrt{2} + 1}$;

$|H(j\omega)| = \frac{1}{\sqrt{(1-\omega^2)^2 + 2\omega^2}} = \frac{1}{\sqrt{1+\omega^4}}$

$|H(j\omega)|_{max}$ occurs when $\omega = 0$

$H(0) = \frac{1}{1} \Rightarrow \phi(0) = 0^0$

15.6 $H(s) = \frac{V_2}{V_1} = \frac{R}{R + sL + 1/sC} = \frac{2s}{s^2 + 2s + 100}$

$|H(j\omega)| = \frac{R}{\sqrt{R^2 + (\omega L - 1/\omega C)^2}}$

$|H(j\omega)|_{max}$ occurs when

$\omega L - 1/\omega C = 0 \Rightarrow \omega_0 = 1/\sqrt{LC} = \underline{10 \text{ rad/s}}$

$|H(j\omega_0)| = 1 \Rightarrow \phi(\omega_0) = 0^0$

Plot of amplitude and phase responses are similiar to those in Prob. 15.3 with $\omega_0 = 10$ and $|H(j\omega)|_{max} = 1$.

15.7 $|H(j\omega)|^2 = \frac{36\omega^2}{(4-\omega^2)^2+\omega^2} = \frac{36}{1+\left(\frac{4-\omega^2}{\omega}\right)^2}$

$|H(j0)| = |H(j\infty)| = 0$; $\Rightarrow$ band pass

$|H(j\omega)|_{max} = |H(j2)| = 6 \therefore \omega_0 = 2$ rad/s

To find ω_{c1} and ω_{c2} set

$\frac{36}{2} = \frac{36}{1+\left(\frac{4-\omega^2}{\omega}\right)^2} \therefore \frac{4-\omega^2}{\omega} = \pm 1$:

$\omega^2 \pm \omega - 4 = 0 \Rightarrow \omega_{c1} = 1.56$ rad/s;

$\omega_{c2} = 2.56$ rad/s; $B = \omega_{c2} - \omega_{c1} = 1$ rad/s

15.8 $|H(j\omega)|^2 = \frac{36\omega^4}{(4-\omega^2)+\omega^2} = \frac{36}{1+\left(\frac{16-7\omega^2}{\omega^4}\right)}$

$|H(j0)| = 0$; $|H(j\infty)| = 6$; $\Rightarrow$ high pass

$|H(j\omega)|_{max} = |H(j\infty)| = 6$

$|H(j\omega_c)| = \left(\frac{6}{\sqrt{2}}\right)^2 = \frac{36}{1+\frac{16-7\omega^2}{\omega^4}}$

$\therefore 1 = \frac{16-7\omega^2}{\omega^4}$: $\omega^4 + 7\omega^2 - 16 = 0$

$\omega_c = 1.347$ rad/s.

15.9 $|H(j\omega)|^2 = \frac{36(4-\omega^2)^2}{(4-\omega^2)^2+\omega^2} = \frac{36}{1+\left(\frac{\omega}{4-\omega^2}\right)^2}$

$|H(j2)| = 0 \Rightarrow \omega_0 = 2$ rad/s

$|H(j0)| = |H(j\infty)| = 6 \Rightarrow$ band reject

ω_{c1} and ω_{c2} satisfy

$\frac{36}{2} = \frac{36}{1+\left(\frac{\omega}{4-\omega^2}\right)^2}$: $\pm 1 = \frac{\omega}{4-\omega^2}$

$\therefore \omega^2 \pm \omega - 4 = 0$: $\omega_{c1} = 1.56$ rad/s

$\omega_{c2} = 2.56$ rad/s

15.10 V_3 = left side node voltage

KCL gives

$V_2\left(1+s+\frac{1}{2s}\right) - V_3\left(\frac{1}{2s}\right) = 0$ or

$V_3 = V_2(2s^2+2s+1)$

$V_3\left(1+s+\frac{1}{2s}\right) - V_2\left(\frac{1}{2s}\right) = V_1$ or

$V_2\left[\frac{(2s^2+2s+1)(2s^2+2s+1)}{2s} - \frac{1}{2s}\right] = V_1$

$H(s) = \frac{V_2}{V_1} = \frac{2s}{(2s^2+2s+1)^2-1}$

$= \frac{1}{2s^3+4s^2+4s^2+2}$

15.10 cont.

$|H(j\omega)|^2 = \frac{1}{(2-4\omega^2)^2+(4\omega-2\omega^3)^2}$

$= \frac{1/4}{1+\omega^6}$

$|H(j0)| = \frac{1}{2}$; $H(j\infty) = 0$; $\Rightarrow$ low pass

$|H(j\omega)|_{max} = H(j0) = \frac{1}{2}$

$|H(j\omega_c)| = \frac{1/4}{2} = \frac{1/4}{1+\omega^6}$:

$\therefore \omega^6 = 1 \Rightarrow \omega_c = 1$ rad/s.

15.11 V_3 = output voltage of first opamp

V_4 = voltage across 8-Ω resistor

At inverting input of second op amp KCL gives

$\frac{V_2}{4} + \frac{V_3}{2} = 0 \Rightarrow V_3 = -\frac{1}{2}V_2$

At inverting input of first opamp KCL gives

$\frac{V_3}{2} + \frac{s}{4}V_4 = 0 \Rightarrow V_4 = -\frac{2}{s}V_3 = \frac{1}{s}V_2$

KCL at V_4:

$V_4\left(\frac{1}{2}+\frac{1}{8}+\frac{s}{4}+\frac{s}{4}+\frac{1}{4}\right) - \frac{s}{4}V_3 - \frac{1}{4}V_2 = \frac{1}{2}V_1$

$\left[\frac{4s+7}{8s}\left(\frac{1}{s}\right) - \frac{s}{4}\left(-\frac{1}{2}\right) - \frac{1}{4}\right]V_2 = \frac{1}{2}V_1$

$\therefore H(s) = \frac{V_2}{V_1} = \frac{4s}{s^2+2s+7}$

$K = 4$, $B = 2$, $\omega_0^2 = 7$,

$(a=2)$, $G = \frac{K}{a} = \frac{4}{2} = 2$

15.12 $H(j\omega) = \frac{j6\omega}{(4-\omega^2)+j\omega}$.

$H(j\omega)$ is real if $4-\omega^2 = 0$ or $\omega = 2$ rad/s, which is ω_0.

15.13 (a) $\omega_0 = 1/\sqrt{LC} = 1/\sqrt{2 \cdot 2} = 1/2$ rad/s

(b) $\omega_0 = \sqrt{\omega_d^2 + \alpha^2} = \sqrt{(20)^2+(10)^2}$

$= 10\sqrt{5}$ rad/s

(c) $\omega_0 = 1/\sqrt{LC}$; $\alpha = \frac{1}{2RC} \Rightarrow C = \frac{1}{2R\alpha}$

$C = \frac{1}{2(20)(10)} = 2.5$ mF

$\omega_0 = 1/\sqrt{2(2.5\times10^{-3})} = 10\sqrt{2}$ rad/s

15.14 $\underline{V} = \underline{I}g/\underline{Y} = \dfrac{10^{-3}}{\frac{1}{5} + \frac{1}{2s} + 2s}$

$$\underline{V}(j\omega) = \frac{10^{-3}}{\frac{1}{5} + j(2\omega - \frac{1}{2\omega})} = \frac{10^{-3}}{\frac{1}{5} + j\left(\frac{4\omega^2 - 1}{2\omega}\right)}$$

(a) $|\underline{V}(j0.05)| = \dfrac{10^{-3}}{\sqrt{(\frac{1}{5})^2 + \left(\frac{4(0.05)^2 - 1}{2(0.05)}\right)^2}}$

$= \underline{0.101 \text{ mV}}$

(b) $|\underline{V}(j0.5)| = \dfrac{10^{-3}}{\sqrt{(\frac{1}{5})^2 + \left(\frac{4(0.5)^2 - 1}{2(0.5)}\right)^2}}$

$= \underline{5 \text{ mV}}$

(c) $|\underline{V}(j5)| = \dfrac{10^{-3}}{\sqrt{(\frac{1}{5})^2 + \left(\frac{4(5)^2 - 1}{2(5)}\right)^2}}$

$= \underline{0.101 \text{ mV}}$

15.15 $Y(j\omega) = \dfrac{1}{1+j\omega} + \dfrac{j\omega}{10} + 1$

$= \dfrac{(20-\omega^2) + j11\omega}{10(1+j\omega)}$

$$|Y(j\omega)|^2 = \frac{(20-\omega^2)^2 + 121\omega^2}{100(1+\omega^2)}$$

maximum occurs when

$$\frac{d}{d\omega^2}|Y(j\omega)|^2 = 0 = \frac{d}{d\omega^2}(100)|Y(j\omega)|^2$$

$$(100)\frac{d}{d\omega^2}|Y(j\omega)|^2 = \frac{(2\omega^2+81)(\omega^2+1) - (\omega^4 + 81\omega^2 + 400)(1)}{(1+\omega^2)^2}$$

$$= \frac{\omega^4 + 2\omega^2 - 319}{(1+\omega^2)^2}$$

$\therefore \omega^4 + 2\omega^2 - 319 = 0$ and

$\omega_0 = \underline{4.11 \text{ rad/s}}$

To find real part.

$$Y(j\omega) = \frac{[(20-\omega^2) + j11\omega](1-j\omega)}{10(1+j\omega)(1-j\omega)}$$

$$= \frac{20 + 10\omega^2 + j(\omega^3 - 9\omega)}{10(1+\omega^2)}$$

real when $\omega^3 - 9\omega = 0$ or when $\omega = \underline{3 \text{ rad/s}}$

15.16 $Q = \omega_0 RC = R\sqrt{C/L} = 10^3\sqrt{\frac{10^{-5}}{10^{-3}}} = \underline{100}$

15.17 $Q = \omega_0 L/R = \frac{1}{R}\sqrt{\frac{L}{C}} = \frac{1}{20}\sqrt{\frac{0.1}{0.001}} = \underline{\frac{1}{2}}$

15.18 $s^2 + 2s + 101 = 0 \Rightarrow$ poles are $-1 \pm j10$

$\omega_0 \approx \underline{10 \text{ rad/s}}$, $M_1 = 1$, $M_2 \approx 20$, $N = 10$

$\omega_{c1} = N - M_1 = 10 - 1 = \underline{9 \text{ rad/s}}$

$\omega_{c2} = N + M_1 = 10 + 1 = \underline{11 \text{ rad/s}}$

$B = \omega_{c2} - \omega_{c1} = 11 - 9 = \underline{2 \text{ rad/s}}$

$Q = \dfrac{\omega_0}{B} = \dfrac{10}{2} = \underline{5}$

15.19 $\omega_0 = \sqrt{101} = \underline{10.05 \text{ rad/s}}$

B = coefficient of s in denominator $= \underline{2 \text{ rad/s}}$

$Q = \omega_0/B = 10.05/2 = \underline{5.025}$

$$\omega_{c_1, c_2} = \omega_0\left(\pm\frac{1}{2Q} + \sqrt{\left(\frac{1}{2Q}\right)^2 + 1}\right)$$

$\omega_{c1} = \underline{11.1 \text{ rad/s}}$, $\omega_{c2} = \underline{9.1 \text{ rad/s}}$

15.20 $\omega_c = \Omega_c k_f$; $k_f = \dfrac{\omega_c}{\Omega_c}$

$k_f = \dfrac{400\pi}{\sqrt{2}} = 200\pi\sqrt{2} = 888.6$

$C = \dfrac{C'}{k_i k_f} \Rightarrow k_i = \dfrac{C'}{C k_f} = \dfrac{1/2 F}{(2\mu F)(888.6)}$

$= 281.3$

$R = k_i R' = (281.3)(2) = \underline{562.7\,\Omega}$.

15.21 $k_f = \dfrac{\omega_c}{\Omega_c} = \dfrac{2\pi(10^6)}{100} = 2\pi(10^4)$

$k_i = \dfrac{C'}{C k_f} = \dfrac{0.001}{(5\times10^{-9})(2\pi)(10^4)} = \dfrac{10}{\pi}$

$R = k_i R' = \frac{10}{\pi}(20) = \frac{200}{\pi} = 63.66\,\Omega$

$L = \dfrac{k_i}{k_f} L' = \dfrac{10/\pi}{2\pi(10^4)}(0.1) = \underline{5.066\,\mu H}$

$Q = \omega_0 L/R = \dfrac{2\pi(10^6)(5.066\mu H)}{63.66\,\Omega} \approx \underline{\frac{1}{2}}$.

15.22 KCL at inverting input gives

$$\frac{V_1}{R_1} + V_2\left(\frac{1}{R_2} + sC\right) = 0$$

$$H(s) = \frac{V_2}{V_1} = \frac{-1/R_1}{1/R_2 + sC} = \frac{-R_2/R_1}{1 + sR_2C}$$

$H(0) = -R_2/R_1 = -2$, since

$$\frac{4}{2} = \frac{4}{1 + (\omega_c R_2 C)^2} \Rightarrow \omega_c R_2 C = 1$$

$R_2 = 1/\omega_c C = 1/(10^5)(10^{-9}) = \underline{10\text{K}\Omega}$

15.22 cont.

$R_1 = R_2/2 = 10\text{k}\Omega/2 = \underline{5\text{k}\Omega}$

15.23 $k_f = \frac{\omega_c}{\Omega_c} = 2\times10^4$

$k_i = \frac{C'}{Ck_f} = \frac{1}{(10^{-6})(2\times10^4)} = 50$

if $R_1' = \frac{1}{2}\Omega$, $R_2' = 1\Omega$ and $R_3' = 2\Omega$

then $R_1 = k_iR_1' = \underline{25\Omega}$,

$R_2 = k_iR_2' = \underline{50\Omega}$, $R_3 = k_iR_3' = \underline{100\Omega}$

15.24 KCL yields

$\underline{V}_2\left[1 + \frac{1}{Q(1-K)(s+\frac{1}{s})} + \frac{K}{1-K}\right] = \frac{K}{1-K}\underline{V}_1$

$\therefore H(s) = \frac{\underline{V}_2}{\underline{V}_1} = \frac{K(s^2+1)}{s^2+(1/Q)s+1}$; $\omega_0 = 1$ rad/s

$k_f = 10^6$; $k_i = \frac{C'}{Ck_f} = \frac{1/Q(1-K)}{(10^6)(10^{-8})}$

$= \frac{1}{5(0.5)(10^6)(10^{-8})} = 40$

$R' = \frac{1-K}{K} = 1\Omega \rightarrow \underline{40\Omega}$

$L' = Q(1-K) = 2.5\text{H} \rightarrow 2.5\frac{k_i}{k_f} = \underline{100\mu\text{H}}$

15.25 $k_f = 2\pi(10^3)$; $k_i = \frac{C'}{Ck_f} = \frac{2}{2\pi(10^3)(10^{-8})}$

$= 10^5/\pi$

$\frac{1}{2}\Omega \rightarrow \underline{50/\pi\ \text{k}\Omega}$; $2\Omega \rightarrow \underline{\frac{200}{\pi}\text{k}\Omega}$;

$2/5\Omega \rightarrow \underline{\frac{40}{\pi}\text{k}\Omega}$.

15.26 $k_f = 10^3/10^4 = 1/10$

$k_i = C'/Ck_f = 10^{-5}/(5\times10^{-9})(1/10) = 2\times10^4$

$R = k_iR' = (2\times10^4)(10^3) = \underline{20\text{M}\Omega}$

$L = \frac{k_i}{k_f}L' = \frac{2\times10^4}{10^{-1}}(10^{-3}) = \underline{200\text{H}}$

15.27 $k_f = 10^6/10 = 10^5$

$k_i = Lk_f/L' = (10^{-4})(10^5)/10^{-2} = 10^3$

$\frac{1}{3}\Omega \rightarrow \underline{\frac{1}{3}\text{k}\Omega}$; $1\Omega \rightarrow \underline{1\text{k}\Omega}$;

$1\text{F} \rightarrow \frac{1}{(10^5)(10^3)} = \underline{10\text{nF}}$.

15.28 $k_f = 2000$; $k_i = R/R' = 10^4$;

$2\text{H} \rightarrow \frac{10^4}{2000}(2) = \underline{10\text{H}}$,

$1\text{F} \rightarrow \frac{1}{(2000)(10^4)} = \underline{50\text{nF}}$.

15.29 $R' = 1\Omega$, $L' = Q = 5\text{H}$,

$C' = 1/Q = 0.2\text{H}$, $\Omega_0 = 1/\sqrt{LC} = 1$ rad/s

$k_f = 2\pi(10^5)$

$k_i = C'/Ck_f = 0.2/(10^{-9})(2\pi)10^5 = 318.3$

$R = k_iR' = \underline{318.3\Omega}$.

$L = \frac{k_i}{k_f}L' = \frac{318.3}{2\pi(10^5)} = \underline{2.53\text{mH}}$.

15.30 $|H(j\omega)| = 1/\sqrt{\left(\frac{\omega^2-4}{\omega}\right)^2+1}$,

$\alpha(\omega) = 10\log_{10}\left[1+\left(\frac{\omega^2-4}{\omega}\right)^2\right]$,

$\alpha(0.02) = \underline{46\text{dB}}$, $\alpha(1.56) = \underline{3\text{dB}}$

$\alpha(2) = 0\text{ dB}$, $\alpha(2.56) = \underline{3\text{dB}}$

$\alpha(20) = 26\text{dB}$, $\alpha(200) = \underline{46\text{dB}}$.

15.31 $|H(j\omega)| = 1/\sqrt{\left(\frac{\omega^2-10^4}{200\omega}\right)^2+1}$

$\alpha(\omega) = 10\log_{10}\left[1+\left(\frac{\omega^2-10^4}{200\omega}\right)^2\right]$

$\alpha(1) = \underline{34\text{dB}}$, $\alpha(10) = \underline{14\text{ dB}}$

$\alpha(100) = \underline{0\text{dB}}$, $\alpha(1\text{K}) = \underline{14\text{dB}}$

$\alpha(10\text{K}) = \underline{34\text{dB}}$.

15.32 $H(s) = \frac{3}{4+j(10^{-5}\omega - 10^7/\omega)}$

$|H(j\omega)| = 1/\sqrt{(4/3)^2+\left(\frac{10^{-5}\omega^2-10^7}{3\omega}\right)^2}$

$\alpha(\omega) = 10\log_{10}\left[(4/3)^2+\left(\frac{10^{-5}\omega^2-10^7}{3\omega}\right)^2\right]$

$\alpha(10\text{K}) = \underline{50\text{dB}}$, $\alpha(100\text{K}) = \underline{30\text{dB}}$

$\alpha(1\text{M}) = \underline{2.5\text{dB}}$, $\alpha(10\text{M}) = \underline{30\text{dB}}$

$\alpha(100\text{M}) = \underline{50\text{dB}}$.

15.33 $|H(j\omega)| = 1/\sqrt{4(1+\omega^6)}$

$\alpha(\omega) = 10\log_{10}\left[4(1+\omega^6)\right]$

$\alpha(0) = \underline{6\text{dB}}$, $\alpha(0.\ \ ) = \underline{6\text{dB}}$

$\alpha(1) = \underline{9\text{dB}}$, $\alpha(10) = \underline{66\text{dB}}$

$\alpha(100) = \underline{126\text{dB}}$, $\alpha(1\text{K}) = \underline{186\text{dB}}$

Chapter 16

Transformers

16.1 Mutual Inductance

16.1 In the circuit $L_1 = L_2 = 0.1$ H and $M = 10$ mH. Find v_1 and v_2 if $i_1 = 10$ mA and $i_2 = 0$.

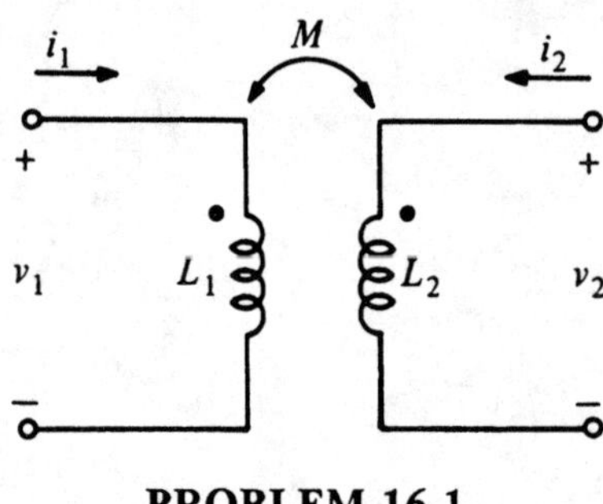

PROBLEM 16.1

16.2 In the circuit in Prob. 16.1, $L_1 = L_2 = 0.1$ H and M = 10 mH. Find v_1 and v_2 if $i_1 = 0$, and $i_2 = 10\sin(100t)$ mA.

16.3 Find $\mathbf{I}_2$ for a sinusoidal voltage of $\mathbf{V}_1 = 26\angle 0°$ V having $\omega = 2$ rad/s if the 2Ω resistor is replaced by the series combination of a 12Ω resistor and a 0.1F capacitor.

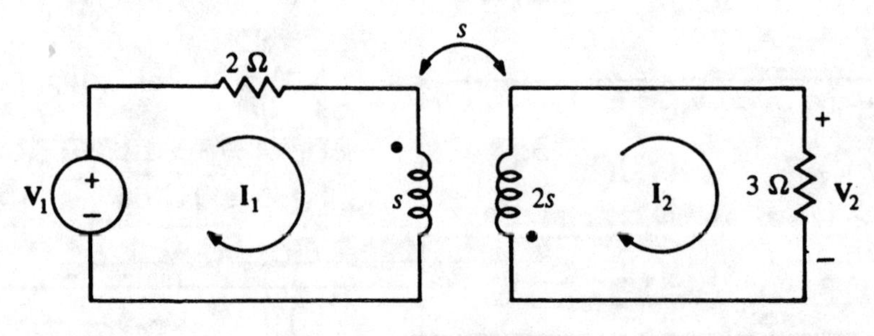

PROBLEM 16.3

16.4 If $N_2 = 500$ turn and $\phi_{21} = 30\mu$Wb when $i_1 = 4$A. Determine v_2 if $i_1 = 5\cos(10t)$ A.

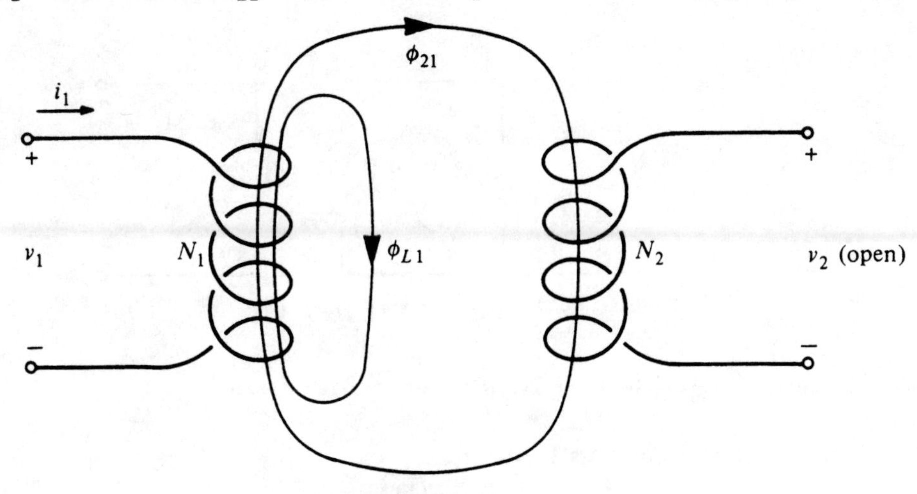

PROBLEM 16.4

16.5 If the inductance measured between terminals a and d is 4H when terminals b and c are connected, and the inductance measured between terminals a and c is 10H when terminals b and d are connected, find the mutual inductance M between the two coils, and the position of the dots.

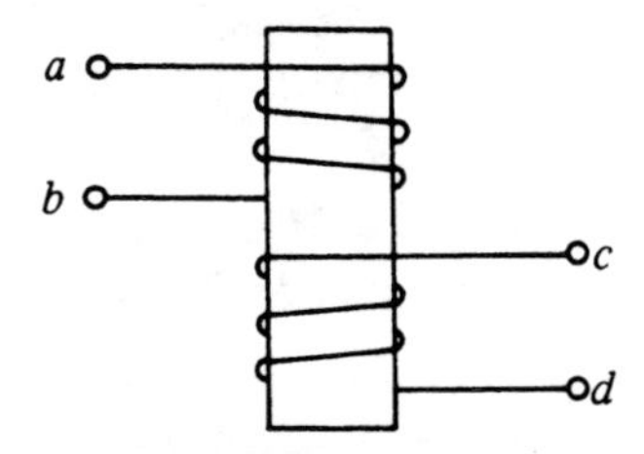

PROBLEM 16.5

16.6 Find v_1 and v_2 if $L_1 = 6$H, $L_2 = 4$H, $M = 2$H, $i_1 = \sin(t)$ A, and $i_2 = -2\cos(2t)$ A.

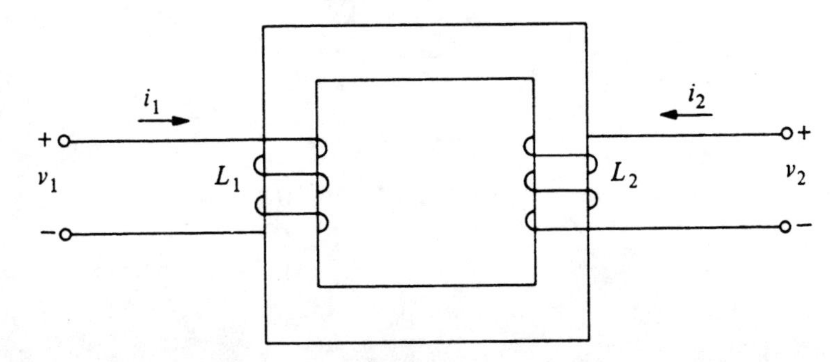

PROBLEM 16.6

16.7 Find v for $t > 0$ across the open circuit, if $i = 4u(t)$ A.

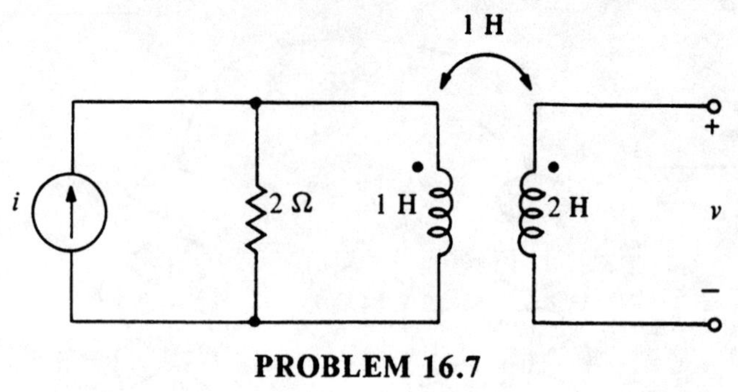

PROBLEM 16.7

16.8 Repeat Prob. 16.7 if $i(t) = 2e^{-t}u(t)$ A.

16.9 Find the phasor currents $\mathbf{I}_1$ and $\mathbf{I}_2$.

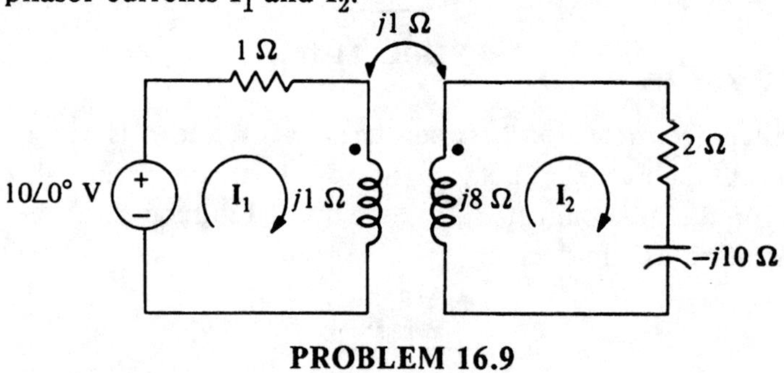

PROBLEM 16.9

16.10 Find v_1 and v_2 if $L_1 = 2$H, $L_2 = 5$H, $M = 3$H, and the currents i_1 and i_2 are changing at the rates -10 A/s and -2 A/s, respectively.

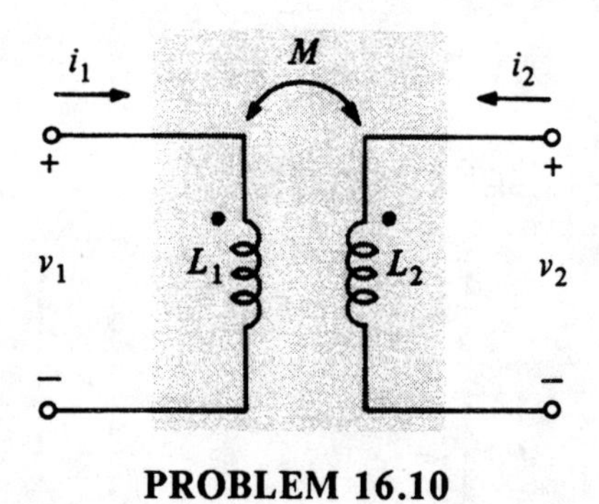

PROBLEM 16.10

16.11 Find v for $t > 0$ if $v_g = 4u(t)$ V.

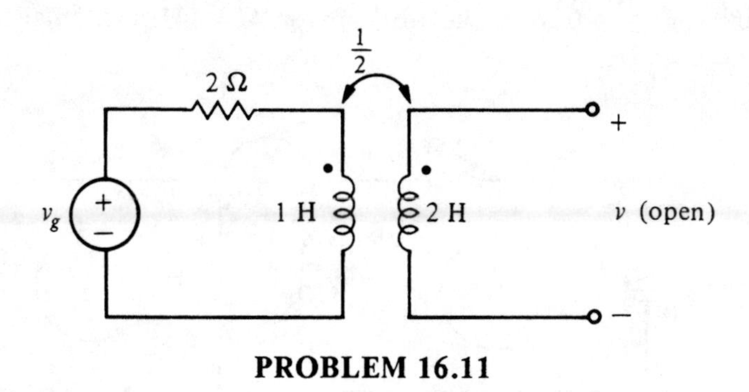

PROBLEM 16.11

16.2 Energy Storage

16.12 Determine the energy stored in Prob. 16.1 at $t = 0$.

16.13 Determine the energy stored in Prob. 16.2 at $t = 0$.

16.14 Find the energy stored in the transformer of Prob. 16.9 at $t = 0$ if the frequency is $\omega = 10$ rad/s.

16.15 (a) Find the energy stored in the transformer at a time when $i_1 = 3$A and $i_2 = 2$A if $L_1 = 1$H, $L_2 = 8$H, and $M = 3$H. (b) Repeat part (a) if one of the dots is moved to another terminal.

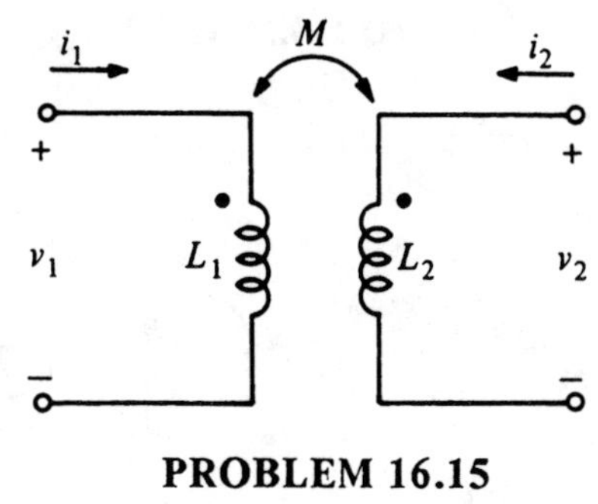

PROBLEM 16.15

16.3 Circuits with Linear Transforms

16.16 Determine i_1 for $t > 0$ in the network, given $M = \frac{1}{2}$H. Assume the circuit is in steady state at $t = 0^-$.

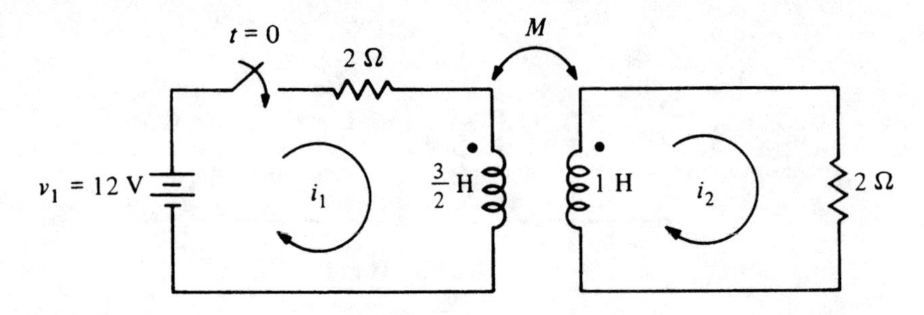

PROBLEM 16.16

16.17 Find the forced response v_2 if $v_1 = 3e^{-t}\cos(3t)$ V.

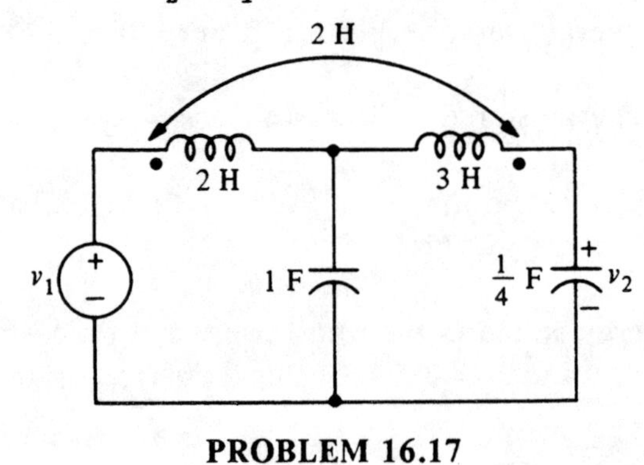

PROBLEM 16.17

16.18 Find the steady-state currents i_1 and i_2 if (a) M = 1H and R = 2Ω, and (b) M = 2H and R = 15/8 Ω.

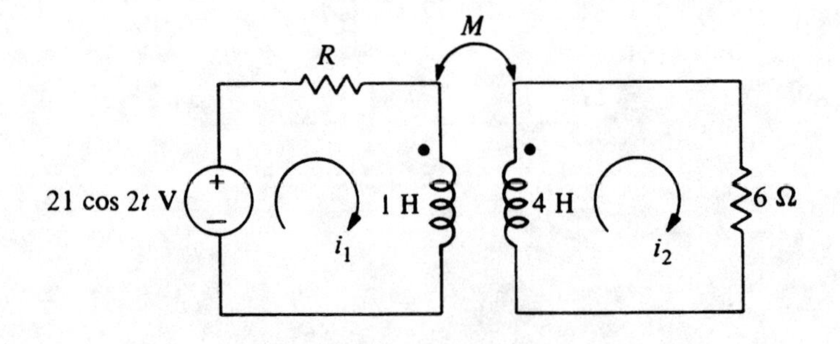

PROBLEM 16.18

16.4 Reflected Impedance

16.19 Given $V_g = 20\angle 0°V$, $\mathbf{Z}_g = 20\Omega$, $L_1 = 1$H, $L_2 = 0.5$H, $M = 0.5$H, and $\omega = 10$ rad/s. If $\mathbf{Z}_2 = -(j10/\omega)\Omega$, find (a) $\mathbf{Z}_{in}$, (b) $\mathbf{I}_1$, (c) $\mathbf{I}_2$, (d) $\mathbf{V}_1$, and (e) $\mathbf{V}_2$.

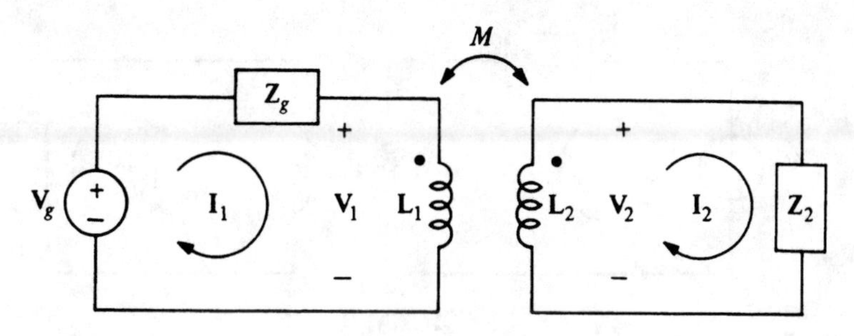

PROBLEM 16.19

16.20 Repeat Prob. 16.19 if the polarity dot is on the lower terminal of the secondary.

16.21 If $\mathbf{Z}_2 = (10 - j10/\omega)\Omega$ in Prob. 16.19, find the frequency for which the reflected impedance is real.

16.22 Find the steady-state current i_1 in Prob. 16.18(a) using reflected impedance.

16.23 Find the steady-state current i_2 in Prob. 16.18(a) by replacing everything in the corresponding phasor circuit except the 6Ω resistor by its Thevenin equivalent circuit.

16.24 Find the steady-state current i_1 using reflected impedance.

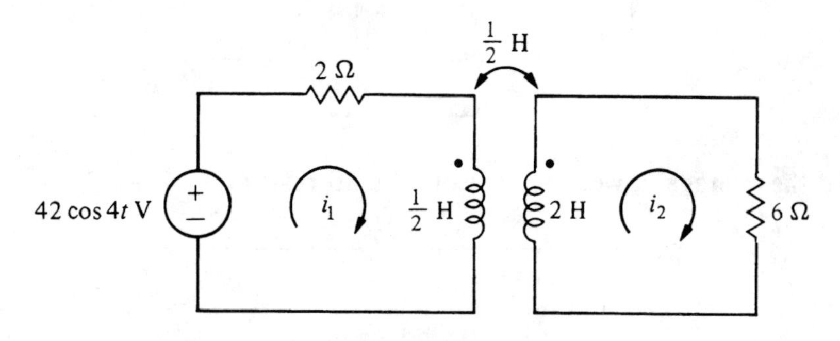

PROBLEM 16.24

16.25 Find the steady-state current i_2 in Prob. 16.24 by replacing everything in the corresponding phasor circuit except the 6Ω resistor by its Thevenin equivalent circuit.

16.5 The Ideal Transformer

16.26 $\mathbf{V}_g = 100\angle 0°$ V, $\mathbf{Z}_g = 100\Omega$, and $\mathbf{Z}_2 = 90\text{k}\Omega$. Find n such that $\mathbf{Z}_1 = \mathbf{Z}_g$, and then find the power delivered to $\mathbf{Z}_2$.

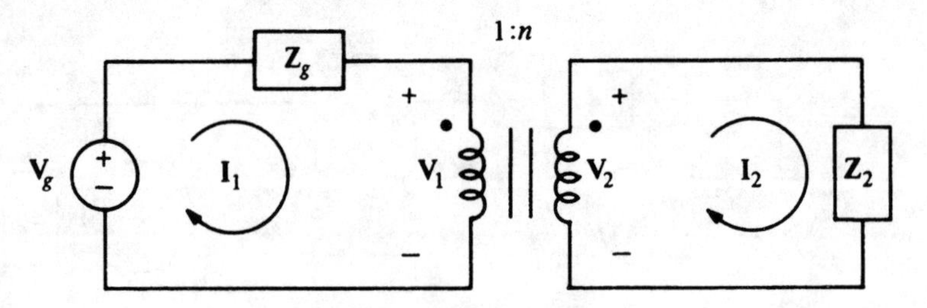

PROBLEM 16.26

16.27 Find the steady-state current i.

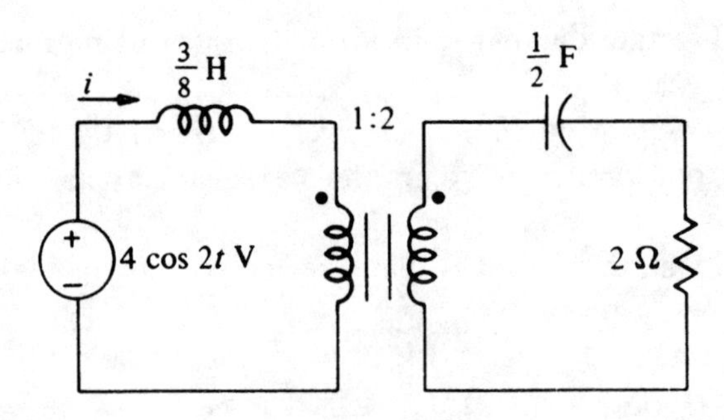

PROBLEM 16.27

16.28 Find the average power delivered to the 8Ω resistor.

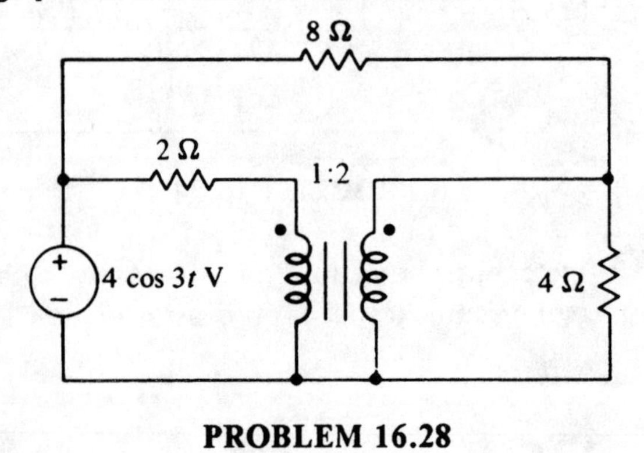

PROBLEM 16.28

16.29 Find $\mathbf{V}_1$, $\mathbf{V}_2$, $\mathbf{I}_1$ and $\mathbf{I}_2$ using the method of reflected impedance.

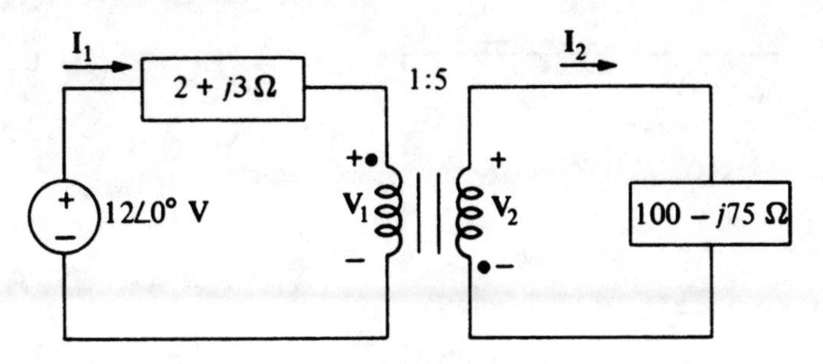

PROBLEM 16.29

16.30 Repeat Prob. 16.29 using the method of the figure below.

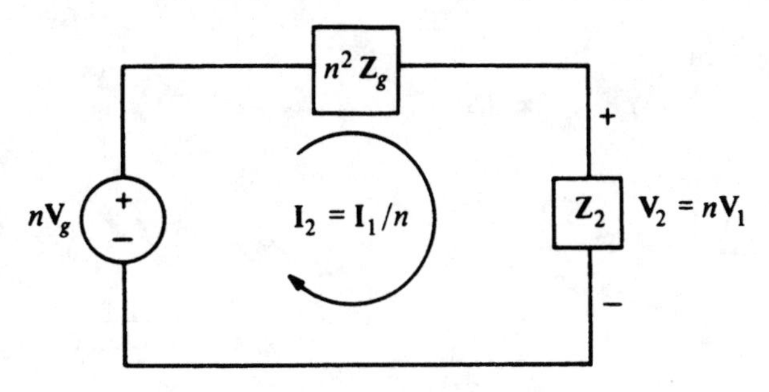

PROBLEM 16.30

16.31 Find the turns ratio n so that the maximum power is delivered to the 10kΩ resistor.

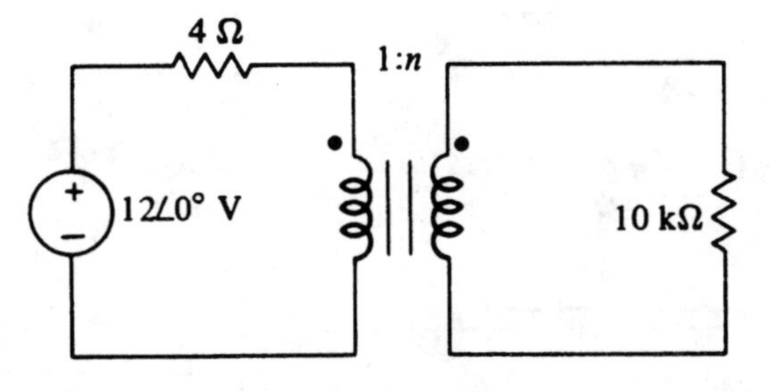

PROBLEM 16.31

16.6 Equivalent Circuits

16.32 Employing the T circuit for the linear transformer in Prob. 16.18(a) determine the current i_2.

16.33 Find $\mathbf{I}_2$ in Prob. 16.9 using the equivalent T circuit.

16.34 Find the equivalent T circuit for Prob. 16.7.

16.1 $\frac{di_1}{dt} = \frac{di_2}{dt} = 0 \therefore v_1 = v_2 = 0$

16.2 $v_1 = L_1\frac{di_1}{dt} + M\frac{di_2}{dt}$

$= (0.1)(0) + (10^{-2})(10^3\cos 100t)$ mA

$= 10\cos 100t$ mV

$v_2 = M\frac{di_1}{dt} + L_2\frac{di_2}{dt}$

$= 0 + (0.1)(10^3\cos 100t$ mA$)$

$= 100\cos 100t$ mV.

16.3 KVL yields

$(12 + s + 10/s)I_1 + sI_2 = V_1$

$sI_1 + (3+2s)I_2 = 0$

$$I_2 = \frac{\begin{vmatrix} 12-j3 & 26 \\ j2 & 0 \end{vmatrix}}{\begin{vmatrix} 12-j3 & j2 \\ j2 & 3+j4 \end{vmatrix}} = \frac{-j52}{52+j39}$$

$= 0.8\angle -126.9^\circ$ A

16.4 $M_{21} = \frac{N_2\phi_{21}}{i_1} = \frac{(500)(30\times10^{-6})}{4}$

$= 3.75$ mH

$v_2 = M_{21}\frac{di_1}{dt} = (3.75m)\frac{d}{dt}(5\cos 10t)$

$= -0.1875\sin 10t$ V

16.5 with b & c connected

$v_{ad} = (L_{ab} + L_{cd} - 2M)\frac{di}{dt} = L_{ad}\frac{di}{dt}$

with b & d connected

$v_{ac} = (L_{ab} + L_{cd} + 2M)\frac{di}{dt} = L_{ac}\frac{di}{dt}$

$\therefore L_{ab} + L_{cd} - 2M = L_{ad} = 4$ H

$L_{ab} + L_{cd} + 2M = L_{ac} = 10$ H

subtracting equations

$4M = 10 - 4 \Rightarrow M = 3/2$ H

16.6 $v_1 = L_1\frac{di_1}{dt} - M\frac{di_2}{dt}$

$= 6\cos t - 2(4\sin 2t)$

$= 6\cos t - 8\sin 2t$ V.

$v_2 = -M\frac{di_1}{dt} + L_2\frac{di_2}{dt} = (-2)\cos t + 4(4\sin 2t)$

$= -2\cos t + 16\sin 2t$ V.

16.7 v_1 = voltage across the primary coil

$v_1 = \frac{di_1}{dt}$; $i = \frac{v_1}{2} + i_1 \Rightarrow \frac{di_1}{dt} + 2i_1 = 2i$

$v = \frac{di_1}{dt}$; $\frac{di_1}{dt} + 2i_1 = 8u(t) = 8$ for $t>0$

$i_1 = A_1e^{-2t} + 4$; $i_1(0^+) = i(0^-) = 0$

$i_1(0) = A_1 + 4 \Rightarrow A_1 = -4$

$v = \frac{d}{dt}(-4e^{-2t} + 4) = 8e^{-2t}$ V for $t>0$

16.8 $\frac{di_1}{dt} + 2i = 4e^{-t}u(t) = 4e^{-t}$ for $t>0$

$i_{1f} = Ae^{-t} \Rightarrow -A + 2A = 4 \Rightarrow A = 4$

$i_1 = A_1e^{-2t} + 4e^{-t}$; $i(0) = i_1(0^-) = 0$

$i_1 = A_1 + 4 \Rightarrow A_1 = -4$

$v = \frac{d}{dt}(4e^{-t} - 4e^{-2t}) = 8e^{-2t} - 4e^{-t}$ V for $t>0$.

16.9 KVL yields

$(1+j)I_1 - jI_2 = 10$

$-jI_1 + (2 + j8 - j10)I_2 = 0$

$$I_1 = \frac{\begin{vmatrix} 10 & -j \\ 0 & 2-j2 \end{vmatrix}}{\begin{vmatrix} 1+j & -j \\ -j & 2-j2 \end{vmatrix}} = \frac{20-j20}{5} = 4-j4\text{ A};$$

$$I_2 = \frac{\begin{vmatrix} 1+j & 10 \\ -j & 0 \end{vmatrix}}{5} = j2\text{ A}.$$

16.10 $v_1 = 2(-10) + 3(-2) = -26$ V

$v_2 = 3(-10) + 5(-2) = -40$ V

16.11 v_1 = voltage across primary coil

$v_1 = \frac{di_1}{dt} \Rightarrow v_g = 2i_1 + \frac{di_1}{dt}$; $v = \frac{1}{2}\frac{di_1}{dt}$

$\frac{di_1}{dt} + 2i_1 = 4u(t) \Rightarrow \frac{di_1}{dt} + 2i_1 = 4$ for $t>0$.

$i_1 = A_1e^{-2t} + 2$; $i(0^+) = i(0^-) = 0$

$i_1(0^-) = A_1 + 2 \Rightarrow A_1 = -2$

$v = \frac{1}{2}\frac{d}{dt}(2 - 2e^{-2t}) = 2e^{-2t}$ V for $t>0$

16.12 $W = \frac{1}{2}L_1i_1^2 + Mi_1i_2 + \frac{1}{2}L_2i_2^2$

at $t=0$, $i_1 = 10$ mA, $i_2 = 0$

$W(0) = \frac{1}{2}(0.1)(10^{-2})^2 = 5\ \mu$J

16.13 at $t=0$ $i_1=0$, $i_2=0$,

$W(0) = \underline{0\ J}$.

16.14 $\mathbf{I}_1 = 4 - j4 = 4\sqrt{2}\angle -45^\circ$

$i_1 = 4\sqrt{2}(10t - 45^\circ)$, $i_1(0) = 4$

$\mathbf{I}_2 = j2$, $i_2 = 2\cos(10t + 90^\circ)$,

$i_2(0) = 0$.

$W(0) = \frac{1}{2}L_1 i_1^2(0) = \frac{1}{2}(\frac{1}{10})(4)^2$

$= \underline{0.8\ J}$ $[\omega L = 10L = 1]$

16.15 (a) $W = \frac{1}{2}L_1 i_1^2 + M i_1 i_2 + \frac{1}{2}L_2 i_2^2$

$= \frac{1}{2}(1)(3)^2 + (3)(3)(2) + \frac{1}{2}(8)(2)^2$

$= \underline{38.5\ J}$

(b) $W = \frac{1}{2}L_1 i_1^2 - M i_1 i_2 + \frac{1}{2}L_2 i_2^2$

$= \underline{2.5\ J}$

16.16 $V_1(s) = (\frac{3}{2}s + 2)\mathbf{I}_1 + \frac{s}{2}\mathbf{I}_2$

$0 = -\frac{s}{2}\mathbf{I}_1 + (s+2)\mathbf{I}_2$ or

$\mathbf{I}_2 = \frac{s}{2(s+2)}\mathbf{I}_1$, substitution

$H(s) = \frac{\mathbf{I}_1}{V_1} = \frac{4(s+2)}{5s^2 + 20s + 16}$

$= \frac{4/5(s+2)}{(s+1.11)(s+2.89)}$

$i_{1f} = H(0)V_1 = \frac{8}{16}(12) = 6$

$i_1 = 6 + A_1 e^{-1.11t} + A_2 e^{-2.89t}$

$i_1(0^+) = i_1(0^-) = 0 = 6 + A_1 + A_2$

From KVL

$\frac{3}{2}\frac{di_1(0^+)}{dt} - \frac{1}{2}\frac{di_2(0^+)}{dt} + 2i_1(0^+) = 12$

$-\frac{1}{2}\frac{di_1(0^+)}{dt} + \frac{di_2(0^+)}{dt} + 2i_2(0^+) = 0$

since $i_1(0^+) = i_2(0^+) = 0$ we obtain

$\frac{di_1(0^+)}{dt} = \frac{2di_2(0^+)}{dt} = \frac{4}{5}(12) = 9.6$

$\frac{di_1(0^+)}{dt} = 9.6 = -1.11A_1 - 2.89A_2$

$\therefore A_1 = -4.34$ & $A_2 = -1.66$

$\underline{i_1 = 6 - 4.34e^{-1.11t} - 1.66e^{-2.89t}\ A}$

16.17 Let i_1 and i_2 be clockwise mesh currents then

$V_1 = (2s + 1/s)\mathbf{I}_1 - (2s + \frac{1}{s})\mathbf{I}_2$

$0 = -(2s + 1/s)\mathbf{I}_1 + (3s + \frac{1}{s} + \frac{4}{s})\mathbf{I}_2$

$V_2 = \frac{4}{s}\mathbf{I}_2$

$\therefore \mathbf{I}_2 = \frac{sV_1}{s^2+4}$, $V_2 = \frac{4V_1}{s^2+4}$,

$s = -1 + j3$, $V_1 = 3$,

$V_2 = \frac{4(3)}{(-1+j3)^2 + 4} = 1.66\angle 123.7^\circ$

$v_{2f} = \underline{1.66e^{-t}\cos(2t + 123.7^\circ)\ V}$

16.18 $21 = (R + j2)\mathbf{I}_1 - j2M\mathbf{I}_2$

$0 = -j2M\mathbf{I}_1 + (6 + j8)\mathbf{I}_2$

(a)
$$\mathbf{I}_1 = \frac{\begin{vmatrix} 21 & -j2(1) \\ 0 & 6+j8 \end{vmatrix}}{\begin{vmatrix} 2+j2 & -j2 \\ -j2 & 6+j8 \end{vmatrix}} = \frac{21(6+j8)}{j28}$$

$= 7.5\angle -36.87^\circ$

$i_1 = \underline{7.5\cos(2t - 36.87^\circ)}$ A

$$\mathbf{I}_2 = \frac{\begin{vmatrix} 2+j2 & 21 \\ -j2 & 0 \end{vmatrix}}{j28} = \frac{j42}{j28} = 1.5$$

$i_2 = \underline{1.5\cos 2t\ A}$

(b)
$$\mathbf{I}_1 = \frac{\begin{vmatrix} 21 & -j4 \\ 0 & 6+j8 \end{vmatrix}}{\begin{vmatrix} 15/8+j2 & -j4 \\ -j4 & 6+j8 \end{vmatrix}} = \frac{21(6+j8)}{11.25+j27}$$

$= 7.18\angle -14.25^\circ$

$i_1 = \underline{7.18\cos(2t - 14.25^\circ)\ A}$

$$\mathbf{I}_2 = \frac{\begin{vmatrix} 15/8+j2 & 21 \\ -j4 & 0 \end{vmatrix}}{11.25+j27} = \frac{j84}{11.25+j27}$$

$= 2.87\angle 22.62^\circ$

$i_2 = \underline{2.87\cos(2t + 22.62^\circ)\ A}$

16.19 $Z_1 = j\omega L_1 + \dfrac{\omega^2 M^2}{Z_2 + j\omega L_2}$

$= j(10 - \frac{25}{4}) = j3.75\ \Omega$

(a) $Z_{in} = Z_g + Z_1 = \underline{20 + j3.75\ \Omega}$

(b) $I_1 = V_g / Z_{in} = \dfrac{20}{20 + j3.75} = 0.97 - j0.18$

$= \underline{0.983\angle -10.62^\circ \text{ A}}$

(c) $I_2 = \dfrac{j\omega M}{Z_2 + j\omega L_2} I_1 = \dfrac{j5(0.97 - j0.18)}{-j + j5}$

$= 1.21 - j0.23 = \underline{1.23\angle -10.6^\circ \text{A}}$

(d) $V_1 = Z_1 I_1 = (j3.75)(0.97 - j0.18)$

$= \underline{3.686\angle 79.38^\circ \text{ V}}$

(e) $V_2 = Z_2 I_2 = (-j)(1.23\angle -10.62^\circ)$

$= \underline{0.983\angle -100.62^\circ \text{V}}$

16.20 Z_1 is the same as in Prob. 16.19

(a) $Z_{in} = \underline{20 + j3.75\ \Omega}$

(b) $I_1 = \underline{0.983\angle -10.62^\circ \text{ A}}$

(c) $I_2 = \dfrac{(-j\omega M)}{Z_2 + j\omega L_2} = \underline{1.23\angle 169.4^\circ \text{ A}}$

(d) $V_1 = \underline{3.686\angle 79.38^\circ} \text{ V}$

(e) $V_2 = Z_2 I_2 = \underline{0.983\angle 79.38^\circ \text{V}}$

16.21 $Z_R = \dfrac{\omega^2 M^2}{Z_2 + j\omega L_2}$,

$Z_2 + j\omega L_2 = 10 - j\,10/\omega + j\frac{\omega}{2}$

Z_R is real when $Z_2 + j\omega L_2$ is real.

$\therefore\ \frac{\omega}{2} - \frac{10}{\omega} = 0 \Rightarrow \omega^2 = 20$

$\underline{\omega = \sqrt{20} \text{ rad/s}}$

16.22 $Z_1 = j(2)(1) + \dfrac{(2)^2(1)^2}{6 + j(2)(4)} = 0.24 + j1.68$

$Z_{in} = R + Z_1 = 2.24 + j1.68\ \Omega$

$I_1 = V_g / Z_{in} = \dfrac{21}{2.24 + j1.68} = 7.5\angle -36.9^\circ$

$i_1 = \underline{7.5\cos(2t - 36.87^\circ) \text{ A}}$

16.23 open circuit secondary:

$21 = (2 + j2) I_1 \Rightarrow I_1 = \dfrac{21}{2 + j2}$;

$V_{oc} = j2 I_1 = \frac{21}{2}(1 + j) \text{ V}$

short circuit secondary:

$\left.\begin{aligned} 21 &= (2 + j2) I_1 - j2 I_{sc} \\ 0 &= -j2 I_1 + j8 I_{sc} \end{aligned}\right\} I_{sc} = \dfrac{21}{8 + j6} \text{ A}$

$Z_{th} = \dfrac{V_{oc}}{I_{sc}} = 1 + j7\ \Omega$

From the Thevenin circuit.

$I_2 = \dfrac{\frac{21}{2}(1 + j)}{6 + 1 + j7} = 1.5\angle 0^\circ \text{A}$

$i_2 = \underline{1.5\cos 2t \text{ A}}$

16.24 $Z_R = \dfrac{\omega^2 M^2}{Z_2 + j\omega L_2} = \dfrac{(4)^2(\frac{1}{2})^2}{6 + j(4)(2)} = \dfrac{2}{3 + j4}$

$Z_1 = j(4)(\tfrac{1}{2}) + \dfrac{2}{3 + j4} = \dfrac{6(-1 + j)}{3 + j4}$

$I_1 = \dfrac{V_g}{Z_1 + Z_g} = \dfrac{42(3 + j4)}{j14} = 15\angle -36.9^\circ \text{ A}$

$i_1 = \underline{15\cos(4t - 36.9^\circ) \text{ A}}$

16.25 open circuit secondary:

$42 = (2 + j2) I_1 \Rightarrow I_1 = \dfrac{42}{2 + j2}$

$V_{oc} = j2 I_1 = 21(1 + j) \text{ V}$

short circuit secondary:

$\left.\begin{aligned} 42 &= (2 + j2) I_1 - j2 I_{sc} \\ 0 &= -j2 I_1 + j8 I_{sc} \end{aligned}\right\} I_{sc} = \dfrac{21}{4 + j3} \text{ A}$

$Z_{th} = \dfrac{V_{oc}}{I_{sc}} = 1 + j7\ \Omega$

From the Thevenin circuit

$I_2 = \dfrac{21(1 + j)}{6 + 1 + j7} = 3\angle 0^\circ \text{A}$

$i_2 = \underline{3\cos 4t \text{ A}}$.

16.26 $Z_1 = \dfrac{Z_2}{n^2} = \dfrac{(90 \times 10^3)}{n^2}$;

$Z_g = Z_1 \Rightarrow 100 = \dfrac{90 \times 10^3}{n^2} \Rightarrow \underline{n = 30}$

$I_1 = \dfrac{V_g}{Z_g + Z_1} = \dfrac{V_g}{2Z_g} = \dfrac{100}{2(100)} = 0.5 \text{ A}$

$I_2 = -\dfrac{I_1}{n} = -\dfrac{0.5}{30} = -\dfrac{1}{60} \text{ A}$

$P_{LOAD} = |I_2|^2 (90 \times 10^3) = \underline{25 \text{W}}$.

16.27 $Z_1 = \frac{Z_2}{n^2} = \frac{2-j}{(2)^2} = \frac{1}{2} - j\frac{1}{4}\ \Omega$

$$I_1 = \frac{V_g}{Z_g + Z_1} = \frac{4}{j\,6/8 + \frac{1}{2} - j\,1/4} = 4 - j4$$

$$= 4\sqrt{2}\,\angle{-45^\circ}$$

$$i_1 = \underline{4\sqrt{2}\cos(2t - 45^\circ)\ \text{A}}$$

16.28

$4\angle 0^\circ$, 2, 8, 2:1, V_1, V_2, I_1, I_2, 4

KVL: $4 = 2I_1 + V_1 = 2(2I_2) + \frac{V_2}{2}$ or

$$V_2 + 8I_2 = 8$$

KCL: $\frac{V_2 - 4}{8} - I_2 + \frac{V_2}{4} = 0$ or

$$3V_2 - 8I_2 = 4$$

Adding, $4V_2 = 12 \Rightarrow V_2 = 3\text{V}$

$$P_{8\Omega} = \frac{1}{2}\left|\frac{V_2 - 4}{8}\right|^2 (8) = \underline{1/16\ \text{W}}$$

16.29 $Z_1 = \frac{Z_2}{n^2} = \frac{100 - j75}{(5)^2} = 4 - j3\ \Omega$

$$I_1 = \frac{V_g}{Z_g + Z_1} = \frac{12}{2 + j3 + 4 - j3} = \underline{2\angle 0^\circ \text{A}}$$

$$I_2 = -\frac{I_1}{n} = \underline{-2/5\,\angle 0^\circ}\ \text{A}$$

$$V_1 = Z_1 I_1 = (4 - j3)\,2 = \underline{10\angle{-36.87^\circ}\,\text{V}}$$

$$V_2 = -nV_1 = \underline{50\angle 143.13^\circ\ \text{V}}$$

16.30 $V_2 = \frac{-Z_2}{Z_2 + n^2 Z_g}(nV_g)$, voltage division

$$= \frac{-(100 - j75)(5)(12)}{100 - j75 + 25(2 + j3)}$$

$$= \underline{50\angle 143.1^\circ\ \text{V}}$$

$$V_1 = -\frac{V_2}{n} = \underline{10\angle{-36.87^\circ}\ \text{V}}$$

$$I_2 = \frac{-nV_g}{n^2 Z_g + Z_2} = \frac{5(12)}{25(2+j3) + 100 - j75}$$

$$= \underline{2/5\,\angle 180^\circ\ \text{A}}$$

$$I_1 = -I_2 n = \underline{2\angle 0^\circ \text{A}}$$

16.31 $Z_g = \frac{Z_2}{n^2} \Rightarrow 4 = \frac{10000}{n^2}$

$$n^2 = \frac{10^4}{4} \Rightarrow n = \underline{50}$$

16.32

$21\angle 0^\circ$, 2, j0, j6, I_1, $j2$, I_2, 6

KVL: $21 = (2 + j2)I_1 - j2I_2$

$$0 = -j2I_1 + (6 + j2 + j6)I_2$$

$$I_2 = \frac{\begin{vmatrix} 2+j2 & 21 \\ -j2 & 0 \end{vmatrix}}{\begin{vmatrix} 2+j2 & -j2 \\ -j2 & 6+j8 \end{vmatrix}} = \frac{j42}{j28} = 1.5\angle 0^\circ \text{A}$$

$$i_2 = \underline{1.5\cos 2t\ \text{A}}$$

16.33

$10\angle 0^\circ$ V, 1, j0, j7, I_1, $j1$, I_2, 2, $-j10$

KVL: $10 = (1 + j)I_1 - jI_2$

$$0 = -jI_2 + (2 + j7 - j10 + j)I_2$$

$$I_2 = \frac{\begin{vmatrix} 1+j & 10 \\ -j & 0 \end{vmatrix}}{\begin{vmatrix} 1+j & -j \\ -j & 2-j2 \end{vmatrix}} = \frac{j10}{4+1} = \underline{j2}$$

16.34

i, 2Ω, 1H, 1H

Chapter 17

Fourier Series

17.1 The Trigonometric Fourier Series

17.1 Find the trigonometric series for the function $f(x)$ with a period of 4. where

$$f(x) = \begin{cases} 2, & -2 < x \leq 0 \\ x, & 0 < x < 2 \end{cases}$$

17.2 Find the trigonometric Fourier series for the function $f(x)$ of period 2π. where

$$f(x) = x^2, \ -\pi < x < \pi$$

17.3 Find the trigonometric fourier series for the function $f(x)$ of period 2π. where

$$f(x) = \begin{cases} 3\pi + 2x, & -\pi < x < 0 \\ \pi + 2x, & 0 < x < \pi \end{cases}$$

17.4 Find the trigonometric fourier series for the function $f(x)$ of period $1/4$ where

$$f(x) = 3\cos 8\pi x, \ 0 < x < 1/4$$

17.2 Symmetry Properties

17.5 Find the trigonometric Fourier series for the function

$$f(x) = \begin{cases} 0, & 0 < x < 1/2 \\ 1, & 1/2 < x < 1 \end{cases}$$

where f(x) is an odd function with a period $T = 2$.

17.6 Find the trigonometric Fourier series for the function

$$f(x) = x(2 - x), \ 0 < x < 2,$$

where f(x) is an even function with period $T = 4$.

17.7 Find the trigonometric Fourier series for the function

$$\begin{aligned} f(x) &= -1, & -1 < x < -1/2 \\ &= 1, & -1/2 < x < 1/2 \\ &= -1, & 1/2 < x < 1 \end{aligned}$$

where f(x) is an even function of period $T = 2$.

17.8 Find the trigonometric Fourier series for the function

$$\begin{aligned} f(x) &= -1, & 1- < x < 0 \\ &= 1, & 0 < x < 1 \end{aligned}$$

where f(x) is an odd function of period $T = 2$.

17.3 Response to Periodic Excitations

17.9 Find the forced component of the voltage across the inductor if $v(t) = f(t)$ from Prob. 17.2.

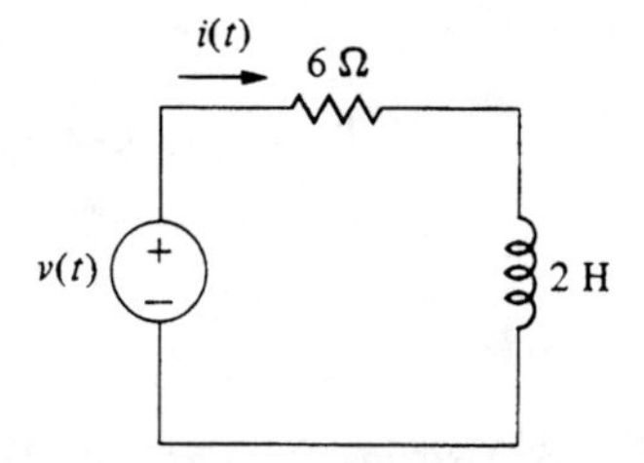

PROBLEM 17.9

17.10 Find the forced component of the voltage across the capacitor if $v(t) = f(x)$ from Prob. 17.6.

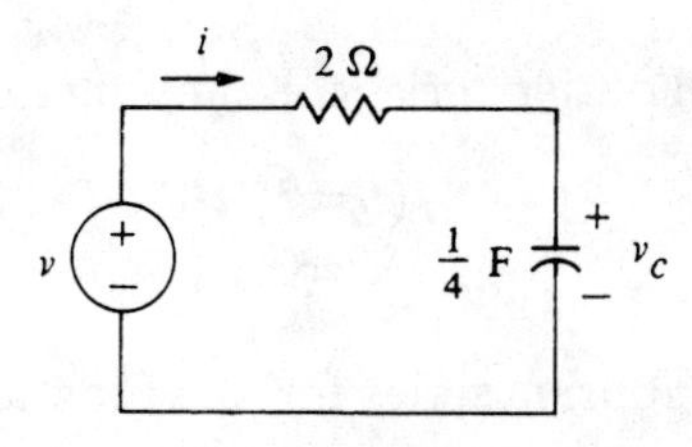

PROBLEM 17.10

17.11 Find the forced component of the current in Prob. *17.9* if

$$\begin{aligned} v(t) &= 1 + t, \quad -1 < t < 0 \\ &= 1 - t, \quad 0 < t < 1 \end{aligned}$$

where $v(t)$ is a periodic function of period $T = 2$.

17.4 Average Power and RMS Values

17.12 Find the average power delivered by the source in Prob. 17.9.

17.13 Find the average power delivered by the source in Prob. 17.10.

17.14 Find the average power delivered by the source in Prob. 17.11.

17.15 Find the forced rms voltage across the inductor in Prob. 17.9.

17.16 Find the forced rms voltage across the capacitor in Prob. 17.10.

17.17 Find the forced rms current in the circuit of Prob. 17.11.

17.5 The Exponential Fourier Series

17.18 Find the exponential Fourier series for the current

$$\begin{aligned} i &= I_m, \quad -1 < t < 1 \\ &= 0, \quad 1 < t < 3 \end{aligned}$$

where $i(t)$ is periodic with period $T = 4$.

17.19 Find the rms value for i in Prob. 17.18.

17.20 Find the exponential Fourier series for the function $f(t)$ where

$$\begin{aligned} f(t) &= e^{-t} - 1, \quad -1 < t < 0 \\ &= e^{t} - 1, \quad 0 < t < 1 \\ f(t+2) &= f(t) \end{aligned}$$

17.21 Find the exponential Fourier series for the function $f(x)$ where

$$\begin{aligned} f(x) &= x \quad 0 < x < 2 \\ f(x+2) &= f(x) \end{aligned}$$

17.22 Find the exponential Fourier series for the function $f(x)$ where

$$\begin{aligned} f(x) &= 2\cos 2\pi x \quad 0 < x < 1 \\ f(x+1) &= f(x) \end{aligned}$$

17.23 Find the exponential Fourier series for the function $f(t)$ where

$$\begin{aligned} f(t) &= 2, \quad 0 < t < 3 \\ &= 0, \quad 3 < t < 4 \end{aligned}$$

17.24 Find the exponential Fourier series for the function of Prob. 17.1.

17.25 Find the exponential Fourier series for the function of Prob. 17.5.

17.6 Frequency Spectra

17.26 Plot the discrete amplitude spectra for Prob. 17.1.

17.27 Plot the discrete amplitude and phase spectra for Prob. 17.2.

17.28 Plot the discrete amplitude and phase spectra for Prob. 17.3.

17.29 Plot the discrete amplitude and phase spectra for Prob. 17.4.

17.30 Plot the discrete amplitude and phase spectra for Prob. 17.21.

17.31 Plot the discrete amplitude spectra for Prob. 17.22.

17.1 $T=4,\ \omega_0=\frac{2\pi}{T}=\frac{\pi}{2}$

$$a_0=\frac{2}{4}\left[\int_{-2}^{0}2\,dx+\int_0^2 x\,dx\right]=\underline{3}$$

$$a_n=\frac{2}{4}\left[\int_{-2}^{0}2\cos\frac{n\pi x}{2}\,dx+\int_0^2 x\cos\frac{n\pi x}{2}\,dx\right]$$

$$=\frac{1}{2}\left[\frac{2\sin\frac{n\pi x}{2}}{n\pi/2}\Bigg|_{-2}^{0}\right]+\frac{1}{2}\left[\frac{\cos n\pi x/2}{(n\pi/2)^2}+\frac{x\sin n\pi x/2}{n\pi/2}\right]_0^2$$

$$=\underline{\frac{2}{n^2\pi^2}\left[(-1)^n-1\right]}$$

$$b_n=\frac{1}{2}\left[\int_{-2}^{0}2\sin\frac{n\pi x}{2}\,dx+\int_0^2 x\sin\frac{n\pi x}{2}\,dx\right]$$

$$=\frac{1}{2}\left\{\frac{-2\cos\frac{n\pi x}{2}}{n\pi/2}\Bigg|_{-2}^{0}+\left[\frac{\sin n\pi x/2}{(n\pi/2)^2}-\frac{x\cos n\pi x/2}{n\pi/2}\right]_0^2\right\}$$

$$=\underline{-\frac{2}{n\pi}}$$

17.2 $T=2\pi,\ \omega_0=\frac{2\pi}{T}=1$

$$a_0=\frac{2}{2\pi}\int_{-\pi}^{\pi}x^2\,dx=\frac{1}{\pi}\frac{x^3}{3}\Bigg|_{-\pi}^{\pi}=\underline{\frac{2\pi^2}{3}}$$

$$a_n=\frac{1}{\pi}\int_{-\pi}^{\pi}x^2\cos nx\,dx$$

$$=\frac{1}{\pi}\left[\frac{2x\cos nx}{n^2}+\frac{n^2x^2-2}{n^3}\sin nx\right]_{-\pi}^{\pi}$$

$$=\underline{\frac{4(-1)^n}{n^2}}$$

$$b_n=\frac{1}{\pi}\int_{-\pi}^{\pi}x^2\sin nx\,dx$$

$$=\frac{1}{\pi}\left[\frac{2x\sin nx}{n^2}-\frac{n^2x^2-2}{n^3}\cos nx\right]_{-\pi}^{\pi}$$

$$=\underline{0}$$

17.3 $T=2\pi,\ \omega_0=\frac{2\pi}{T}=1$

$$a_0=\frac{1}{\pi}\left[\int_{-\pi}^{0}(3\pi+2x)\,dx+\int_0^{\pi}(\pi+2x)\,dx\right]$$

17.3 cont. $a_0=2\pi+2\pi=\underline{4\pi}$

$$a_n=\frac{1}{\pi}\left[\int_{-\pi}^{0}(3\pi+2x)\cos nx\,dx+\int_0^{\pi}(\pi+2x)\cos nx\,dx\right]$$

$$=\frac{1}{\pi}\left[\frac{-3\pi\sin nx}{n}\Bigg|_{-\pi}^{0}+2\left(\frac{\cos nx}{n^2}+\frac{x}{n}\sin nx\right)\Bigg|_{-\pi}^{0}-\frac{\pi}{n}\sin nx\Bigg|_0^{\pi}+2\left(\frac{\cos nx}{n^2}+\frac{x}{n}\sin nx\right)\Bigg|_0^{\pi}\right]$$

$$=\underline{0}$$

$$b_n=\frac{1}{\pi}\left[\int_{-\pi}^{0}(3\pi+2x)\sin nx\,dx+\int_0^{\pi}(\pi+2x)\sin nx\,dx\right]$$

$$=\frac{1}{\pi}\left\{\frac{-3\pi\cos nx}{n}\Bigg|_{-\pi}^{0}+2\left(\frac{\sin nx}{n^2}-\frac{x}{n}\cos nx\right)\Bigg|_{-\pi}^{0}-\frac{\pi\cos nx}{n}\Bigg|_0^{\pi}+2\left(\frac{\sin nx}{n^2}-\frac{x}{n}\cos nx\right)\Bigg|_0^{\pi}\right\}$$

$$=\underline{-\frac{2}{n}\left[1+(-1)^n\right]}$$

$\underline{b_n=0,\ n\text{ odd}}$; $\underline{b_n=-\frac{4}{n},\ n\text{ even}}$

17.4 $\omega_0=\frac{2\pi}{T}=\frac{2\pi}{(1/4)}=8\pi$

$$a_0=\frac{2}{T}\int_0^{1/4}3\cos 8\pi x\,dx$$

$$=24\,\frac{\sin 8\pi x}{8\pi}\Bigg|_0^{1/4}=\underline{0}$$

$$a_n=8\int_0^{1/4}3\cos 8\pi x\cos 8n\pi x\,dx$$

$a_1=\underline{3}$; $a_n=\underline{0},\ n\neq 1$

$$b_n=8\int_0^{1/4}3\cos 8\pi x\sin 8n\pi x\,dx$$

$$=\underline{0}$$

$f(x)=\underline{3\cos 8\pi x}$

17.5 $T=2, \omega_0 = 2\pi/2 = \pi$

$f(x)$ odd $\Rightarrow a_n = \underline{0}$

$$b_n = \frac{4}{2}\int_{1/2}^{1} \sin n\pi t\, dt$$
$$= 2\left[-\frac{1}{n\pi}\cos n\pi t\right]_{1/2}^{1}$$
$$= -\frac{2}{n\pi}\left[\cos n\pi - \cos\frac{n\pi}{2}\right]$$
$$= \underline{-\frac{2}{n\pi}\left[(-1)^n - \cos\frac{n\pi}{2}\right]}$$

17.6 $\omega_0 = \frac{2\pi}{4} = \frac{\pi}{2}$

$$a_0 = \frac{4}{4}\int_0^2 x(2-x)dx$$
$$= x^2 - \frac{x^3}{3}\Big|_0^2 = \frac{4}{3}$$
$$a_n = \frac{4}{4}\int_0^2 (2x - x^2)\cos\frac{n\pi x}{2}dx$$
$$= \left\{2\left[\frac{4\cos\frac{n\pi x}{2}}{n^2\pi^2} + \frac{2x}{n\pi}\sin\frac{n\pi x}{2}\right] - \left[\frac{\frac{n^2\pi^2 x^2}{4} - 2}{n^3\pi^3/8}\sin\frac{n\pi x}{2} + \frac{4x}{n^2\pi^2}\cos\frac{n\pi x}{2}\right]\right\}\Bigg|_0^2$$
$$= \frac{-8}{n^2\pi^2}\left[1 + (-1)^n\right]$$

$a_n = \underline{0, n \text{ odd}}; \; a_n = \underline{-\frac{16}{n^2\pi^2}, n \text{ even}}$

$b_n = \underline{0}$ since $f(x)$ is even.

17.7 $T=2, \omega_0 = \frac{2\pi}{2} = \pi$

$$a_0 = \frac{4}{2}\left[\int_0^{1/2} dx + \int_{1/2}^{1}(-1)dx\right] = \underline{0}$$
$$a_n = 2\left[\int_0^{1/2}\cos n\pi x\, dx - \int_{1/2}^{1}\cos n\pi x\, dx\right]$$
$$= 2\left[\frac{\sin n\pi x}{n\pi}\Big|_0^{\frac{1}{2}} - \frac{\sin n\pi x}{n\pi}\Big|_{\frac{1}{2}}^{1}\right]$$
$$= \underline{\frac{4}{n\pi}\sin\frac{n\pi}{2}}$$

17.7 Cont.

$b_n = 0$ since $f(x)$ is even

17.8 $T=2, \omega_0 = 2\pi/2 = \pi$

$a_n = 0$ since $f(x)$ is odd.

$$b_n = \frac{4}{2}\int_0^1 (1)\sin n\pi x\, dx$$
$$= 2\,\frac{-\cos n\pi x}{n\pi}\Big|_0^1 = \frac{2}{n\pi}\left[1 - (-1)^n\right]$$

$b_n = \underline{0, n \text{ even}}; \; b_n = \underline{\frac{4}{n\pi}, n \text{ odd.}}$

17.9 $v = \frac{\pi^2}{3} + \sum_{n=1}^{\infty}\frac{4(-1)^n}{n^2}\cos nt$

$$V_n = \frac{4(-1)^n}{n^2}\angle 0°, \quad \omega_n = n$$
$$Z(j\omega_n) = 6 + j2n$$
$$I_n = \frac{4(-1)^n}{n^2(6+j2n)}$$
$$V_{Ln} = j2n I_n = \frac{j4(-1)^n}{n(3+jn)}$$
$$= \frac{4(-1)^n}{n\sqrt{9+n^2}}\angle 90° - \tan^{-1}\frac{n}{3}$$
$$V_{Lo} = 0$$
$$v_L(t) = \sum_{n=1}^{\infty} v_{Ln}(t)$$
$$= \underline{4\sum_{n=1}^{\infty}\frac{(-1)^n\cos(nt + 90° - \tan^{-1}\frac{n}{3})}{n\sqrt{n^2+9}}}$$

17.10 $v = \frac{2}{3} - \frac{4}{\pi^2}\sum_{n=1}^{\infty}\frac{1}{n^2}\cos n\pi t$

$$V_0 = \frac{2}{3}; \; V_n = -\frac{4}{n^2\pi^2}, \; \omega_n = n\pi$$
$$Z(j\omega_n) = 2 - j\frac{4}{n\pi}; \; V_{co} = V_0 = \frac{2}{3}$$

Voltage division:

$$V_{cn} = \frac{-j4/\omega_n}{2 - j\frac{4}{\omega_n}}V_n = \frac{-j2}{n\pi - j2}V_n$$
$$= \frac{8}{n^2\pi^2\sqrt{n^2\pi^2+4}}\angle\theta_n$$
$$\theta_n = 90° - \tan^{-1}\frac{-2}{n\pi}$$
$$= 90° + \tan^{-1}\frac{2}{n\pi}$$
$$v_c = v_0 + \sum_{n=1}^{\infty} v_{cn}$$

17.10 cont.

$$v_c = \frac{2}{3} + \frac{8}{\pi^2}\sum_{n=1}^{\infty} \frac{\cos(n\pi t + \theta_n)}{n^2\sqrt{n^2\pi^2+4}}$$

17.11 $T = 2,\ \omega_o = 2\pi/2 = \pi$

$b_n = 0$ since v is even

$$a_o = \frac{4}{2}\int_0^1 (1-t)dt = 1$$

$$a_n = \frac{4}{2}\int_0^1 (1-t)\cos n\pi t\,dt$$

$$= 2\left[\frac{\sin n\pi t}{n\pi} - \left(\frac{\cos n\pi t}{n^2\pi^2} + \frac{t\sin n\pi t}{n\pi}\right)\right]_0^1$$

$$= \frac{2}{n^2\pi^2}\left[1-(-1)^n\right]$$

$a_n = 0$, n even

$a_n = \frac{4}{n^2\pi^2}$, n odd

$$v = \frac{1}{2} + \sum_{n=1}^{\infty} \frac{4\cos(2n-1)\pi t}{(2n-1)^2\pi^2}$$

$\omega_n = (2n-1)\pi$; $Z(j\omega_n) = 6 + j2\omega_n$

$$I_o = \frac{V_o}{6} = \frac{1}{12}$$

$$I_n = \frac{V_n}{Z(j\omega_n)} = \frac{4/(2n-1)^2\pi^2}{6 + j2(2n-1)\pi}$$

$$|I_n| = \frac{2}{(2n-1)^2\pi^2\sqrt{9+(2n-1)^2\pi^2}}$$

$$i = \frac{1}{12} + \sum_{n=1}^{\infty} |I_n|\cos[(2n-1)\pi t + \theta_n]$$

where $\theta_n = -\tan^{-1}\frac{(2n-1)\pi}{3}$

17.12 $P = P_{dc} + \sum_{n=1}^{\infty} P_n$

$$P_{dc} = V_{dc}I_{dc} = \frac{\pi^2}{3}\left[\frac{1}{6}\left(\frac{\pi^2}{3}\right)\right] = \frac{\pi^4}{54}\ W$$

$$|I_n| = \frac{2}{n^2\sqrt{n^2+9}}$$

$\theta_n = -\tan^{-1}\frac{n}{3}$, n even

$= 180° - \tan^{-1}\frac{n}{3}$, n odd

$\phi_n = 0$, n even

$= 180°$, n odd

17.12 cont. $|V_n| = \frac{4}{n^2}$

$$\phi_n - \theta_n = \tan^{-1}\frac{n}{3}$$

$$P_n = |V_n||I_n|\left(\tfrac{1}{2}\right)\cos(\phi_n - \theta_n)$$

$$= \frac{4}{n^4\sqrt{n^2+9}}\cos\left(\tan^{-1}\frac{n}{3}\right)$$

$$= \frac{4}{n^4\sqrt{n^2+9}}\ \frac{3}{\sqrt{n^2+9}}$$

$$P = \frac{\pi^4}{54} + \sum_{n=1}^{\infty} \frac{12}{n^4(n^2+9)}\ W$$

17.13 $I_{dc} = I_o = 0 \Rightarrow P_{dc} = 0$

$$V_n = \frac{4}{n^2\pi^2}\angle 180°$$

$$I_n = \frac{V_n}{Z_n} = \frac{(4/n^2\pi^2)\angle 180°}{2 - j\frac{4}{n\pi}}$$

$$= \frac{2}{n\pi\sqrt{n^2\pi^2+4}}\angle 180° + \tan^{-1}\frac{2}{n\pi}$$

$$\phi_n - \theta_n = \tan^{-1}\frac{2}{n\pi}$$

$$P_n = \frac{1}{2}\,\frac{4}{n^2\pi^2}\,\frac{\cos(\phi_n - \theta_n)}{n\pi\sqrt{n^2\pi^2+4}}$$

$$\cos(\phi_n - \theta_n) = \frac{n\pi}{\sqrt{n^2\pi^2+4}}$$

$$P = \sum_{n=1}^{\infty} \frac{2}{n^2\pi^2(n^2\pi^2+4)}\ W$$

17.14 $P_{dc} = V_{dc}I_{dc} = \frac{1}{2}\,\frac{1}{12} = \frac{1}{24}$

$$V_n = \frac{4}{(2n-1)^2\pi^2}\angle 0°$$

$$P_n = \frac{1}{2}\,\frac{4}{(2n-1)^2\pi^2}|I_n|\cos\theta_n$$

$$\cos\theta_n = \frac{3}{\sqrt{(2n-1)^2\pi^2+9}}$$

$$P_n = \frac{12}{(2n-1)^4\pi^4[(2n-1)^2\pi^2+9]}$$

$$P = \frac{1}{24} + \sum_{n=1}^{\infty} P_n\ W$$

17.15 $V_{rms}^2 = V_{dc}^2 + \sum_{n=1}^{\infty} V_{nrms}^2$

$V_{dc} = 0$

17.15 Cont. $V_{n\,rms}^2 = \frac{1}{2}V_n^2$

$$V_{rms}^2 = \sum_{n=1}^{\infty} \frac{1}{2} \frac{16}{n^2(n^2+9)}$$

$$\underline{V_{rms} = 4\sqrt{\sum_{n=1}^{\infty} \frac{1}{2n^2(n^2+9)}}}$$

17.16 $V_{dc} = \frac{2}{3}$

$$V_n = \frac{8}{n^2\pi^2\sqrt{n^2\pi^2+4}}$$

$$V_{rms}^2 = \frac{4}{9} + \sum_{n=1}^{\infty} \frac{1}{2} \frac{64}{n^4\pi^4(n^2\pi^2+4)}$$

$$\underline{V_{rms} = \left\{\frac{4}{9} + \frac{32}{\pi^4}\sum_{n=1}^{\infty} \frac{1}{n^4(n^2\pi^2+4)}\right\}^{1/2}}$$

17.17 $I_{dc} = \frac{1}{12}$ A

$$\frac{I_n^2}{2} = \frac{1}{2}\frac{4}{(2n-1)^4\pi^4[9+(2n-1)^2\pi^2]}$$

$$I_{rms}^2 = \frac{1}{144}$$

$$\underline{+\frac{2}{\pi^4}\sum_{n=1}^{\infty}\frac{1}{(2n-1)^4[9+(2n-1)^2\pi^2]}}$$

17.18 $\omega_0 = \frac{2\pi}{4} = \frac{\pi}{2}$, $T = 4$

$$c_0 = \frac{1}{4}\int_{-1}^{1} I_m\,dt = \underline{\frac{1}{2}I_m}$$

$$c_n = \frac{1}{4}\int_{-1}^{1} I_m e^{-jn\pi t/2}\,dt$$

$$= -\frac{I_m}{j2n\pi}\left[e^{-j\frac{n\pi}{2}} - e^{j\frac{n\pi}{2}}\right]$$

$$\underline{= \frac{I_m}{n\pi}\sin\frac{n\pi}{2}, \quad n \neq 0}$$

17.19 $I_{rms} = \left[|c_0|^2 + \sum_{n=-\infty, n\neq 0}^{\infty} |c_n|^2\right]^{1/2}$

$$I_{rms}^2 = \frac{I_m^2}{4} + \sum_{n=-\infty, n\neq 0}^{\infty} \frac{I_m^2}{n^2\pi^2}\sin^2\frac{n\pi}{2}$$

$\sin\frac{n\pi}{2} = 0$ for n even

$\sin(2n-1)\frac{\pi}{2} = (-1)^{n+1}$

$$\underline{I_{rms} = I_m\left[\frac{1}{4} + \frac{1}{\pi^2}\sum_{n=-\infty}^{\infty}\frac{1}{(2n-1)^2}\right]^{\frac{1}{2}}}$$

17.19 Cont

$$\underline{I_{rms} = I_m\left[\frac{1}{4} + \frac{2}{\pi^2}\sum_{n=1}^{\infty}\frac{1}{(2n-1)^2}\right]^{\frac{1}{2}}}$$

17.20 $T = 2$, $\omega_0 = \frac{2\pi}{2} = \pi$

$$c_0 = \frac{1}{2}\left[\int_{-1}^{0}(e^{-t}-1)\,dt + \int_0^1 (e^t-1)\,dt\right]$$

$$= \int_0^1 (e^t-1)\,dt = e^t - t\Big|_0^1$$

$$= e - 1 - 1 = \underline{e-2}$$

$$c_n = \frac{1}{2}\left[\int_{-1}^{0}(e^{-t}-1)e^{-jn\pi t}\,dt\right.$$

$$\left.+\int_0^1 (e^t-1)e^{-jn\pi t}\,dt\right]$$

$$= \frac{1}{2}\left[\frac{e^{-(jn\pi+1)t}}{-(jn\pi+1)} - \frac{e^{-jn\pi t}}{jn\pi}\Big|_{-1}^{0}\right.$$

$$\left.-\frac{e^{-(jn\pi-1)t}}{jn\pi-1} + \frac{e^{-jn\pi t}}{jn\pi}\Big|_0^1\right]$$

$$= \frac{1}{2}\left[\frac{-1+e(-1)^n}{jn\pi+1} - \frac{e(-1)^n-1}{jn\pi-1}\right]$$

$$\underline{= \frac{e(-1)^n-1}{1+n^2\pi^2}}$$

17.21 $T = 2$, $\omega_0 = \pi$

$$c_0 = \frac{1}{2}\int_0^2 x\,dx = \underline{1}$$

$$c_n = \frac{1}{2}\int_0^2 x e^{-jn\pi x}\,dx$$

$$= \frac{1}{2(-jn\pi)^2}e^{-jn\pi x}(-jn\pi x-1)\Big|_0^2$$

$$= \frac{-1}{2n^2\pi^2}\left[e^{-j2n\pi}(-j2n\pi-1)-(-1)\right]$$

$$e^{-j2n\pi} = 1$$

$$c_n = \underline{\frac{j}{n\pi}}, \quad n \neq 0$$

17.22 $T = 1$, $\omega_0 = 2\pi$

$$c_0 = \int_0^1 2\cos 2\pi x\,dx$$

$$= \frac{1}{\pi}\sin 2\pi x\Big|_0^1 = \underline{0}$$

$$c_n = \int_0^1 2\cos 2\pi x\, e^{-j2n\pi x}\,dx$$

17.22 Cont.

$$c_n = \frac{2e^{-j2n\pi x}}{(2\pi)^2 + (-j2n\pi)^2} \cdot \left[-j2n\pi \cos 2\pi x + 2\pi \sin 2\pi x\right]_0^1$$

$c_n = \underline{0}$ for $n^2 \neq 1$

$$c_1 = 2\int_0^1 \cos 2\pi x \, e^{-j2\pi x} dx$$

$$= 2\int_0^1 \cos^2 2\pi x \, dx - j2\int_0^1 \sin 2\pi x \cos 2\pi x \, dx$$

$$= \int_0^1 (1 + \cos 4\pi x) dx - j\int_0^1 \sin 4\pi x \, dx$$

$$= x + \frac{\sin 4\pi x}{4\pi} + j\frac{\cos 4\pi x}{4\pi}\Big|_0^1 = \underline{1}$$

$$c_{-1} = c_1^* = \underline{1}$$

$$f(x) = e^{j2\pi x} + e^{-j2\pi x}$$

$$= \underline{2\cos 2\pi x}$$

17.23 $T = 4, \ \omega_o = \pi/2$

$$c_o = \frac{1}{4}\int_0^3 2\,dt = \underline{\frac{3}{2}}$$

$$c_n = \frac{1}{4}\int_0^3 2e^{-jn\pi t/2} dt$$

$$= \frac{1}{2} \frac{e^{-jn\pi t/2}}{-jn\pi/2}\Big|_0^3$$

$$= \frac{1}{-jn\pi}\left[e^{-j3n\pi/2} - 1\right]$$

$$= \underline{\frac{1 - e^{-j3n\pi/2}}{jn\pi}}$$

17.24 $T = 4, \omega_o = \pi/2$

$$c_o = \frac{1}{4}\left[\int_{-2}^0 2dx + \int_0^2 x dx\right]$$

$$= \frac{1}{4}\left\{2x\Big|_{-2}^0 + \frac{x^2}{2}\Big|_0^2\right]$$

$$= \frac{1}{4}[4 + 2] = \underline{\frac{3}{2}}$$

17.24 cont.

$$c_n = \frac{1}{4}\left[\int_{-2}^0 2e^{-j\frac{n\pi x}{2}} dx + \int_0^2 x e^{-j\frac{n\pi x}{2}} dx\right]$$

$$= \frac{1}{4}\left\{\frac{2e^{-jn\pi x/2}}{-jn\pi/2}\Big|_{-2}^0 + \frac{e^{-jn\pi x/2}}{(-jn\pi/2)^2}\left(-j\frac{n\pi x}{2} - 1\right)\Big|_0^2\right\}$$

$$= \frac{1}{-jn\pi}\left[1 - e^{jn\pi}\right] - \frac{1}{n^2\pi^2}\left[e^{-jn\pi}(-jn\pi - 1) - (-1)\right]$$

$$= \frac{1}{n^2\pi^2}\left[e^{-jn\pi} - 1\right] - \frac{1}{jn\pi}$$

$$= \underline{\frac{1}{n^2\pi^2}\left[(-1)^n - 1)\right] - \frac{1}{jn\pi}}$$

17.25 $T = 2, \ \omega_o = \frac{2\pi}{2} = \pi$

$f(x)$ odd $\Rightarrow a_n = 0, \ n = 0, 1, 2, 3, \cdots$

$$c_o = \frac{a_o}{2} = \underline{0}; \ c_n = \frac{a_n - jb_n}{2}$$

$$c_n = -\frac{j}{2}b_n = -\frac{j}{2}\frac{4}{2}\int_{1/2}^1 \sin n\pi x dx$$

$$= -j\int_{1/2}^1 (1)\sin n\pi x \, dx$$

$$= -j\frac{-\cos n\pi x}{n\pi}\Big|_{1/2}^1$$

$$= \frac{j}{n\pi}\left[\cos n\pi - \cos\frac{n\pi}{2}\right]$$

$$= \underline{\frac{j}{n\pi}\left[(-1)^n - \cos\frac{n\pi}{2}\right]}$$

17.26 From Prob. 17.24

$$|c_o| = \frac{3}{2}$$

$$|c_n|^2 = \left[\frac{(-1)^n - 1}{n^2\pi^2}\right]^2 + \left[\frac{1}{n\pi}\right]^2$$

$$= \frac{[(-1)^n - 1]^2 + n^2\pi^2}{n^4\pi^4}$$

$$|c_n| = \underline{\frac{1}{n^2\pi^2}\sqrt{[(-1)^n - 1]^2 + n^2\pi^2}}$$

17.27 $c_0 = \frac{a_0}{2} = \frac{1}{2}\,\frac{2\pi^2}{3} = \frac{\pi^2}{3}$

$|c_0| = \underline{\frac{\pi^2}{3}}, \ \phi_0 = \underline{0}$

$c_n = \frac{1}{2}(a_n - jb_n) = \frac{1}{2}\,\frac{4(-1)^n}{n^2}$

$= \frac{2(-1)^n}{n^2}$

$|c_n| = \underline{\frac{2}{n^2}}$

$\phi_n = \underline{0}, \ n$ even

$= \underline{180°}, \ n$ odd, $n>0$

17.28 $c_0 = \frac{a_0}{2} = \frac{1}{2}(4\pi) = 2\pi$

$|c_0| = \underline{2\pi}, \ \phi_0 = \underline{0}$

$c_n = \frac{1}{2}(a_n - jb_n) = -\frac{j}{2}b_n$

$= \frac{j}{n}[1+(-1)^n]$

$c_n = \underline{0}$ for n odd

$c_n = j\frac{2}{n} \Rightarrow |c_n| = \underline{\frac{2}{|n|}}, \ n$ even

$\phi_n = \underline{90°}, \ n$ even $+ \ n>0$

$= \underline{-90°}, \ n$ even $+ \ n<0$

17.29 $c_0 = \frac{a_0}{2} = 0$

$c_n = \frac{1}{2}(a_n - jb_n)$

$= 0$ for $n^2 \neq 1$

$c_1 = c_{-1} = 3$

$|c_1| = |c_{-1}| = 3$

$\phi_n = \underline{0}$

17.30 $c_0 = 1 \Rightarrow |c_0| = \underline{1}, \ \phi_0 = \underline{0}$

$c_n = \frac{j}{n\pi} \Rightarrow |c_n| = \frac{1}{|n|\pi}$

$\phi_n = \underline{90°}, \ n>0$

$= \underline{-90°}, \ n<0$

17.31 $c_0 = 0$

$c_n = 0$ for $n^2 \neq 1$

17.31 cont.

$c_1 = c_{-1} = 1$

$|c_1| = |c_{-1}| = \underline{1}$

$\phi_1 = \phi_{-1} = \underline{0}$

Chapter 18

Fourier Transforms

18.2 Development of the Fourier Transform

18.1 Find the Fourier transform of the function

$$\begin{aligned} f(t) &= 2(t+1), & -1 \le t \le 0 \\ &= 2(-t+1), & 0 \le t \le 1 \\ &= 0, & \text{elsewhere} \end{aligned}$$

18.2 Find the Fourier transform of the function

$$\begin{aligned} f(t) &= \cos \pi t, & -1 < t < 1 \\ &= 0, & \text{elsewhere} \end{aligned}$$

18.3 Find the Fourier transform of the function

$$\begin{aligned} f(t) &= \cos^2 \pi t, & -1 < t < 1 \\ &= 0, & \text{elsewhere} \end{aligned}$$

18.4 Find the Fourier transform of the function

$$\begin{aligned} f(t) &= e^t, & 0 < t < 10 \\ &= 0, & \text{elsewhere} \end{aligned}$$

18.5 Find the Fourier transform of the function

$$\begin{aligned} f(t) &= t, & 0 < t < 1 \\ &= 2, & 1 < t < 3 \\ &= 0, & \text{elsewhere} \end{aligned}$$

18.6 Find the Fourier transform of the function

$$f(t) = \cos \pi t u(t) - \cos \pi t u(t) u(t-10)$$

18.7 Find the Fourier transform of the function

$$f(t) = \cos \pi t u(t+5) - \cos \pi t u(t) u(t-5)$$

18.8 Find the Fourier transform of the function

$$\begin{aligned} f(t) &= -1, \quad -1 < t < 0 \\ &= 1, \quad 0 < t < 1 \\ &= 0, \quad \text{elsewhere} \end{aligned}$$

18.9 Find the Fourier transform of the function

$$\begin{aligned} f(t) &= -1, \quad -1 < t < -1/2 \\ &= -1, \quad -1/2 < t < 1/2 \\ &= -1, \quad 1/2 < t < 1 \\ &= 0, \quad \text{elsewhere} \end{aligned}$$

18.10 Find the Fourier transform of the function

$$\begin{aligned} f(t) &= 2\cosh t, \quad -2 < t < 2 \\ &= 0, \quad \text{elsewhere} \end{aligned}$$

18.11 Find the Fourier transform of the function

$$\begin{aligned} f(t) &= 2\sinh t, \quad -2 < t < 2 \\ &= 0, \quad \text{elsewhere} \end{aligned}$$

18.3 Fourier Transform Properties

18.12 Solve Prob. 18.8 using the properties of even and odd functions.

18.13 Solve Prob. 18.9 using the properties of even and odd functions.

18.14 Using the properties of even and odd functions find the Fourier transform of the function

$$\begin{aligned} f(t) &= |t|, \quad -2 < t < 2 \\ &= 0, \quad \text{elsewhere} \end{aligned}$$

18.15 Using the properties of even and odd functions find the Fourier transform of the function

$$\begin{aligned} f(t) &= t, \quad -2 < t < 2 \\ &= 0, \quad \text{elsewhere} \end{aligned}$$

18.16 Using the properties of even and odd functions find the Fourier transform of the function

$$\begin{aligned} f(t) &= t^2, \quad -2 < t < 2 \\ &= 0, \quad \text{elsewhere} \end{aligned}$$

18.17 Using the properties of even and odd functions find the Fourier transform of the function

$$\begin{aligned} f(t) &= t|t|, \quad -2 < t < 2 \\ &= 0, \quad \text{elsewhere} \end{aligned}$$

18.4 Fourier Transform Operations

18.18 Find the Fourier transform of $f(t) = (5e^{-2t} + 4e^{-3t})u(t)$.

18.19 Find the Fourier transform of $f(t) = 6e^{-3(t+6)}u(t+6)$.

18.20 Find the Fourier transform of $f(t) = 2e^{-t/6}u(t)$.

18.21 Find the Fourier transform of $f(t) = (d/dt)2e^{-3t}\ u(t)$.

18.22 Find the Fourier transform of

$$\begin{aligned} f(t) &= t, \quad -1 < t < 1 \\ &= 0, \quad \text{elsewhere} \end{aligned}$$

18.23 Find the inverse Fourier transform of $F(\omega) = \frac{\sin\omega}{\omega}e^{-j\omega/2}$.

18.24 Find the inverse Fourier trasform of $F(\omega) = \frac{-\omega^2}{1+j\omega}$

18.5 Network Functions

18.25 If $x(t)$ is the input and $y(t)$ is the output find the network function using Fourier transforms of

$$4y'' + 6y' + y = 2x$$

18.26 If $x(t) = e^{-(t-2)}u(t-2)$ in Prob. 18.25 find $Y(j\omega)$.

18.27 If $x(t)$ is the input and $y(t)$ is the output find the network function using Fourier transforms of

$$y''' + 3y' = 4x$$

18.28 If $x(t) = 1$, $-1/2 < t < 1/2$ in Prob. 18.27 find $Y(j\omega)$.

18.6 Parseval's Equation for Fourier Transforms

18.29 Use Parseval's theorm and Prob. 18.1 to evaluate the integral

$$\int_0^\infty \frac{(1-\cos\omega)^2}{\omega^4}d\omega$$

18.30 The input voltage to the lowpass filter shown is

$$v_i = Ke^{-at}u(t)$$

where $a > 0$. Find the 1-Ω energy w_i at the input and the 1-Ω energy w_o at the output.
(b) If $R = 1\text{k}\Omega$, $C = 1\mu\text{F}$, $a = 10^3$, and $K = 10$, find the ratio w_o/w_i.
(c) Repeat (b) if the filter is the ideal one with transfer function $H(s)$, where

$$\begin{aligned} |H(j\omega)| &= 1, \quad -10^3 < t < 10^3 \\ &= 0, \quad \text{elsewhere} \end{aligned}$$

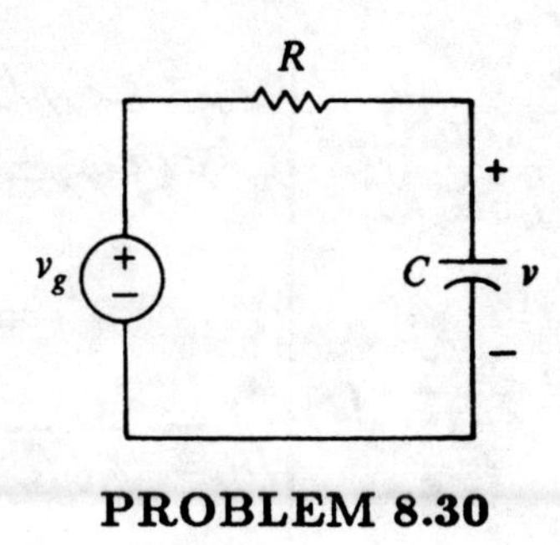

PROBLEM 8.30

18.1 $F(j\omega) = \int_{-1}^{0} 2(t+1)e^{-j\omega t}dt + \int_{0}^{1} 2(-t+1)e^{-j\omega t}dt$

$$= 2\left[\frac{e^{-j\omega t}}{-\omega^2}(-j\omega t-1) + \frac{e^{-j\omega t}}{-j\omega}\right]_{-1}^{0} + 2\left[\frac{e^{-j\omega t}}{\omega^2}(-j\omega t-1) + \frac{e^{-j\omega t}}{-j\omega}\right]_{0}^{1}$$

$$= 2\left[\frac{1}{\omega^2} + \frac{e^{j\omega}}{\omega^2}(j\omega-1) + \frac{1}{-j\omega} + \frac{e^{j\omega}}{j\omega} + \frac{e^{-j\omega}}{\omega^2}(-j\omega-1) + \frac{1}{\omega^2} + \frac{1}{j\omega}\right]$$

$$= \frac{4}{\omega^2}(1-\cos\omega)$$

18.2

$F(j\omega) = \int_{-1}^{1}\cos\pi t\, e^{-j\omega t}dt$

$$= \frac{e^{-j\omega t}}{\pi^2-\omega^2}\left[j\omega\cos\pi t + \pi\sin\pi t\right]_{-1}^{1}$$

$$= \frac{1}{\pi^2-\omega^2}\left[e^{-j\omega}(-j\omega) - e^{j\omega}(-j\omega)\right]$$

$$= \frac{j\omega}{\pi^2-\omega^2}\left[e^{j\omega} - e^{-j\omega}\right]$$

$$= \frac{-2\omega}{\pi^2-\omega^2}\sin\omega$$

18.3

$F(j\omega) = \int_{-1}^{1}\cos^2\pi t\, e^{-j\omega t}dt$

$$= \frac{1}{2}\int_{-1}^{1}(1+\cos 2\pi t)e^{-j\omega t}dt$$

$$= \frac{1}{2}\left\{\frac{e^{-j\omega t}}{-j\omega} + \frac{e^{-j\omega t}}{4\pi^2-\omega^2}\left[-j\omega\cos 2\pi t + 2\pi\sin 2\pi t\right]\right\}\Big|_{-1}^{1}$$

$$= \frac{1}{-2j\omega}\left[e^{-j\omega} - e^{j\omega}\right] + \frac{e^{-j\omega}(-j\omega) - e^{j\omega}(-j\omega)}{2[4\pi^2-\omega^2]}$$

18.3 Cont.

$$F(j\omega) = \frac{\sin\omega}{\omega} - \frac{\omega\sin\omega}{4\pi^2-\omega^2}$$

$$= \frac{2\sin\omega}{4\pi^2-\omega^2}\left[\frac{2\pi^2}{\omega} - \omega\right]$$

18.4 $F(j\omega) = \int_{0}^{10} e^{t}e^{-j\omega t}dt$

$$= \frac{e^{-(j\omega-1)t}}{-j\omega+1}\Big|_{0}^{10}$$

$$= \frac{e^{-10(j\omega-1)}}{-j\omega+1}$$

18.5

$F(j\omega) = \int_{0}^{1} te^{-j\omega t}dt + \int_{1}^{3} 2e^{-j\omega t}dt$

$$= \frac{e^{-j\omega t}}{-\omega^2}(-j\omega t-1)\Big|_{0}^{1} + \frac{2e^{-j\omega t}}{-j\omega}\Big|_{1}^{3}$$

$$= \frac{e^{-j\omega}}{-\omega^2}(-j\omega-1) - \frac{1}{\omega^2} + \frac{2e^{-j3\omega}}{-j\omega} + \frac{2e^{-j\omega}}{j\omega}$$

$$= \frac{e^{-j\omega}(j\omega+1)-1}{\omega^2} + \frac{2(e^{-j3\omega}-e^{-j\omega})}{-j\omega}$$

18.6 $F(j\omega) = \int_{0}^{10}(\cos\pi t)e^{-j\omega t}dt$

$$= \frac{e^{-j\omega t}}{\pi^2-\omega^2}\left[-j\omega\cos\pi t + \pi\sin\pi t\right]_{0}^{10}$$

$$= \frac{j\omega}{\pi^2-\omega^2}\left(1 - e^{-j10\omega}\right)$$

$$= \frac{2\omega e^{-j5\omega}}{\omega^2-\pi^2}\left(\frac{e^{j5\omega} - e^{-j5\omega}}{2j}\right)$$

$$= \frac{2\omega e^{-j5\omega}\sin 5\omega}{\omega^2-\pi^2}$$

18.7 $F(j\omega) = \int_{-5}^{5}(\cos\pi t)e^{-j\omega t}dt$

$$= \frac{e^{-j\omega t}}{\pi^2-\omega^2}\left[-j\omega\cos\pi t + \pi\sin\pi t\right]_{-5}^{5}$$

$$= \frac{j\omega}{\pi^2-\omega^2}\left(e^{-j5\omega} - e^{j5\omega}\right)$$

$$= \frac{2\omega}{\omega^2-\pi^2}\sin 5\omega$$

18.8 $F(j\omega)=\int_{-1}^{0}-e^{-j\omega t}dt+\int_{0}^{1}e^{-j\omega t}dt$

$$=\left.\frac{e^{-j\omega t}}{j\omega}\right|_{-1}^{0}+\left.\frac{e^{-j\omega t}}{-j\omega}\right|_{0}^{1}$$

$$=\frac{1}{j\omega}\left(1-e^{j\omega}-e^{-j\omega}+1\right)$$

$$=\underline{\frac{2-2\cos\omega}{j\omega}}$$

18.9

$$F(j\omega)=\int_{-1}^{-1/2}-e^{-j\omega t}dt+\int_{-1/2}^{1/2}e^{-j\omega t}dt-\int_{1/2}^{1}e^{-j\omega t}dt$$

$$=\frac{1}{j\omega}\left[\left.e^{-j\omega t}\right|_{-1}^{-1/2}-\left.e^{-j\omega t}\right|_{-1/2}^{1/2}+\left.e^{-j\omega t}\right|_{1/2}^{1}\right]$$

$$=\frac{2\left(e^{j\omega/2}-e^{-j\omega/2}\right)-\left(e^{j\omega}-e^{-j\omega}\right)}{j\omega}$$

$$=\underline{\frac{4\sin\frac{\omega}{2}-2\sin\omega}{\omega}}$$

18.10 $F(j\omega)=\int_{-2}^{2}(e^{t}+e^{-t})e^{-j\omega t}dt$

$$=\left.\frac{e^{(1-j\omega)t}}{1-j\omega}-\frac{e^{-(1+j\omega)t}}{1+j\omega}\right|_{-2}^{2}$$

$$=\frac{e^{2(1-j\omega)}-e^{-2(1-j\omega)}}{1-j\omega}-\frac{e^{-2(1+j\omega)}-e^{2(1+j\omega)}}{1+j\omega}$$

$$=e^{2}\left(\frac{e^{-j2\omega}}{1-j\omega}+\frac{e^{j2\omega}}{1+j\omega}\right)-e^{-2}\left(\frac{e^{j2\omega}}{1-j\omega}+\frac{e^{-j2\omega}}{1+j\omega}\right)$$

$$=2\,\mathrm{Re}\left[e^{2}\left(\frac{e^{j2\omega}}{1+j\omega}\right)-e^{-2}\left(\frac{e^{-j2\omega}}{1+j\omega}\right)\right]$$

$$=\frac{2}{1+\omega^{2}}\Big[e^{2}\,\mathrm{Re}\left[(1-j\omega)(\cos 2\omega+j\sin 2\omega)\right]-e^{-2}\,\mathrm{Re}\left[(1-j\omega)(\cos 2\omega-j\sin 2\omega)\right]\Big]$$

$$=\frac{2}{1+\omega^{2}}\left[e^{2}(\cos 2\omega+\omega\sin 2\omega)-e^{-2}(\cos 2\omega-\omega\sin 2\omega)\right]$$

18.10 Cont.

$$F(j\omega)=\frac{4\omega}{1+\omega^{2}}\left[\frac{e^{2}-e^{-2}}{2}\frac{\cos 2\omega}{\omega}+\frac{e^{2}+e^{-2}}{2}\sin 2\omega\right]$$

$$=\underline{\frac{4}{1+\omega^{2}}\left(\sinh 2\cos 2\omega+\omega\cosh 2\sin 2\omega\right)}$$

18.11 $F(j\omega)=\int_{-2}^{2}(e^{t}-e^{-t})e^{-j\omega t}dt$

$$=\left.\frac{e^{(1-j\omega)t}}{1-j\omega}+\frac{e^{-(1+j\omega)t}}{1+j\omega}\right|_{-2}^{2}$$

$$=\frac{e^{2}e^{-j2\omega}}{1-j\omega}+\frac{e^{-2}e^{-j2\omega}}{1+j\omega}-\frac{e^{-2}e^{j2\omega}}{1-j\omega}-\frac{e^{2}e^{j2\omega}}{1+j\omega}$$

$$=\left[\frac{e^{2}e^{-j2\omega}}{1-j\omega}-\frac{e^{2}e^{j2\omega}}{1+j\omega}\right]+\left[\frac{e^{-2}e^{-j2\omega}}{1+j\omega}-\frac{e^{-2}e^{j2\omega}}{1-j\omega}\right]$$

$$=2j\,\mathrm{Im}\left[\frac{e^{2}e^{-j2\omega}}{1-j\omega}+\frac{e^{-2}e^{-j2\omega}}{1+j\omega}\right]$$

$$=\frac{2j\,\mathrm{Im}}{1+\omega^{2}}\Big[e^{2}(1+j\omega)(\cos 2\omega-j\sin 2\omega)+e^{-2}(1-j\omega)(\cos 2\omega-j\sin 2\omega)\Big]$$

$$=\underline{\frac{4j}{1+\omega^{2}}\left(\omega\cos 2\omega\sinh 2-\sin 2\omega\cosh 2\right)}$$

18.12 $f(t)$ is odd.

$$F(j\omega)=-2j\int_{0}^{\infty}f(t)\sin\omega t\,dt$$

$$=-2j\int_{0}^{1}\sin\omega t\,dt$$

$$=-2j\left[-\frac{\cos\omega t}{\omega}\right]_{0}^{1}$$

$$=\frac{-2j}{\omega}(-\cos\omega+1)$$

$$=\underline{\frac{2(1-\cos\omega)}{j\omega}}$$

18.13 $f(t)$ is even.

$$F(j\omega) = 2\int_0^\infty f(t)\cos\omega t\,dt$$

$$= 2\left[\int_0^{1/2}\cos\omega t\,dt - \int_{1/2}^{1}\cos\omega t\,dt\right]$$

$$= 2\left[\frac{\sin\omega t}{\omega}\bigg|_0^{1/2} - \frac{\sin\omega t}{\omega}\bigg|_{1/2}^{1}\right]$$

$$= \frac{2}{\omega}\left[\sin\frac{\omega}{2} - \sin\omega + \sin\frac{\omega}{2}\right]$$

$$= \frac{4\sin\frac{\omega}{2} - 2\sin\omega}{\omega}$$

18.14 $f(t)$ is even.

$$F(j\omega) = 2\int_0^3 t\cos\omega t\,dt$$

$$= 2\left[\frac{1}{\omega^2}\cos\omega t + \frac{t}{\omega}\sin\omega t\right]_0^3$$

$$= 2\left[\frac{\cos 3\omega - 1}{\omega^2} + \frac{3\sin 3\omega}{\omega}\right]$$

18.15 $f(t)$ is odd.

$$F(j\omega) = -j2\int_0^3 t\sin\omega t\,dt$$

$$= -j2\left[\frac{1}{\omega^2}\sin\omega t - \frac{t}{\omega}\cos\omega t\right]_0^3$$

$$= -j2\left[\frac{\sin 3\omega}{\omega^2} - \frac{3\cos 3\omega}{\omega}\right]$$

18.16 $f(t)$ is even.

$$F(j\omega) = 2\int_0^2 t^2\cos\omega t\,dt$$

$$= 2\left[\frac{2t\cos\omega t}{\omega^2} + \frac{\omega^2 t^2 - 2}{\omega^3}\sin\omega t\right]_0^2$$

$$= 2\left[\frac{4\cos 2\omega}{\omega^2} + \frac{4\omega^2 - 2}{\omega^3}\sin 2\omega\right]$$

18.17 $f(t)$ is odd.

$$F(j\omega) = -j2\int_0^2 t^2\sin\omega t\,dt$$

$$= \frac{-j2}{\omega^3}\Big[-\omega^2 t^2\cos\omega t + 2\cos\omega t + 2\omega t\sin\omega t\Big]_0^2$$

18.17 Cont.

$$F(j\omega) = -\frac{j2}{\omega^3}\Big[-4\omega^2\cos 2\omega + 2\cos 2\omega + 4\omega\sin 2\omega - 2\Big]$$

$$= \frac{j4}{\omega^3}\Big[1 + (2\omega^2 - 1)\cos 2\omega - 2\omega\sin 2\omega\Big]$$

18.18 Since $e^{-at}u(t) \longleftrightarrow \frac{1}{a + j\omega}$,

$$F(j\omega) = \frac{5}{2 + j\omega} + \frac{4}{3 + j\omega}$$

18.19 From $f(t - \tau) \longleftrightarrow F(j\omega)e^{-j\omega\tau}$,

$$F(j\omega) = \frac{6e^{j6\omega}}{3 + j\omega}$$

18.20
$$F(j\omega) = \frac{2}{\frac{1}{6} + j\omega} = \frac{12}{1 + j6\omega}$$

18.21
$$F(j\omega) = j\omega\,\mathcal{F}\left[2e^{-3t}u(t)\right]$$

$$= j\omega\frac{2}{3 + j\omega} = \frac{j2\omega}{3 + j\omega}$$

18.22 Let $f_1(t) = 1,\ -1 < t < 1$
$= 0$, elsewhere

Then $f(t) = t f_1(t)$; by (18.16)

$$F_1(j\omega) = \mathcal{F}[f_1(t)] = \frac{2\sin\omega}{\omega}$$

$$t f_1(t) \longleftrightarrow -\frac{d}{d(j\omega)}F_1(j\omega)$$

or

$$f(t) \longleftrightarrow -\frac{d}{d(j\omega)}\left(\frac{2\sin\omega}{\omega}\right)$$

$$= j\frac{d}{d\omega}\left(\frac{2\sin\omega}{\omega}\right)$$

$$= \frac{j2}{\omega^2}(\omega\cos\omega - \sin\omega)$$

18.23 Let $g(t) = \mathcal{F}^{-1}\left[\frac{\sin\omega}{\omega}\right] = \frac{1}{2},\ -1 < t < 1$
$= 0$, elsewhere

Then
$$f(t) = \mathcal{F}^{-1}\left[\frac{\sin\omega}{\omega}e^{-j\omega/2}\right]$$

$$= g(t - \tfrac{1}{2})$$

$= \frac{1}{2},\ -1 < t - \frac{1}{2} < 1$

$= 0$, elsewhere

Or

$f(t) = \frac{1}{2},\ -\frac{1}{2} < t < \frac{3}{2}$

$= 0$, elsewhere

18.24 $F(j\omega) = \frac{(j\omega)^2}{1+j\omega}$; since $\mathcal{F}^{-1}\left[\frac{1}{1+j\omega}\right] = e^{-t}u(t)$, then

$$f(t) = \frac{d^2}{dt^2}\left[e^{-t}u(t)\right]$$

18.25 $[4(j\omega)^2+6j\omega+1]Y(j\omega) = 2X(j\omega)$

$$H(j\omega) = \frac{Y(j\omega)}{X(j\omega)} = \frac{2}{-4\omega^2+6j\omega+1}$$

18.26 $X(j\omega) = \frac{e^{-j2\omega}}{1+j\omega}$

$$Y(j\omega) = H(j\omega)X(j\omega)$$

$$= \frac{2e^{-j2\omega}}{(-4\omega^2+6j\omega+1)(1+j\omega)}$$

18.27 $[(j\omega)^3+3j\omega]Y(j\omega) = 4X(j\omega)$

$$H(j\omega) = \frac{Y(j\omega)}{X(j\omega)} = \frac{4}{-j\omega^3+3j\omega}$$

18.28 $X(j\omega) = \frac{2}{\omega}\sin\frac{\omega}{2}$

$$Y(j\omega) = H(j\omega)X(j\omega)$$

$$= \frac{4(\frac{2}{\omega}\sin\frac{\omega}{2})}{-j\omega^3+3j\omega}$$

$$= \frac{8\sin\frac{\omega}{2}}{j\omega^2(-\omega^2+3)}$$

18.29 $f(t)$ is an even function defined by $f(t) = 2(-t+1)$, $0 \le t \le 1$, and $F(j\omega) = \frac{4}{\omega^2}(1-\cos\omega)$; For even functions, (18.43) is

$$\frac{1}{\pi}\int_0^\infty |F(j\omega)|^2 d\omega = 2\int_0^\infty f^2(\tau)\,d\tau$$

or

$$\frac{1}{\pi}\int_0^\infty \frac{16}{\omega^4}(1-\cos\omega)^2 d\omega$$

$$= 2\int_0^1 [2(-t+1)]^2 dt$$

or

$$\int_0^\infty \frac{(1-\cos\omega)^2}{\omega^4} d\omega = \frac{\pi}{16}(8)\int_0^1 (-t+1)^2 dt$$

$$= -\frac{\pi}{2}\left.\frac{(-t+1)^3}{3}\right|_0^1 = \frac{\pi}{6}$$

18.30

(a) $W_i = \int_0^\infty K^2 e^{-2at} dt = \frac{K^2}{2a}$

$$V_o(j\omega) = \frac{\frac{1}{j\omega C}}{R+\frac{1}{j\omega C}} \cdot \frac{K}{a+j\omega}$$

$$= \frac{K}{(1+j\omega RC)(a+j\omega)}$$

$$W_o = \frac{1}{\pi}\int_0^\infty \frac{K^2}{(1+\omega^2R^2C^2)(a^2+\omega^2)} d\omega$$

$$= \frac{K^2}{\pi}\int_0^\infty \left(\frac{A}{1+\omega^2R^2C^2} + \frac{B}{a^2+\omega^2}\right) d\omega$$

where $A = \frac{-R^2C^2}{1-a^2R^2C^2}$, $B = \frac{1}{1-a^2R^2C^2}$

$$W_o = \frac{K^2}{\pi}\left[\frac{B}{RC}\tan^{-1}\omega RC + \frac{D}{a}\tan^{-1}\frac{\omega}{a}\right]_0^\infty$$

$$= \frac{K^2}{\pi}\left[\frac{B\pi}{2RC} + \frac{D\pi}{2a}\right] = \frac{K^2}{2a(1+aRC)}$$

(b) $\frac{W_o}{W_i} = \frac{K^2}{2a(1+aRC)} \cdot \frac{2a}{K^2}$

$$= \frac{1}{1+aRC} = \frac{1}{1+1} = \frac{1}{2}$$

(c) $W_o = \frac{1}{\pi}\int_0^{10^3} \frac{K^2}{a^2+\omega^2} d\omega$

$$= \frac{K^2}{a\pi}\tan^{-1}\left.\frac{\omega}{a}\right|_0^{10^3} = \frac{K^2}{4(10^3)}$$

$$\frac{W_o}{W_i} = \frac{K^2}{4(10^3)} \cdot \frac{2(10^3)}{K^2} = \frac{1}{2}$$

Chapter 19

Laplace Transforms

19.1 Definition

19.1 Find the Laplace transform $F(s)$ of

$$f(t) = 3\cos 10t u(t)$$

19.2 Find the Laplace transform $F(s)$ of

$$f(t) = 4\sin 5t u(t)$$

19.3 Find the Laplace transform $F(s)$ of

$$f(t) = 2e^{-6t}u(t)$$

19.4 Find the Laplace transform $F(s)$ of $df(t)/dt$ if

$$f(t) = 3\cos 10t u(t)$$

19.5 Find the Laplace transform $F(s)$ of $df(t)/dt$ if

$$f(t) = (5-t)u(t) - (5-t)u(t-5)$$

19.6 Find the Laplace transform $F(s)$ of

$$f(t) = u(t)$$

19.2 Results Using Linearity

19.7 Find the Laplace transform $F(s)$ of

$$f(t) = (6\cos 3t + 2\sin 3t)u(t)$$

19.8 Find the inverse Laplace transform $f(t)$ of

$$F(s) = \frac{2}{s} + \frac{3}{s+3}$$

19.9 Find the Laplace transform $F(s)$ of

$$f(t) = (2e^{-2t} + 4e^{-4t})u(t)$$

19.10 Find the inverse Laplace transform $f(t)$ of

$$F(s) = \frac{1}{s^2} + \frac{s}{s^2 + 9}$$

19.3 Translation Theorems

19.11 Find the Laplace transform $F(s)$ of

$$f(t) = e^{-5t}u(t)$$

19.12 Find the Laplace transform $F(s)$ of

$$f(t) = \sin 5(t - 3)u(t - 3)$$

19.13 Find the Laplace transform $F(s)$ of

$$f(t) = 3u(t) - 3u(t - 6) + 2u(t - 7) - 2u(t - 9)$$

19.14 Find the inverse Laplace transform $f(t)$ of

$$F(s) = \frac{s}{(s + 2)^2 + 4} = \frac{s}{s^2 + 4s + 8}$$

19.15 Find the inverse Laplace transform $f(t)$ of

$$F(s) = \frac{e^{-3s}}{s^2}$$

19.16 Find the inverse Laplace transform $f(t)$ of

$$F(s) = \frac{4}{4s + 8}$$

19.4 Convolution

19.17 Find $f * g$ if

$$\begin{aligned} f(t) &= u(t) \\ g(t) &= -u(t - 3) \end{aligned}$$

19.18 Find $f * g$ if

$$\begin{aligned} f(t) &= u(t) - u(t - 4) \\ g(t) &= tu(t) - tu(t - 2) \end{aligned}$$

19.19 Find $f * g$ if

$$\begin{aligned} f(t) &= 4 \cos 4tu(t) \\ g(t) &= u(t) - u(t - 5) \end{aligned}$$

19.20 Find the inverse Fourier transform of

$$F(s) = \frac{1}{(s + 3)(s + 5)}$$

19.5 Impulse Function

19.21 Find the Fourier transform $F(s)$:

$$f(t) = \sum_{n=0}^{\infty} \delta(t-n)$$

19.22 Evaluate

$$\int_a^{\infty} e^{-3t}\delta(t-2)dt$$

for (a) $a = 3$ (b) $a = 1$.

19.23 Find $f * g$ if

$$\begin{aligned} f(t) &= \delta(t-3) \\ g(t) &= \quad g(t) \end{aligned}$$

19.6 Inverse Transforms

19.24 Find the inverse Laplace transform $f(t)$ of

$$F(s) = \frac{11s^2 - 10s + 11}{(s^2+1)(s^2-2s+5)}$$

19.25 Find the inverse Laplace transform $f(t)$ of

$$F(s) = \frac{s^3 + 2s^2 + 4s + 5}{(s+1)^2(s+2)^2}$$

19.26 Find the inverse Laplace transform $f(t)$ of

$$F(s) = \frac{2 - 2e^{-2s} - 4se^{-2s} + 2s^2e^{-2s}}{s^3}$$

19.27 Find the inverse Laplace transform $f(t)$ of

$$F(s) = \frac{s}{(s^2+1)^2}$$

19.7 Differential Theorems

19.28 Find the Laplace transform $F(s)$ of

$$f(t) = t^3e^{-3t}u(t)$$

19.8 Integrodifferntial Equations

19.29 Use Laplace transforms to solve for $y(t)$ for $t > 0$

$$y'' + 2y' + y = 3te^{-t}$$

$$y(0) = 4, \quad y'(0) = 2$$

19.30 Use Laplace transforms to solve for $y(t)$ for $t > 0$

$$y'' - 4y' + 4y = 4\cos 2t$$

$$y(0) = 2, \quad y'(0) = 5$$

19.31 Use Laplace transforms to solve for $y(t)$ for $t > 0$

$$\int_0^t y(\tau)\sin(t-\tau)d\tau = y(t) + \sin t - \cos t$$

19.32 Solve for $x(t)$ for $t > 0$

$$\begin{aligned} x'' + x &= f(t) \\ f(t) &= 1, \quad 0 \le t < \pi/2 \\ &= 0, \quad t > \pi/2 \\ x(0) &= 0, \quad x' = 1 \end{aligned}$$

19.33 Use Laplace transforms to solve for $x(t)$ for $t > 0$

$$x' = t + \int_0^t x(t-\tau)\cos\tau d\tau$$

$$x(0) = 4$$

19.1 $F(s) = \int_0^\infty 3(\cos 10t)e^{-st}dt$

$= \frac{3e^{-st}}{s^2+100}\left[-s\cos 10t + 10\sin 10t\right]_0^\infty$

$= 0 - \frac{-3s}{s^2+100} = \underline{\frac{3s}{s^2+100}}$

19.2 $F(s) = \int_0^\infty 4(\sin 5t)e^{-st}dt$

$= \frac{4e^{-st}}{s^2+25}\left[-s\sin 5t - 5\cos 5t\right]_0^\infty$

$= 0 - \frac{-20}{s^2+25} = \underline{\frac{20}{s^2+25}}$

19.3 $F(s) = \int_0^\infty 2e^{-6t}e^{-st}dt$

$= 2\int_0^\infty e^{-(s+6)t}dt$

$= \frac{2e^{-(s+6)t}}{-(s+6)}\Big]_0^\infty = \underline{\frac{2}{s+6}}$

19.4 From Prob. 19.1, $F(s) = \frac{3s}{s^2+100}$

Also $f(0) = 3$;

$\mathcal{L}\left[\frac{df}{dt}\right] = sF(s) - f(0)$

$= \underline{\frac{3s}{s^2+100} - 3}$

19.5 $F(s) = \int_0^\infty (5-t)[u(t)-u(t-5)]e^{-st}dt$

$= \int_0^5 (5-t)e^{-st}dt$

$= \frac{5e^{-st}}{-s} - \frac{e^{-st}}{s^2}(-5t-1)\Big]_0^5$

$= \frac{e^{-5s}-1}{s^2} + \frac{5}{s}$

$f(0) = 5$;

$\mathcal{L}\left[\frac{df}{dt}\right] = s\left[\frac{e^{-5s}-1}{s^2} + \frac{5}{s}\right] - 5$

$= \underline{\frac{e^{-5s}-1}{s}}$

19.6 $F(s) = \int_0^\infty e^{-st}dt = \frac{e^{-st}}{-s}\Big]_0^\infty = \underline{\frac{1}{s}}$

19.7 $F(s) = 6\mathcal{L}[\cos 3t] + 2\mathcal{L}[\sin 3t]$

$= 6\left(\frac{s}{s^2+9}\right) + 2\left(\frac{3}{s^2+9}\right)$

$= \underline{6\,\frac{s+1}{s^2+9}}$

19.8 $f(t) = \mathcal{L}^{-1}\left[\frac{2}{s}\right] + \mathcal{L}^{-1}\left[\frac{3}{s+2}\right]$

$= 2u(t) + 3e^{-2t}u(t)$

$= \underline{(2+3e^{-2t})u(t)}$

19.9 $F(s) = \mathcal{L}[2e^{-2t}u(t)] + \mathcal{L}[4e^{-4t}u(t)]$

$= \underline{\frac{2}{s+2} + \frac{4}{s+4}}$

19.10 $f(t) = \mathcal{L}^{-1}\left[\frac{1}{s^2}\right] + \mathcal{L}^{-1}\left[\frac{s}{s^2+9}\right]$

$= \underline{(t + \cos 3t)u(t)}$

19.11 $F(s) = \mathcal{L}[e^{-5t}u(t)]$

$\mathcal{L}[u(t)] = \frac{1}{s}$

$\therefore F(s) = \underline{\frac{1}{s+5}}$

19.12 $F(s) = e^{-3s}\mathcal{L}[\sin 5t] = \underline{\frac{5e^{-3s}}{s^2+25}}$

19.13 $F(s) = 3\mathcal{L}[u(t)] - 3e^{-6s}\mathcal{L}[u(t)]$
$+ 2e^{-7s}\mathcal{L}[u(t)] - 2e^{-9s}\mathcal{L}[u(t)]$

$= \underline{\frac{1}{s}\left[3 - 3e^{-6s} + 2e^{-7s} - 2e^{-9s}\right]}$

19.14 $f(t) = \mathcal{L}^{-1}\left[\frac{s}{(s+2)^2+4}\right]$

$= \mathcal{L}^{-1}\left[\frac{s+2}{(s+2)^2+2^2} - \frac{2}{(s+2)^2+2^2}\right]$

$= \underline{e^{-2t}[\cos 2t - \sin 2t]u(t)}$

19.15 $f(t) = \mathcal{L}^{-1}\left[\frac{e^{-3s}}{s^2}\right]$;

$\mathcal{L}^{-1}\left[\frac{1}{s^2}\right] = t\,u(t)$;

$\underline{f(t) = (t-3)u(t-3)}$

19.16 $\mathcal{L}^{-1}[F(as)] = \frac{1}{a} f\left(\frac{t}{a}\right)$

$$\mathcal{L}^{-1}\left[\frac{4}{4s+8}\right] = \frac{1}{4}(4e^{-8t/4}) = \underline{e^{-2t}u(t)}$$

19.17 $f*g = \mathcal{L}^{-1}[F(s)\,G(s)]$

$$= \mathcal{L}^{-1}\left[\left(\frac{1}{s}\right)\left(-\frac{e^{-3s}}{s}\right)\right] = \mathcal{L}^{-1}\left[-\frac{e^{-3s}}{s^2}\right]$$

$$\underline{= -(t-3)u(t-3)}$$

19.18 $F(s) = \frac{1-e^{-4s}}{s}$

$$G(s) = \mathcal{L}^{-1}[tu(t) - (t-2)u(t-2) - 2u(t-2)]$$

$$= \frac{1}{s^2}(1-e^{-2s}) - \frac{2e^{-2s}}{s}$$

$$f(t)*g(t) = \mathcal{L}^{-1}[F(s)\,G(s)]$$

$$= \mathcal{L}^{-1}\left[\frac{(1-e^{-4s})(1-e^{-2s})}{s^3} - \frac{(1-e^{-4s})(2e^{-2s})}{s^2}\right]$$

$$= \mathcal{L}^{-1}\left[\frac{1}{s^3}(1-e^{-2s}-e^{-4s}+e^{-6s}) - \frac{2}{s^2}(e^{-2s}-e^{-6s})\right]$$

$$= \frac{t^2}{2}u(t) - \frac{(t-2)^2}{2}u(t-2) - \frac{(t-4)^2}{2}u(t-4)$$

$$\underline{+\frac{(t-6)^2}{2}u(t-6) - 2(t-2)u(t-2) + 2(t-6)u(t-6)}$$

19.19 $F(s) = \frac{4s}{s^2+16}$; $G(s) = \frac{1}{s} - \frac{e^{-5s}}{s}$

$$f*g = \mathcal{L}^{-1}\left[\frac{4}{s^2+16} - \frac{4e^{-5s}}{s^2+16}\right]$$

$$\underline{= \sin 4t\,u(t) - \sin 4(t-5)\,u(t-5)}$$

19.20 $f(t) = \mathcal{L}^{-1}\left[\frac{1}{s+3}\cdot\frac{1}{s+5}\right]$

$$= e^{-3t}u(t) * e^{-5t}u(t)$$

$$= \int_0^t e^{-3\tau}e^{-5(t-\tau)}\,d\tau$$

$$= e^{-5t}\int_0^t e^{2\tau}\,d\tau$$

$$= e^{-5t}\left[\frac{1}{2}e^{2\tau}\right]_0^t$$

$$\underline{= \frac{1}{2}\left(e^{-3t} - e^{-5t}\right)u(t)}$$

19.21 $F(s) = \int_0^\infty e^{-st}\left[\sum_{n=0}^\infty \delta(t-n)\right]dt$

$$= \sum_{n=0}^\infty \int_0^\infty e^{-st}\,\delta(t-n)\,dt$$

$$\underline{= \sum_{n=0}^\infty e^{-sn}}$$

19.22 (a) $\int_3^{10} e^{-3t}\,\delta(t-2)\,dt = \underline{0}$

$(t \neq 2$ on $3 < t < 10)$

(b) $\int_1^{10} e^{-3t}\,\delta(t-2)\,dt = e^{-3(2)} = \underline{e^{-6}}$

19.23 $F(s) = e^{-3s}$

$$f(t)*g(t) = \mathcal{L}^{-1}[e^{-3s}G(s)]$$

$\underline{= g(t-3)}$; (Assuming $g(t)=0$, $t<0$)

19.24 $F(s) = \frac{11s^2-10s+11}{(s^2+1)(s^2-2s+5)}$

$$= \frac{As+B}{s^2+1} + \frac{Cs+D}{s^2-2s+5}$$

$$11s^2-10s+11 = (As+B)(s^2-2s+5) + (Cs+D)(s^2+1)$$

s^3: $0 = A + C$

s^2: $11 = -2A + B + D$

s : $-10 = 5A - 2B + C$

s^0: $11 = 5B + D$

which yields

$A = -2$, $B = 1$, $C = 2$, $D = 6$

$$F(s) = \frac{-2s+1}{s^2+1} + \frac{2s+6}{(s-1)^2+4}$$

$$= \frac{-2s}{s^2+1} + \frac{1}{s^2+1} + \frac{2(s-1)}{(s-1)^2+4} + \frac{8}{(s-1)^2+4}$$

$$\underline{f(t) = -2\cos t + \sin t + 2e^t(\cos 2t + 2\sin 2t)}$$

19.25

$$\frac{s^3+2s^2+4s+5}{(s+1)^2(s+2)^2}=\frac{A}{(s+1)^2}+\frac{B}{s+1}+\frac{C}{(s+2)^2}+\frac{D}{s+2}$$

$$A=\left.\frac{s^3+2s^2+4s+5}{(s+2)^2}\right|_{s=-1}=2$$

$$C=\left.\frac{s^3+2s^2+4s+5}{(s+1)^2}\right|_{s=-2}=-3$$

$$s^3+2s^2+4s+5=2(s+2)^2+B(s+1)(s^2+4s+4)$$
$$-3(s+1)^2+D(s+2)(s^2+2s+1)$$

s^3: $1=B+D$

s^0: $5=8+4B-3+2D$

which yields $B=-1, D=2$

$$F(s)=\frac{2}{(s+1)^2}-\frac{1}{s+1}-\frac{3}{(s+2)^2}+\frac{2}{s+2}$$

$$f(t)=(2t-1)e^{-t}+(-3t+2)e^{-2t}$$

19.26

$$F(s)=\frac{2}{s^3}-\frac{2e^{-2s}}{s^3}-\frac{4e^{-2s}}{s^2}+\frac{2e^{-2s}}{s}$$

$$f(t)=t^2u(t)-(t-2)^2u(t-2)$$
$$-4(t-2)u(t-2)+2u(t-2)$$

or $f(t)=t^2,\ 0<t<2$

$=6,\ t>2$

19.27 $F(s)=\frac{s}{s^2+1}\cdot\frac{1}{s^2+1}$

$$f(t)=\left[\mathcal{L}^{-1}\left(\frac{s}{s^2+1}\right)\right]*\left[\mathcal{L}^{-1}\left(\frac{1}{s^2+1}\right)\right]$$

$$=\cos t*\sin t$$

$$=\int_0^t\cos\tau\sin(t-\tau)\,d\tau$$

$$=\int_0^t\cos\tau[\sin t\cos\tau-\cos t\sin\tau]\,d\tau$$

$$=\sin t\left[\frac{\tau}{2}+\frac{1}{4}\sin 2\tau\right]-\cos t\left[\frac{1}{2}\sin^2\tau\right]\Bigg|_{\tau=0}^{t}$$

$$=\sin t\left(\frac{t}{2}+\frac{1}{4}\sin 2t\right)-\frac{1}{2}\cos t\sin^2 t$$

$$=\frac{1}{2}t\sin t\ u(t)$$

19.28 $F(s)=(-1)^3\frac{d^3}{ds^3}\left(\frac{1}{s+3}\right)$

$$=-\frac{d^2}{ds^2}\left[-(s+3)^{-2}\right]$$

$$=-\frac{d}{ds}\left[2(s+3)^{-3}\right]$$

$$=\frac{6}{(s+3)^4}$$

19.29 $s^2Y(s)-4s-2+2[sY(s)-4]+Y(s)$

$$=\frac{3}{(s+1)^2};$$

$$Y(s)=\frac{4s+10+\frac{3}{(s+1)^2}}{s^2+2s+1}$$

$$=\frac{4s+10}{(s+1)^2}+\frac{3}{(s+1)^4}$$

$$=\frac{4(s+1)+6}{(s+1)^2}+\frac{3}{(s+1)^4}$$

$$=\frac{4}{s+1}+\frac{6}{(s+1)^2}+\frac{3}{(s+1)^4}$$

$$y(t)=\left(4e^{-t}+6te^{-t}+\frac{1}{2}t^3e^{-t}\right)u(t)$$

19.30 $s^2Y(s)-2s-5-4[sY(s)-2]+4Y(s)$

$$=\frac{4s}{s^2+4}$$

$$Y(s)=\frac{2s-3+\frac{4s}{s^2+4}}{s^2-4s+4}$$

$$=\frac{(2s-3)(s^2+4)+4s}{(s-2)^2(s^2+4)}=\frac{A}{(s-2)^2}+\frac{B}{s-2}+\frac{Cs+D}{s^2+4}$$

$$A=\left.\frac{(2s-3)(s^2+4)+4s}{s^2+4}\right|_{s=2}=2$$

$$(2s-3)(s^2+4)+4s=2(s^2+4)+B(s-2)(s^2+4)$$
$$+(Cs+D)(s^2-4s+4)$$

s^3: $2=B+C$

s^2: $-3=2-2B-4C+D$

s^0: $-12=8-8B+4D$

which yields

$B=2,\ C=0,\ D=-1$

19.30 Cont.

$$Y(s) = \frac{2}{(s-2)^2} + \frac{2}{s-2} - \frac{1}{s^2+4}$$

$$y(t) = \left[2(t+1)e^{2t} - \tfrac{1}{2}\sin 2t\right]u(t)$$

19.31 $\mathcal{L}\left[\int_0^t y(\tau)\sin(t-\tau)d\tau\right]$

$$= Y(s) + \frac{1-s}{s^2+1}$$

$$Y(s)\frac{1}{s^2+1} = Y(s) + \frac{1-s}{s^2+1}$$

$$Y(s)\left[1-(s^2+1)\right] = 1-s$$

$$Y(s) = \frac{s-1}{s^2} = \frac{1}{s} - \frac{1}{s^2}$$

$$y(t) = (1-t)\,u(t)$$

19.32 $x'' + x = u(t) - u(t-\pi/2)$

$$s^2X(s) - 1 + X(s) = \frac{1}{s} - \frac{1}{s}e^{-\pi s/2}$$

$$X(s) = \frac{\frac{1}{s}+1}{s^2+1} - \frac{\frac{1}{s}e^{-s\pi/2}}{s^2+1}$$

$$= \frac{s+1}{s(s^2+1)} - \frac{1}{s(s^2+1)}e^{-s\pi/2}$$

$$\frac{1}{s(s^2+1)} = \frac{A}{s} + \frac{Bs+C}{s^2+1}$$

$$A = \left.\frac{1}{s^2+1}\right|_{s=0} = 1$$

$$1 = s^2+1+(Bs+C)s$$

s^2: $0 = 1+B$, $B=-1$

s: $0 = C$

$$X(s) = \frac{1}{s^2+1} + \frac{1}{s(s^2+1)} - \frac{1}{s(s^2+1)}e^{-s\pi/2}$$

$$= \frac{1}{s^2+1} + \frac{1}{s} - \frac{s}{s^2+1} - \left[\frac{1}{s} - \frac{s}{s^2+1}\right]e^{-s\pi/2}$$

$$x(t) = (\sin t + 1 - \cos t)u(t) - \left[1 - \cos(t - \tfrac{\pi}{2})\right]u(t-\tfrac{\pi}{2})$$

19.32 Cont.

$$x(t) = (\sin t + 1 - \cos t)\,u(t) - (1 - \sin t)\,u(t - \tfrac{\pi}{2})$$

19.33 $sX(s) - 4 = \frac{1}{s^2} + \frac{s}{s^2+1}X(s)$

$$X(s)\left[s - \frac{s}{s^2+1}\right] = \frac{1}{s^2} + 4$$

$$X(s) = \frac{(\frac{1}{s^2}+4)(s^2+1)}{s(s^2+1)-s}$$

$$= \frac{(4s^2+1)(s^2+1)}{s^5}$$

$$= \frac{4}{s} + \frac{5}{s^3} + \frac{1}{s^5}$$

$$x(t) = \left(4 + \tfrac{5}{2}t^2 + \tfrac{1}{24}t^4\right)u(t)$$

Chapter 20

Laplace Transform Applications

20.1 Applications

20.1 Use the describing equations and Laplace transforms to find i for > 0 if $i(0) = 4$ A and $v(0) = 6$V.

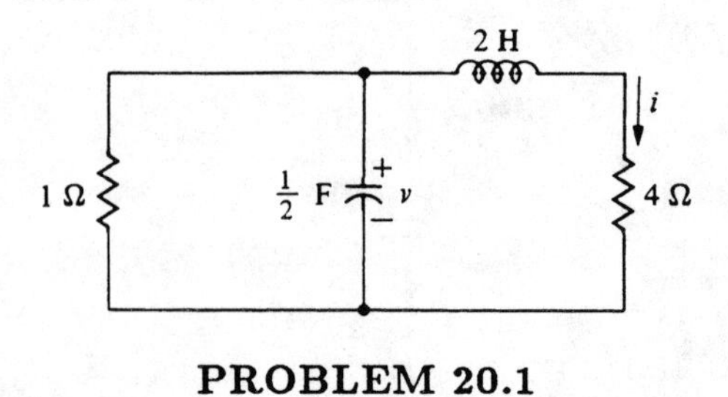

PROBLEM 20.1

20.2 Use the describing equations and Laplace transforms to find i for $t > 0$ if $L = 3$H and the circuit is in steady-state at $t = 0^-$.

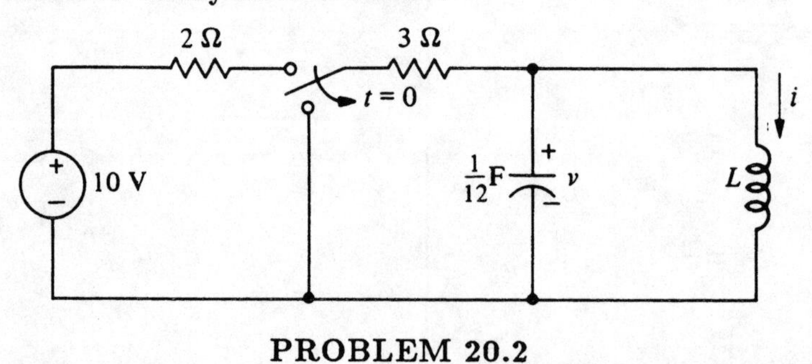

PROBLEM 20.2

20.3 Use the describing equations and Laplace transforms to find i for $t > 0$ if the circuit is in steady-state at $t = 0^-$.

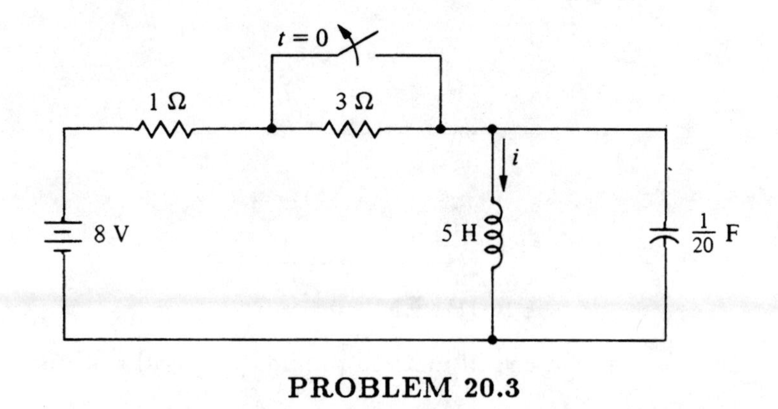

PROBLEM 20.3

20.4 Use the describing equations and Laplace transforms to find v for $t > 0$ if $i_g = 2u(t)$A.

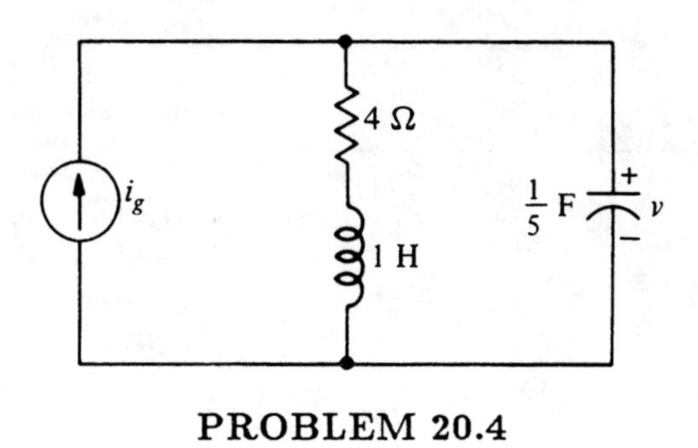

PROBLEM 20.4

20.2 Transformed Circuits

20.5 Use the Laplace transformed circuit method to find $i(t)$, $t > 0$, if $v(0) = 4$V and $i(0) = 10$A.

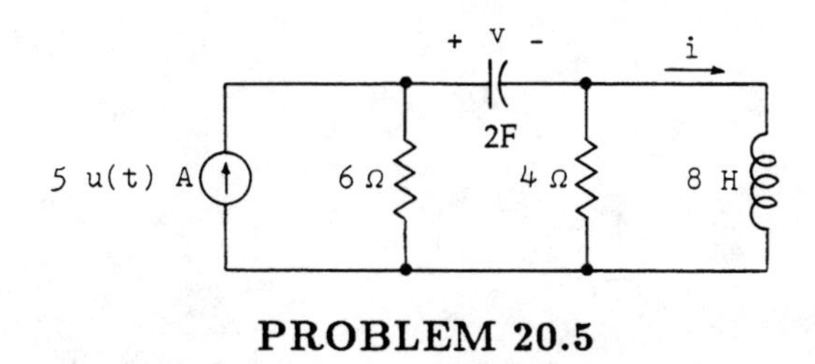

PROBLEM 20.5

20.6 Use the Laplace transformed circuit method to find $i(t)$, $t > 0$, if $v(0) = 4$V and $i(0) = 2$A.

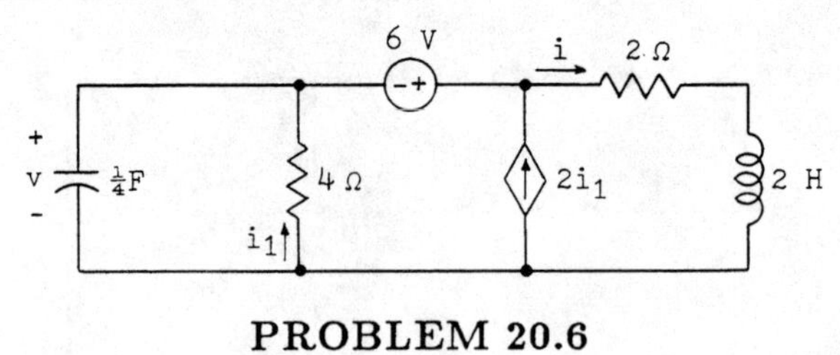

PROBLEM 20.6

20.7 Use the Laplace transformed circuit method to find v, if $v_1(0) = 4\text{V}$ and $v_2(0) = 2\text{V}$.

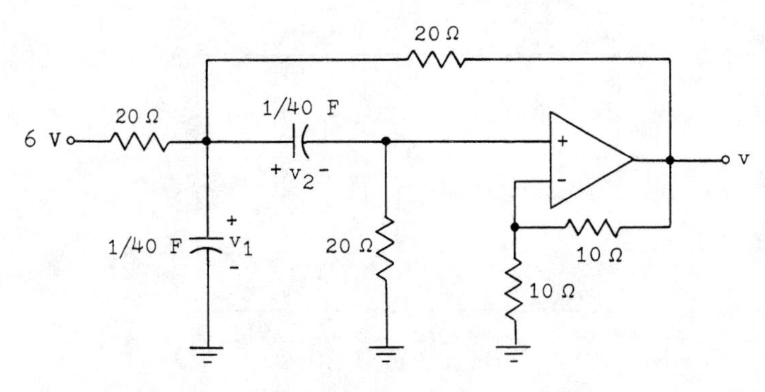

PROBLEM 20.7

20.3 Network Functions

20.8 Find $\mathbf{H}(s)$ in Prob 20.5 if i is the output and the current source is the input.

20.9 For $R = 20\Omega$, $L = 2\text{H}$, and $C = 1/80\text{F}$ find $\mathbf{H}(s)$ if v is the output and v_g is the input.

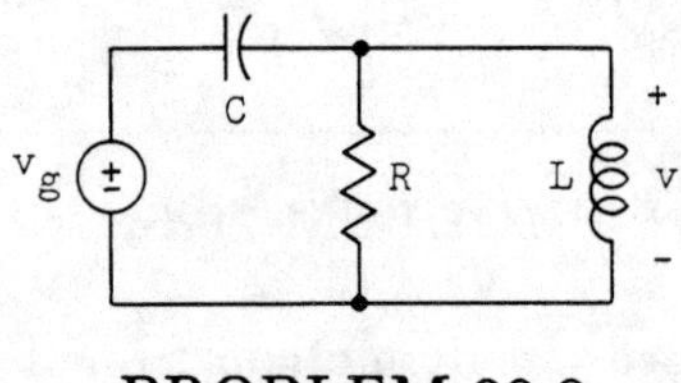

PROBLEM 20.9

20.10 For $R = 7\Omega$, $L = 8\text{H}$ and $C = 1/56\text{F}$ find $\mathbf{H}(s)$ in Prob. 20.9.

20.11 Find $\mathbf{H}(s)$ in Prob. 20.7 if v is the output and the 6V DC source is the input.

20.4 Step and Impulse Responses

20.12 Find the impulse and step responses for Prob. 20.9.

20.13 Find the impulse and step responses for Prob. 20.10.

20.14 Find the impulse and step responses for Prob. 20.8.

20.15 Find the impulse and step responses for Prob. 20.11.

20.5 Stability

20.16 Find $\frac{\mathbf{V}_2(s)}{\mathbf{V}_g(s)}$ and use the result to determine μ so that the network shown is (a) absolutely stable, (b) conditionally stable, and (c) unstable.

20.17 Determine if the following are network functions of absolutely stable, conditionally stable, or unstable circuits, and state the reasons.

$$(a) \quad \frac{s}{s^3+6s^2+11s+6} \quad (b) \quad \frac{s}{s^3+4s^2+s-6}$$

20.18 Determine if the following are network functions of absolutely stable, conditionally stable, or unstable circuits, and state the reasons.

$$(a) \quad \frac{s^3}{(s^2+1)^2(s+4)} \quad (b) \quad \frac{s}{s^3+2s^2+s+2}$$

20.19 Determine if the following are network functions of absolutely stable, conditionally stable, or unstable circuits, and state the reasons.

$$(a) \quad \frac{s^4}{(s+1)(s+3)} \quad (b) \quad \frac{s^4}{s^4+4s^2+3}$$

20.6 Initial and Final Value Theorems

20.20 Find the initial and final values using the initial and final value theorems were applicable of

$$(a) \quad F(s) = \frac{3s^2}{s^2+5s+6} \quad (b) \quad F(s) = \frac{3s+2}{s^2+5s+6}$$

20.21 If $\mu = 1$ in Prob. 20.16 and $v_g = 4\text{V}$, $v_1 = 2\text{V}$, and $v_3 = 0\text{V}$, find the initial value $v_2(0^+)$ and the final value $\lim_{t \to \infty} v_2(t)$ using the initial and final value theorems.

20.22 Find the initial and final values using the initial and final value theorems were applicable of

$$F(s) = \frac{3}{s^3 + 4s^2 + s - 6}$$

20.23 Find the initial and final values using the initial and final value theorems were applicable of

$$F(s) = \frac{s^2 + s + 1}{s^2 + 7s + 12}$$

20.7 Steady-State Sinusoidal Response

20.24 Find the sinusoidal steady-state response for the transfer function in Prob. 20.8 if the current source is $10\cos 2tu(t)$.

20.25 Find the sinusoidal steady-state response for the transfer function in Prob. 20.9 if v_g is $4\sin 3tu(t)$.

20.26 Find the sinusoidal steady-state response for the transfer function in Prob. 20.11 if v_{in} is $3\sin 6tu(t)$.

20.8 Bode Plots

20.27 Find the straight-line approximation to the Bode magnitude plot for the function

$$\mathbf{H}(s) = \frac{100(s+10)}{(s+1)(s+100)}$$

20.28 Find the straight-line approximation to the Bode magnitude plot for the function

$$\mathbf{H}(s) = \frac{.1s}{(s+1)(s+10)}$$

20.9 Quadratic Factors

20.29 Find ς_k, ω_k, ω_{max}, and M_{max} for

$$\mathbf{H}(s) = \frac{3}{100+4s+s^2}$$

20.30 Find ς_k, ω_k, ω_{max}, and M_{max} for

$$\mathbf{H}(s) = \frac{9}{36+s+s^2}$$

20.1 Node equation:

$$\frac{v}{1} + \frac{1}{2}\frac{dv}{dt} + i = 0$$

Loop equation:

$$2\frac{di}{dt} + 4i - v = 0$$

$$V(s) + \frac{1}{2}[sV(s) - 6] + I(s) = 0 \quad (1)$$

$$2[sI(s) - 4] + 4I(s) - V(s) = 0 \quad (2)$$

$$V(s) = (2s+4)I(s) - 8 \quad (2)$$

$$(\tfrac{1}{2}s+1)[(2s+4)I(s) - 8] - 3 + I(s) = 0 \quad (1)$$

$$I(s) = \frac{8(\frac{1}{2}s+1) + 3}{(\frac{1}{2}s+1)(2s+4)+1} = \frac{4s+11}{s^2+4s+5}$$

$$= \frac{4(s+2)}{(s+2)^2+1} + \frac{3}{(s+2)^2+1}$$

$$i(t) = e^{-2t}(4\cos t + 3\sin t) \text{ A}$$

20.2 At $t=0^-$, $v=0$, $i = \frac{10}{2+3} = 2$ A

$\therefore v(0^+) = 0$, $i(0^+) = 2$ A

Node equation ($t>0$):

$$\frac{1}{12}\frac{dv}{dt} + \frac{v}{3} + i = 0$$

Loop equation:

$$3\frac{di}{dt} = v$$

$$\frac{1}{12}[sV(s) - 0] + \frac{V(s)}{3} + I(s) = 0 \quad (1)$$

$$V(s) = 3[sI(s) - 2]$$

$$(\tfrac{s}{12} + \tfrac{1}{3})(3)[sI(s) - 2] + I(s) = 0 \quad (1)$$

$$I(s) = \frac{6(\frac{s}{12}+\frac{1}{3})}{3s(\frac{s}{12}+\frac{1}{3})+1}$$

$$= \frac{2(s+4)}{(s+2)^2} = \frac{2(s+2)+4}{(s+2)^2}$$

$$= \frac{2}{s+2} + \frac{4}{(s+2)^2}$$

$$i(t) = 2e^{-t} + 4te^{-2t}$$

$$= (2+4t)e^{-2t} \text{ A}$$

20.3 $v_c(0^-) = v_c(0^+) = 0$

$$i(0^-) = i(0^+) = \frac{8}{1} = 8 \text{ A}$$

Node equation ($t>0$):

$$\frac{v_c - 8}{4} + i + \frac{1}{20}\frac{dv_c}{dt} = 0$$

Loop equation:

$$5\frac{di}{dt} = v_c$$

$$\frac{V_c(s)}{4} - \frac{2}{s} + I(s) + \frac{1}{20}[sV_c(s) - 0] = 0$$

$$5[sI(s) - 8] = V_c(s)$$

$$(\tfrac{1}{4} + \tfrac{s}{20})(5)[sI(s) - 8] + I(s) = \frac{2}{s}$$

$$I(s) = \frac{\frac{2}{s} + 40(\frac{1}{4}+\frac{s}{20})}{5s(\frac{1}{4}+\frac{s}{20})+1}$$

$$= \frac{8(s^2+5s+1)}{s(s+1)(s+4)} = \frac{2}{s} + \frac{8}{s+1} - \frac{2}{s+4}$$

$$i(t) = 2 + 8e^{-t} - 2e^{-4t} \text{ A}$$

20.4 $v(0^-) = v(0^+) = i_L(0^-) = i_L(0^+) = 0$

Nodal equation:

$$i_L + \frac{1}{5}\frac{dv}{dt} = i_g$$

Loop: $\frac{di_L}{dt} + 4i_L = v$

$$I_L(s) + \frac{1}{5}sV(s) = \frac{2}{s}$$

$$sI_L(s) + 4I_L(s) = V(s) \quad (1)$$

$$(s+4)[\tfrac{2}{s} - \tfrac{s}{5}V(s)] = V(s) \quad (1)$$

$$V(s) = \frac{\frac{2}{s}(s+4)}{1+\frac{s}{5}(s+4)}$$

$$= \frac{10(s+4)}{s(s^2+4s+5)}$$

$$= \frac{8}{s} - \frac{8s+22}{s^2+4s+5}$$

$$= \frac{8}{s} - 8\left[\frac{s+2}{(s+2)^2+1}\right] - \frac{6}{(s+2)^2+1}$$

$$v(t) = 8 - e^{-2t}(8\cos t + 6\sin t) \text{ V}$$

20.5 The transformed circuit is

I_1, 80, I, $\frac{5}{s}$, 6, $\frac{4}{s}$, $\frac{1}{2s}$, 4, $8s$

KVL: $\frac{4}{s} + \frac{1}{2s} I_1 + 4(I_1 - I) + 6\left(I_1 - \frac{5}{s}\right) = 0$

$$-80 + 8sI + 4(I - I_1) = 0 \quad (1)$$

$$I_1 = (2s+1)I - 20 \quad (1)$$

$$\frac{4}{s} + \left(\frac{1}{2s} + 4 + 6\right)\left[(2s+1)I - 20\right] - 4I - \frac{30}{s} = 0$$

$$I = \frac{\frac{30}{s} + 20\left(\frac{1}{2s} + 10\right) - \frac{4}{s}}{\left(\frac{1}{2s} + 10\right)(2s+1) - 4}$$

$$= \frac{400s + 72}{40s^2 + 14s + 1} = \frac{10s + \frac{9}{5}}{\left(s + \frac{1}{10}\right)\left(s + \frac{1}{4}\right)}$$

$$= \frac{\frac{16}{3}}{s + \frac{1}{10}} + \frac{\frac{14}{3}}{s + \frac{1}{4}}$$

$$i(t) = \frac{16}{3} e^{-t/10} + \frac{14}{3} e^{-t/4} \text{ A}$$

20.6 Transformed circuit:

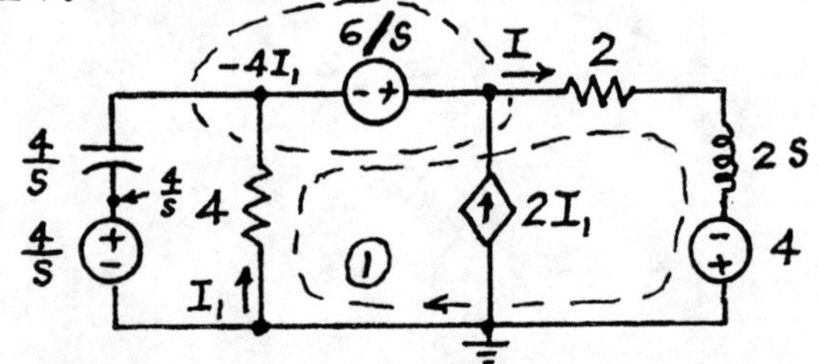

KCL equation (supernode):

$$\frac{-4I_1 - \frac{4}{s}}{\frac{4}{s}} - I_1 - 2I_1 + I = 0 \quad (1)$$

KVL equation (Loop 1):

$$-\frac{6}{s} + (2s+2)I - 4 + 4I_1 = 0 \quad (2)$$

From (1) and (2),

$$I = \frac{2s^2 + 11s + 9}{s(s^2 + 4s + 5)} = \frac{9/5}{s} + \frac{\frac{1}{5}s + \frac{19}{5}}{s^2 + 4s + 5}$$

$$= \frac{9/5}{s} + \frac{1}{5}\left[\frac{s+2}{(s+2)^2 + 1}\right] + \frac{17}{5}\left[\frac{1}{(s+2)^2 + 1}\right]$$

$$i(t) = \frac{9}{5} + \frac{1}{5} e^{-2t}(\cos t + 17 \sin t) \text{ A}$$

20.7 Transformed circuit:

20, $\frac{6}{s}$, 20, V_1, $40/s$, $V/2$, $2/s$, V, $\frac{40}{s}$, $\frac{4}{s}$, 20, 10, 10

KCL: $\frac{V_1 - \frac{6}{s}}{20} + \frac{V_1 - \frac{4}{s}}{\frac{40}{s}} + \frac{V_1 - \frac{V}{2} - \frac{2}{s}}{\frac{40}{s}} + \frac{V_1 - V}{20} = 0 \quad (1)$

$$\frac{\frac{V}{2} + \frac{2}{s} - V_1}{\frac{40}{s}} + \frac{V/2}{20} = 0 \quad (2)$$

from which

$$V(s) = \frac{4(s+2)}{s^2 + 4s + 8} = \frac{4(s+2)}{(s+2)^2 + 4}$$

$$v(t) = 4e^{-2t} \cos 2t \text{ V}$$

20.8 Let $I = 1$, then $V_L = 8s$,

$$I_c = \frac{V_L}{4} + I = 2s + 1,$$

$$V_{6\Omega} = \frac{1}{2s}(2s+1) + 8s = 8s + 1 + \frac{1}{2s}$$

$$I_g = \frac{8s + 1 + \frac{1}{2s}}{6} + 2s + 1 = \frac{20s + 7 + \frac{1}{2s}}{6}$$

$$H = \frac{I}{I_g} = \frac{6}{20s + 7 + \frac{1}{2s}} = \frac{12s}{40s^2 + 14s + 1}$$

20.9 Let $V = 1$, then $I_c = \frac{1}{20} + \frac{1}{2s}$,

$$V_c = \frac{\frac{1}{20} + \frac{1}{2s}}{\frac{s}{80}} = \frac{4}{s} + \frac{40}{s^2}$$

$$V_g = V_c + 1 = 1 + \frac{4}{s} + \frac{40}{s^2}$$

$$H = \frac{V}{V_g} = \frac{1}{1 + \frac{4}{s} + \frac{40}{s^2}} = \frac{s^2}{s^2 + 4s + 40}$$

20.10 By voltage division,

$$H = \frac{V}{V_g}$$

$$= \frac{\frac{56s}{8s+7}}{\frac{56s}{8s+7} + \frac{56}{s}}$$

$$= \frac{s^2}{s^2 + 8s + 7}$$

20.11

Let $V = 1$; then

$$I_1 = \frac{V/2}{20} = \frac{1}{40}$$

$$V_1 = \frac{40}{s}I_1 + \frac{V}{2} = \frac{1}{s} + \frac{1}{2}$$

$$I_2 = \frac{V_1 - V}{20} = \frac{\frac{1}{s}+\frac{1}{2}-1}{20} = \frac{1}{20s} - \frac{1}{40}$$

$$I_3 = \frac{V_1}{40/s} = \frac{s}{40}\left(\frac{1}{s}+\frac{1}{2}\right) = \frac{1}{40} + \frac{s}{80}$$

$$I_4 = I_1 + I_2 + I_3 = \frac{1}{40} + \frac{1}{20s} - \frac{1}{40} + \frac{1}{40} + \frac{s}{80}$$

$$= \frac{s}{80} + \frac{1}{40} + \frac{1}{20s}$$

$$V_g = 20I_4 + V_1 = \frac{s}{4} + \frac{1}{2} + \frac{1}{s} + \frac{1}{s} + \frac{1}{2}$$

$$\frac{V}{V_g} = \frac{1}{\frac{s}{4} + 1 + \frac{2}{s}} = \frac{4s}{s^2 + 4s + 8}$$

20.12 From 20.9, $H(s) = \dfrac{s^2}{s^2 + 4s + 40}$

or

$$H(s) = 1 - \frac{4(s+10)}{s^2+4s+40}$$

$$= 1 - \frac{4(s+2) + 32}{(s+2)^2 + 36}$$

$$h(t) = \mathcal{L}^{-1}[H(s)]$$

$$= \delta(t) - 4e^{-2t}\left(\cos 6t - \frac{4}{3}\sin 6t\right)$$

$$r(t) = \mathcal{L}^{-1}\left[\frac{H(s)}{s}\right]$$

$$= \mathcal{L}^{-1}\left[\frac{s}{s^2+4s+40}\right]$$

$$= \mathcal{L}^{-1}\left[\frac{(s+2) - 2}{(s+2)^2 + 36}\right]$$

$$= e^{-2t}\left(\cos 6t - \frac{1}{3}\sin 6t\right)$$

20.13 From 20.10, $H(s) = \dfrac{s^2}{s^2 + 8s + 7}$

$$h(t) = \mathcal{L}^{-1}[H(s)] = \mathcal{L}^{-1}\left[1 - \frac{8s+7}{s^2+8s+7}\right]$$

$$= \mathcal{L}^{-1}\left[1 + \frac{1/6}{s+1} - \frac{49/6}{s+7}\right]$$

$$= \delta(t) + \frac{1}{6}e^{-t} - \frac{49}{6}e^{-7t}$$

$$r(t) = \mathcal{L}^{-1}\left[\frac{H(s)}{s}\right] = \mathcal{L}^{-1}\left[\frac{s}{(s+1)(s+7)}\right]$$

$$= \mathcal{L}^{-1}\left[\frac{-1/6}{s+1} + \frac{7/6}{s+7}\right] = \frac{1}{6}\left(-e^{-t} + 7e^{-7t}\right)$$

20.14 From 20.8,

$$H(s) = \frac{12s}{40s^2+14s+1} = \frac{0.3s}{(s+0.1)(s+0.25)}$$

$$= \frac{-0.2}{s+0.1} + \frac{0.5}{s+0.25}$$

$$h(t) = -\frac{1}{5}e^{-t/10} + \frac{1}{2}e^{-t/4}$$

$$\frac{H(s)}{s} = \frac{0.3}{(s+0.1)(s+0.25)} = \frac{2}{s+\frac{1}{10}} - \frac{2}{s+\frac{1}{4}}$$

$$r(t) = 2\left(e^{-t/10} - e^{-t/4}\right)$$

20.15 From 20.11,

$$H(s) = \frac{4s}{s^2+4s+8} = \frac{4(s+2) - 8}{(s+2)^2 + 4}$$

$$h(t) = e^{-2t}(4\cos 2t - 4\sin 2t)$$

$$\frac{H(s)}{s} = \frac{4}{(s+2)^2+4}$$

$$r(t) = 2e^{-2t}\sin 2t$$

20.16 The poles of the network function are

$$s_{1,2} = \frac{\mu - 5 \pm \sqrt{(5-\mu)^2 - 12}}{2}$$

(a) If $\mu < 5$, the poles are in the left-half plane, and the network is stable.

(b) If $\mu = 5$, the poles are simple $j\omega$-axis poles, and the network is conditionally stable.

(c) If $\mu > 5$, the poles are in the right-half plane and the network is unstable.

20.17 (a) Denominator $= (s+1)(s+2)(s+3)$ Poles in LHP. Stable absolutely.

(b) Den $= (s-1)(s+2)(s+3)$ One RHP pole $(s=1)$. Unstable.

20.18 (a) Double poles on $j\omega$-axis. (at $\pm j1$). Unstable.

(b) $H(s) = \dfrac{s}{(s+2)(s^2+1)}$

Simple $j\omega$-axis poles($\pm j1$). Conditionally stable.

20.19 (a) Double pole at ∞. Unstable.

(b) $H = \dfrac{s^4}{(s^2+1)(s^2+3)}$.

Poles are simple $j\omega$-axis poles (at $\pm j1$, $\pm j\sqrt{3}$). Conditionally stable.

20.20 (a) F(s) not a proper fraction; Initial value theorem not applicable.

$f(\infty) = \lim_{s\to 0} sF(s) = 0 =$ final value

(b) $f(0^+) = \lim_{s\to\infty} sF(s)$

$$= \lim_{s\to\infty} \frac{3s^2+2s}{s^2+5s+6} = 3$$

$$f(\infty) = \lim_{s\to 0} \frac{3s^2+2s}{s^2+5s+6} = 0$$

20.21 $V_2(s) = H(s)V_1(s) = H(s)\frac{4}{s}$

$$= \frac{12}{s(s^2+4s+3)}$$

$$v_2(0^+) = \lim_{s\to\infty} sV_2(s)$$

$$= \lim_{s\to\infty} \frac{12}{s^2+4s+3} = 0$$

$$v_2(\infty) = \lim_{s\to 0} \frac{12}{s^2+4s+3}$$

$$= \frac{12}{3} = 4$$

20.22 $f(0^+) = \lim_{s\to\infty} \dfrac{3s}{s^3+4s^2+s-6}$

$= 0 =$ initial value.

Final value theorem not applicable because F(s) has a RHP pole $(s=1)$.

20.23 Initial value theorem not applicable because F(s) is not a proper fraction.

$$f(\infty) = \lim_{s\to 0} \frac{s(s^2+s+1)}{s^2+7s+12}$$

$= 0 =$ final value

20.24 $H(s) = \dfrac{12s}{40s^2+14s+1}$

$$I = H(s)\frac{10s}{s^2+4} = \frac{120s^2}{(s^2+4)(40s^2+14s+1)}$$

$$= \frac{As+B}{(s^2+4)} + I_1$$

$$\frac{120s^2}{40s^2+14s+1} = As+B+(s^2+4)I_1$$

Let $s = j2$:

$$\frac{-480}{-160+j28+1} = j2A+B \text{ ; or}$$

$$j2A+B = \frac{96}{13(41)}(159+j28)$$

$$\therefore\ A = \frac{96(28)}{13(41)(2)} = \frac{96(14)}{13(41)} = \frac{96(14)}{533}$$

$$B = \frac{96(159)}{13(41)} = \frac{96(159)}{533}$$

$$I_{ss} = \frac{As+B}{s^2+4}\text{ ; } i_{ss} = A\cos 2t + \frac{B}{2}\sin 2t$$

$$i_{ss} = \frac{96}{533}\left(14\cos 2t + \frac{159}{2}\sin 2t\right)$$

20.25 $H(s) = \dfrac{s^2}{s^2+4s+40}$

$$V = H(s)\frac{12}{s^2+9}$$

$$= \frac{12s^2}{(s^2+9)(s^2+4s+40)} = \frac{As+B}{s^2+9} + V_1$$

$$As+B+(s^2+9)V_1 = \frac{12s^2}{s^2+4s+40}$$

20.25 Cont.

Let $s = j3$:

$$j3A + B = \frac{-108}{-9 + j12 + 40} = \frac{-108(31 - j12)}{1105}$$

$$A = \frac{108(12)}{3(1105)} = 0.39; \quad B = \frac{-108(31)}{1105} = -3.03$$

$$v_{ss} = \mathcal{L}^{-1}\left[\frac{As + B}{s^2 + 9}\right] = \underline{0.39\cos 3t - 1.01 \sin 3t}$$

20.26 $V(s) = H(s)\,\dfrac{18}{s^2 + 36}$

$$V(s) = \frac{72s}{(s^2+36)(s^2+4s+8)} = \frac{As+B}{s^2+36} + V_1(s)$$

$$As + B = \left.\frac{72s}{s^2+4s+8}\right|_{s=j6}$$

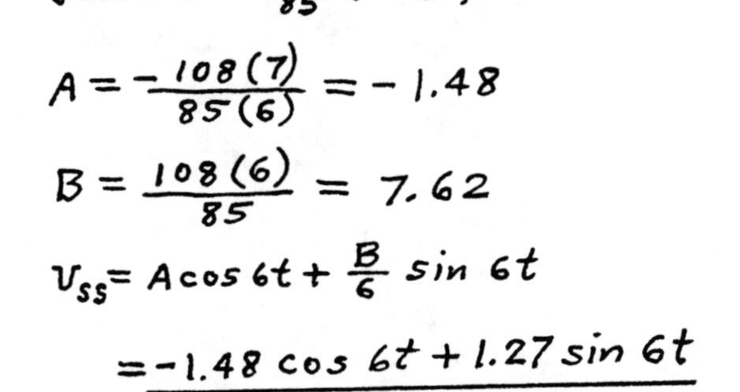

$$j6A + B = \frac{108}{85}(6 - j7)$$

$$A = -\frac{108(7)}{85(6)} = -1.48$$

$$B = \frac{108(6)}{85} = 7.62$$

$$v_{ss} = A\cos 6t + \frac{B}{6}\sin 6t$$

$$= \underline{-1.48\cos 6t + 1.27\sin 6t}$$

20.27 $H(s) = \dfrac{10\left(1 + \frac{s}{10}\right)}{(1+s)\left(1 + \frac{s}{100}\right)}$

$20 \log 10 = 20$

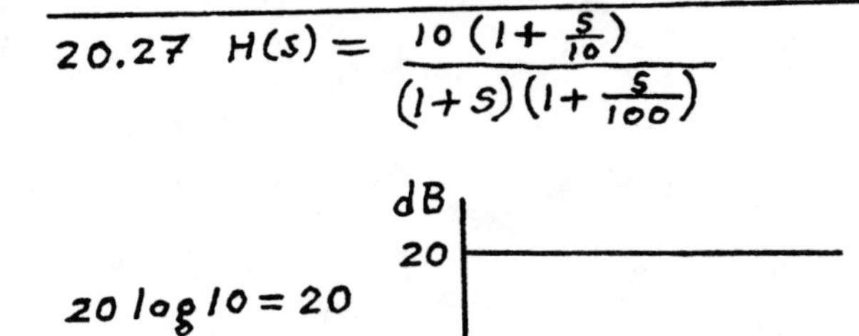

$20 \log \left|1 + \frac{j\omega}{10}\right|$

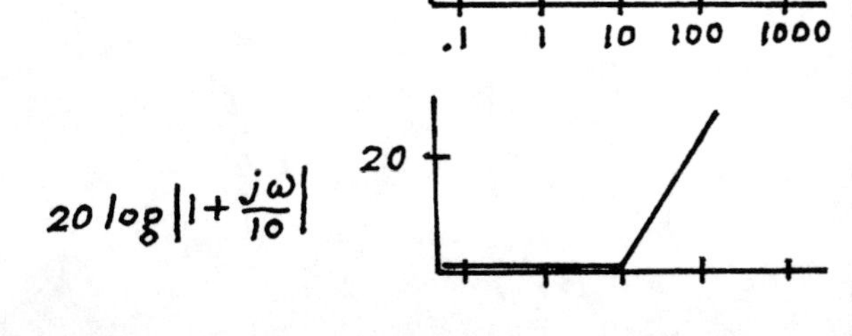

$-20 \log |1 + j\omega|$

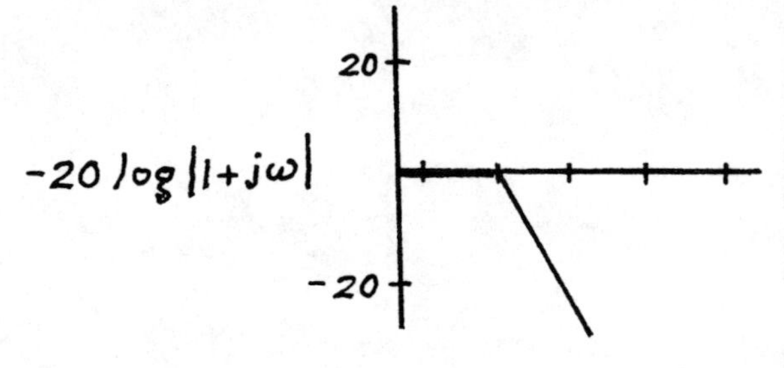

20.27 Cont.

$-20 \log \left|1 + \frac{j\omega}{100}\right|$

Sum

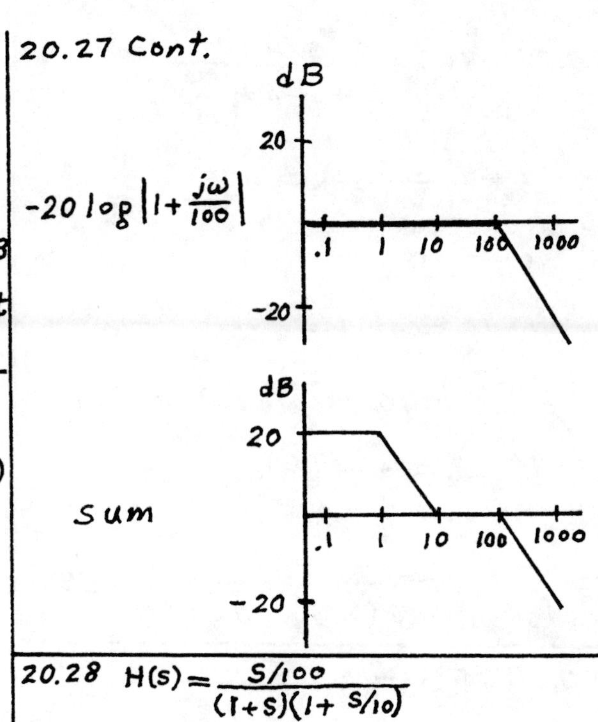

20.28 $H(s) = \dfrac{s/100}{(1+s)(1+s/10)}$

$20 \log |j\omega/100|$

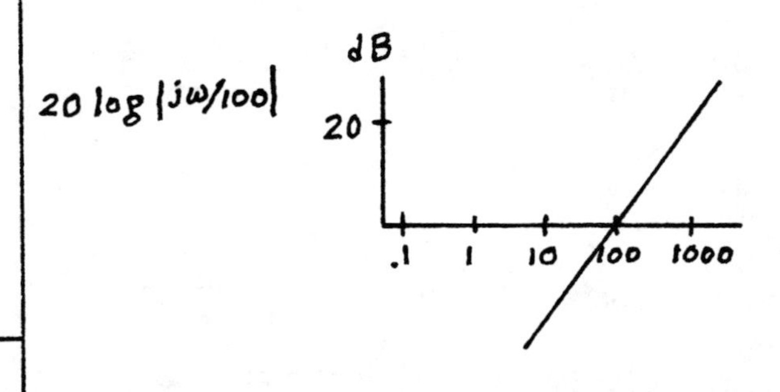

$-20 \log |1 + j\omega|$

$-20 \log |1 + j\omega/10|$

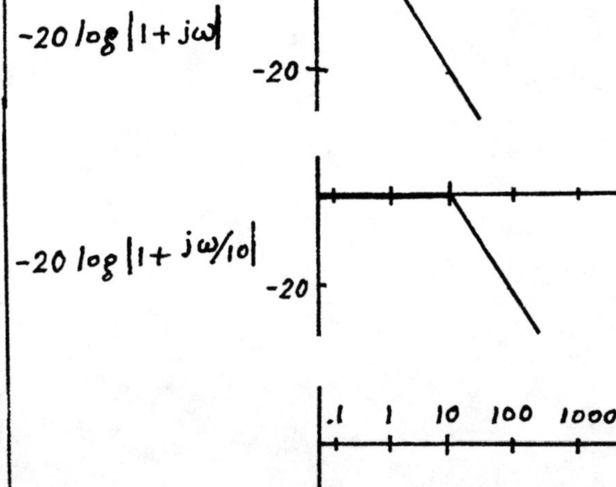

Sum

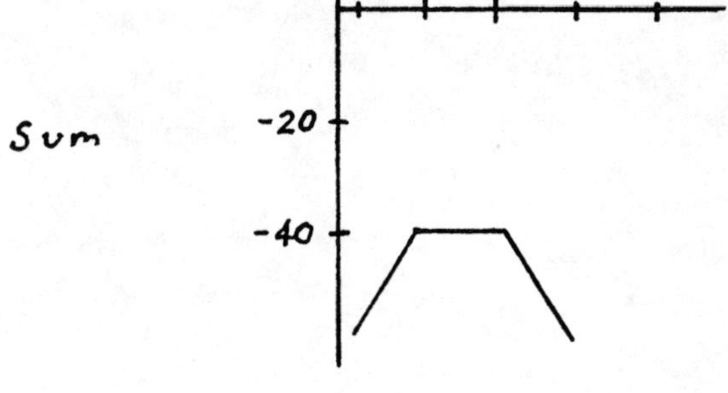

20.29 $F(s) = \dfrac{3/100}{1 + \frac{4}{100}s + \frac{s^2}{10^2}}$

$\zeta_K = \frac{4}{100}\left(\frac{10}{2}\right) = \underline{0.2}$

$\omega_K = \underline{10}$

$\omega_{max} = \omega_K \sqrt{1 - 2\zeta_K^2}$

$= 10\sqrt{1 - 2(0.04)} = \underline{9.6}$

$M_{max} = \dfrac{1}{2\zeta_K\sqrt{1 - \zeta_K^2}}$

$= \dfrac{1}{2(0.2)\sqrt{1 - 0.04}} = \underline{2.55}$

20.30 $F(s) = \dfrac{\frac{1}{4}}{1 + \frac{s}{36} + \frac{s^2}{6^2}}$

$\zeta_K = \frac{1}{36}\left(\frac{6}{2}\right) = \frac{1}{12} = \underline{0.0833}$

$\omega_K = \underline{6}$

$\omega_{max} = 6\sqrt{1 - 2\left(\frac{1}{12}\right)^2} = \underline{5.96}$

$M_{max} = \dfrac{1}{2\left(\frac{1}{12}\right)\sqrt{1 - \left(\frac{1}{12}\right)^2}} = \underline{6.02}$